Edited by
Seymour J. Garte

Molecular Environmental Biology

LEWIS PUBLISHERS
Boca Raton Ann Arbor London Tokyo

Library of Congress Cataloging-in-Publication Data

Molecular environmental biology / by Seymour J. Garte, editor.
p. cm.
Includes bibliographical references and index.
ISBN 0-87371-631-0
1. Genetic toxicology. 2. Molecular biology. 3. Molecular epidemiology. 4. Pollution—Environmental aspects. I. Garte, Seymour J.
RA1224.3.M64 1993
615.9′02′0157488—dc20 93-32995
CIP

Lewis Publishers is an imprint of CRC Press

International Standard Book Number 0-87371-631-0
Library of Congress Card Number 93-32995
Printed in the United States of America 1 2 3 4 5 6 7 8 9 0
Printed on acid-free paper

Dr. Seymour Garte is Professor of Environmental Medicine and Deputy Director of the Nelson Institute of Environmental Medicine at New York University Medical Center. He is the Program Director for Environmental Carcinogenesis in the Kaplan Comprehensive Cancer Center at NYU. Dr. Garte also serves on the Research Committee of the Health Effects Institute, and on the editorial boards of *Carcinogenesis* and *Journal of Cancer Research and Clinical Oncology*. He has served on the Scientific Advisory Board of the Chemical Industry Institute of Toxicology, the American Association for Cancer Research State Legislative Committee, U.S. Congressman Benjamin Gillman's Environmental Advisory Committee, and on numerous National Institutes of Health review panels. Dr. Garte is the recipient of research grants from the National Cancer Institute and the National Institute of Environmental Health Sciences. He is Director of the Training Program in Environmental Oncology, and the Director of the Laboratory of Molecular Oncology, where his research interests are in the molecular mechanisms of environmental carcinogenesis. His current research focuses on the role of oncogenes in chemical- and radiation-induced carcinogenesis, and in molecular epidemiology of environmental toxicology. Dr. Garte has a strong commitment to the application of molecular biology to problems in the environmental health sciences, and environmental biology in general.

PREFACE

Molecular Environmental Biology is the first book that illustrates molecular biological approaches to diverse issues of environmental biology. A panel of international experts has contributed representative chapters on the application of molecular methods and concepts to wildlife management, ecology, pollution control and remediation, and environmental health (including environmental toxicology, risk assessment, and epidemiology). While the book's contents cover a wide range of subject matter, the unifying theme is that the tools of molecular biology (such as gene cloning, DNA hybridization, sequencing, polymerase chain reaction, etc.) have made it possible for scientists to make rapid and sometimes stunning progress on environmental questions that were previously not even open for exploration.

The book is intended for students or established scientists in all fields of environmental biology and/or environmental medicine who are interested in the molecular revolution that is occurring in all fields of biological science. Each chapter presents a readily accessible review of the most current literature and knowledge in these rapidly evolving disciplines. Readers will find information on new techniques and new applications of established molecular methodology that should stimulate new research ideas, collaborations, and progress.

Molecular Environmental Biology

CONTENTS

1

Introduction and Overview

Seymour J. Garte

Nelson Institute of Environmental Medicine, New York University Medical Center, New York, NY

Both molecular biology and environmental science are relatively new areas of scientific research. Neither was found in the academic curriculum of a generation or two ago, but both have become highly popular fields of study within the past two decades. Apart from their modernity, molecular biology and environmental science may seem to have little in common. While molecular biology began as a highly specialized subspecialty of cell biology, environmental science is not in fact even a single science but an amalgam of diverse disciplines with disparate methodologies, underlying concepts, and histories. Some of these disciplines, such as ecology and natural history, are quite old and well established. More recently, the boundaries of environmental science have expanded beyond the range of these established fields, an expansion fueled to a large degree by societal and political influences outside the realm of science. There are now recognized scientific specialties, many with their own societies and journals, in environmental chemistry, environmental toxicology, pollution control technology, hazardous waste management, risk assessment, exposure assessment, and environmental medicine (or health sciences).

0-87371-631-0/94/$0.00+$.50

While the initial impetus for research in and development of these new disciplines may have been partially extrascientific, many of them have grown and matured to the extent of producing a body of basic data and theory to justify their roles as distinct scientific fields. The most successful of the new environmental sciences have taken full advantage of all available concepts and techniques from the cutting edge of the more traditional scientific disciplines — mathematics, physics, engineering, chemistry, biology, and geology.

Political and social forces stemming from a growing public awareness, and then concern regarding the state of the natural human environment, have accelerated both the development and growth of the new fields of scientific inquiry into environmental issues. Along with this development has come a trend toward convergence and interdisciplinary collaboration between scientists from many backgrounds into united efforts toward understanding and mediating the effects of man on the environment and the effects of the environment on man. The latter issue, the province of environmental health science, has especially grown in importance during the past two decades, as evidenced, for example, by the establishment of the National Institute of Environmental Health Sciences of the NIH.

Research in the environmental health sciences, in such fields as environmental toxicology, environmental epidemiology, and occupational health, has expanded tremendously in the past two decades. New fields such as risk assessment are still emerging. With growing recognition that environmental factors have replaced microbiological pathogens as playing the major role in the etiology of important diseases such as cancer, atherosclerosis, and lung disease, many medical schools are including courses in environmental and/or occupational medicine in their curricula. The reality of the potential disaster to human health posed by unrestrained environmental degradation has become tragically obvious in several nations of Eastern Europe where industrial pollution has gone unchecked for decades.

Environmental medicine, like all fields of medicine, may be considered a branch of biology. Other environmental sciences that are biologically based include ecology, wildlife management, forestry, and natural history. All of these biological environmental sciences, like all areas of biology, have recently begun to feel the impact of what may be considered to be the most important technical and conceptual revolution in biology since Darwin — the rise of molecular biology. As the phrase suggests, this new discipline involves the study of biology at a molecular level. Molecular biology has evolved extremely rapidly from its beginnings in the 1950s into a field with its own body of techniques, theory, and terminology. The central molecule of molecular biology is DNA — the genetic molecule. And the central goal of molecular biology is the complete understanding of the structure and function of the genetic apparatus of living cells at the most basic (molecular) level.

Molecular biology was born after the discovery of the structure (and rapidly thereafter the function) of DNA in the 1950s. It was after the development in the 1970s of techniques that allowed for the manipulation of genes, and the ability to study the genetic process at a depth not dreamt of only a decade earlier, that the field of molecular biology exploded with major impact on research in a wide variety of biological sciences. From its origins as the study of the molecular basis of genetics and heredity, molecular biology has more recently exercised enormous influence over diverse areas of biological and medical research. It is difficult in fact to conceive of any field of biology that is not subject to rapid progress by the application of molecular biological methods and ideas.

In the history of science, new fields have usually emerged and then diverged from older ones; rarely, if ever, has a new scientific discipline been born without roots in a previously established area of study. History has also shown that, time and again, major advances in scientific knowledge have occurred at the interface between two previously unconnected fields that are brought together either by the efforts of individuals or by the development of new techniques.

The application of the new molecular techniques has in fact brought exciting and often explosive change to areas of biological research as diverse as taxonomy, cancer biology, virology, evolutionary biology, physiology, medicine, agricultural science, and the subjects of this book — environmental biology and environmental health science. In each of these fields, the new ability to ask questions at the molecular level has enabled scientists to move into areas not previously open to exploration. Techniques such as the measurement of DNA adducts, levels of gene expression, analysis of mutational spectra, and detailed precise knowledge of the effects of environmental stress on the biological macromolecules of cells have added enormous power to scientists seeking to improve the scientific basis for the assessment of exposure to and/or risk from environmental toxicants. The ability to distinguish species and populations on the basis of minute differences in the genetic sequences of specific genes (which of course are responsible for the gross morphometric and anatomical differences used historically) has dramatically altered the study of evolutionary biology, and has had major economic as well as scientific effects on our attempts to manage endangered as well as other wildlife populations. The practical problems associated with environmental remediation have also been affected by molecular biology. Cloning and characterization of bacterial genes that code for degradative enzymes may provide a biological solution to many pollution problems.

The history of molecular biology might be divided into two stages. The first, from 1953 to 1970, was the theoretical phase. Starting with the publication by Watson and Crick of the double helix structure of DNA, the next two decades brought about a tremendous advance in basic knowledge of genes and how they work. During this period, the genetic code was deciphered, the central dogma DNA $\rightarrow$ RNA $\rightarrow$ Protein was drafted, and the mechanisms of DNA replication, gene transcription, and translation were elucidated. The results were often breathtaking and spectacular. For example, when the details of the process of protein synthesis, with the involvement of ribosomes, messenger RNA, transfer RNA, specific transferase enzymes, etc., were finally elaborated, an unprecedented picture emerged of biological complexity, truly beautiful and amazing in its organization and regulation.

All of these discoveries were made using established and newly developed methods of biochemistry and cell biology. Calling this the theoretical phase of molecular biology implies the idea that the achievements of this period led to knowledge of the general functional mechanisms of how genes worked. However, structural and functional analysis of specific genes was beyond the technical ability of the time. With the basic theory of molecular biology (as exemplified by the central dogma) in place, further progress required methodological innovations.

The second stage of molecular biology, which might be termed the practical phase, began in the early 1970s with a rapid succession of technical advances. It must be noted that by stressing the practical aspects of research progress during this period, it should not be construed that basic knowledge of molecular genetics did not make significant advances as well. Introns, regulatory DNA sequences, and other critical structural details of the genetic apparatus were uncovered; but many such advances at the same time had major technological impact. For example, one of the most critical and earliest findings of this stage was the discovery of reverse transcriptase, a viral enzyme that catalyses the synthesis of a complementary DNA strand (cDNA) from RNA. This not only answered a puzzling enigma related to the way RNA viruses function and produced an important exception to the basic theoretical central dogma of molecular biology, but also provided a crucial reagent for the cloning of genes. The discovery of restriction endonucleases, which cleave DNA at specific base pair sequences, the refinement of DNA hybridization techniques on solid matrices, and other technical advances led in the space of 2 or 3 years to the ability to clone specific DNA sequences and to analyze specific genes in detail. Technical progress in this area was so rapid that a number of scientists, including molecular biologists, became alarmed at the possibility that the frenzied pace of research in this area might in fact prove dangerous to human health and/or the environment.

In an unprecedented action, the world's leading molecular biologists in 1975 agreed to a self-imposed moratorium on further gene cloning until sufficient time and data had accumulated to attest to the safety of the new technology. This 2-year moratorium has had, of course, special significance in its social, historical, and ethical contexts; but an important aspect not always fully appreciated is that the research moratorium was made necessary by the revolutionary speed of progress during these years. There was simply no time between publication of major advances for assessment of potential risks, much less any other implications of the latest results. While terms like revolution, breakthrough, and explosion are often misused in the history of science, they may certainly be applied to the burst of recombinant DNA technology and gene cloning during the brief period of the early 1970s.

I have said that this second phase of molecular biology (which persists through the present) was largely technology driven. With the ability to clone and study individual genes, molecular biology moved outside the realm of molecular biologists. The new techniques were quickly applied to research problems in cell biology and biochemistry, physiology, and medicine. For example, research dermatologists cloned and characterized the keratin genes, cancer biologists discovered cellular oncogenes, forsenic pathologists used DNA fingerprinting methodology, neurophysiologists began a major effort to understand the molecular nature of nerve action, botanists developed new molecular technologies for cloning and breeding plants *in vitro* (an enterprise with major economic importance), pharmacologists applied plasmid technology and recombinant DNA to the biosynthesis of drugs and human hormones, and so on.

A mere dozen years ago, the phrases "molecular biology", "gene cloning", and "recombinant DNA" had a slightly magical ring and those skilled in these new arts were in great demand. Today, the majority of graduate students in all fields of biology receive at least some training in molecular techniques, and the tools needed to clone and study genes, manipulate DNA and proteins, analyze specific macromolecular sequences, etc. are extremely widespread. One need no longer be a molecular biologist to use molecular biology, and in fact the vast majority of those who use these methods would not consider themselves to be primarily molecular biologists.

The pace of technical progress in molecular biology continues at a high level. Half a dozen years ago, the automated polymerase chain reaction (PCR) method was developed. This new technique has had immense impact on many fields, and has often replaced gene cloning as a first step in the study of specific genes. With PCR, pathobiologists can sequence the DNA from tiny tissue fragments of preserved museum specimens, cancer researchers can analyze mutation rates in tissues from pathology slides, and scientists can study the sequence of a gene or part of a gene in a tissue or cell culture without first cloning the sequence. The advent of PCR has spurned new journals, and a constant stream of modifications, improvements, and related techniques has enabled scientists to tremendously expand their repertoire of analytical tools.

It is vital for the reader to understand that this book is not a text on molecular biology, just as the scientists who have authored the individual chapters would not all necessarily classify themselves as molecular biologists. Rather, the message we wish to convey is that the *tools* of the molecular biologist are now available and extremely useful for biological scientists in all disciplines, and particularly timely in the areas of environmental biological and medical sciences where the questions are often new or poorly defined and where the potential for major new answers and insights is enormous.

Ecology and evolutionary biology are interrelated disciplines that have historically provided the scientific foundation for a major portion of contemporary environmental biological science. In recent years, molecular biological approaches and techniques have made significant contributions to these fields. Chapter 2 by Bruford and Wayne provides an excellent example of how modern molecular techniques can be used to address questions that were previously unanswerable in population biology of wild animals. What is particularly exciting about this field is the immediate and enormous benefits that have been gained in the conservation and management of our natural world. The unambiguous genetic analysis of individuals, small populations, and subspecies, as described by Bruford and Wayne, has already provided biologists and naturalists with critical information that can be used to assure maximum genetic diversity in captive breeding populations of near extinct species, as just one example. Questions related to phylogenetic distinction and history, interbreeding rates, genetic variability, parentage and reproductive efficiency are all being addressed at a level not previously possible. The new information generated by techniques such as PCR and DNA fingerprinting will result in improved strategies for salvaging and conserving the ecological diversity and richness that is part of this planet's natural environment.

Chapter 3 by Waldman and Wirgin presents a specific example of how molecular biological approaches are being used in the management of one of our most important natural resources — fisheries. Techniques such as mitochondrial DNA analysis, and analysis of repetitive DNA sequences and coding and noncoding single copy sequences, have been applied to differentiate fish stocks on a genetic basis. Waldman and Wirgin illustrate the approach with specific examples and makes very clear the large-scale economic as well as scientific impact that such new technologies may bring.

It has been known for many years that there are vast numbers of unknown species of microorganisms, including bacteria, fungi, algae, etc., that inhabit the natural world, and that many of these species have the capacity to use certain toxic chemical pollutants as carbon and energy sources. The use of natural organisms to aid in the cleanup of the environment (often termed bioremediation or biodegradation) is often an attractive alternative to other disposal methods. How is it possible that chemicals such as industrial solvents, PCBs, even pesticide residues, which are toxic to most living organisms, can serve as a food source for these microorganisms? Toxicologists have long known that toxicity is a relative phenomenon dependent on the metabolic profile of particular organisms. Of course, the existence of enzymes capable of metabolizing or converting any xenobiotic compound is a function of the genetic composition of the species. In order to take full advantage of the biodegrading potential of natural organisms, it is vital to understand the basic molecular mechanisms by which these organisms act.

The study at the molecular genetic level of biodegradation of persistent and toxic chemicals has enormous potential for the future remediation of environmental pollution. In Chapter 4, Yates and Mondello describe the cloning and characterization of the genes responsible for degradation of PCBs and related environmental pollutants. These genes are present (possibly in movable elements) in several new strains of *Pseudomonas* bacteria isolated from the natural environment.

Comparison of the cloned genes that degrade chlorinated environmental pollutants such as PCBs from different strains reveals similarities and differences that lead to the conclusion that soil bacteria possess diverse and complex metabolic pathways for utilization (and destruction) of these chemicals.

In Chapter 5, Zylstra discusses similar strategies for the molecular analysis of aromatic hydrocarbon degradation by soil bacteria. The agents include such environmentally significant chemicals as toluene, cresol, and tricholoethylene, as well as the carcinogenic and well-studied polycyclic aromatic hydrocarbons. Zylstra presents specific examples of the cloning of the genes responsible for biodegradation of these and other toxicants, and also describes the application of such approaches for construction of new biodegradative pathways for bioremediation.

Further understanding of the molecular and biochemical mechanisms involved in bacterial bioremediation of toxic pollution will obviously have dramatic impact, not only on the basic science of environmental microbiology, but also on the technical and economic aspects of pollution control and abatement.

Chapter 6 by Wirgin and Garte presents a novel molecular biological approach to the assessment of environmental pollution status of a natural habitat. Atlantic tomcod inhabit a number of northeastern U.S. and Canadian rivers; the Hudson River, New York population exhibits an unusually high incidence of liver tumors. Using this species as a sentinel, Wirgin and Garte have shown good correlation between levels of expression of the CYP1A (cytochrome P450) gene in livers of these fish and their exposure to hydrocarbon pollutants in water and sediments.

While it might seem overly elaborate to determine the degree of pollution in a river by doing northern blot hybridization, as opposed to simply measuring water and sediment concentrations, the fact is that these molecular techniques provide a sensitive and useful measure of the biologically relevant dose of pollutant from all sources in the river (including prey species) and take into account variations in local concentrations that are missed by water and sediment sampling. The economic and social implications of such new methods for commercial and recreational use of our waterways are very large. We are clearly still in the early stages of taking full advantage of the potential of such approaches.

The idea that nonbiological agents in the environment could play a role in serious and chronic human diseases did not become apparent until modern medicine and technology had helped to eradicate the communicable diseases of the past. With the resulting extension of the human life span, along with major changes in technological and cultural environments, it is now clear that environmental influences play a very important role in human disease etiology. Cancer is an important cause of death in most industrialized countries, and efforts directed at its cure, prevention, and control are subjects of intense research. There is now overwhelming evidence that the preponderance of human cancers owe their origin to nonbiological agents present in the human environment.

The importance of applying new molecular approaches to questions of human disease etiology and epidemiology in general is described in the Chapter 7 by Hemminki. In this review, which may serve as an introduction to molecular epidemiology, Hemminki explores the application of molecular biology to a number of common chronic disease states, most of which are caused by some combination of genetic and environmental factors. One of the major challenges in modern biomedical research into origins of human disease such as cancer, cardiovascular disease, and other chronic diseases is to elucidate the role of specific environmental agents, and/or specific genetic loci, that contribute to disease causation. The molecular biological approaches discussed by Hemminki are beginning to provide solutions to some of these difficult but critical questions.

Perhaps the most widespread development of molecular epidemiology has been in the field of environmental carcinogenesis. Specific applications of molecular biological approaches to this research are discussed in the remaining chapters by Strickland and Groopman, O'Neill et al., and Cosma and Garte. Chapter 8 by Santella and Perera represents a good introduction by one of the pioneering groups in the field. In this chapter, Santella and Perera review the various uses of molecular biological markers in cancer risk assessment, including measurement of internal dose of carcinogens, the analysis of biologically effective dose, determination of individual differences in susceptibility to environmental carcinogens, etc. The explosion of research into biomarkers has resulted in a number of new experimental strategies, a renaissance of collaborations between epidemiologists and laboratory scientists, and a number of new journals and volumes.[1] As exciting discoveries on the basic molecular mechanisms of carcinogenesis related to oncogenes and tumor suppressor genes continue to emerge, these new findings are being exploited wherever possible in epidemiological inquiries into the origin, prevention, and intervention of cancer in human populations.

In relation to human health, the term "environment" encompasses more than is generally the case in other environmental studies. Environmental medicine must include knowledge of the effects of human exposure to agents from all external sources — including diet, smoking, medications, occupational situations, and indoor air — in addition to the more commonly defined environmental sources such as outdoor air, ground water, and toxic dump sites. Perhaps the most significant of these environmental sources for important chronic diseases such as cancer and heart disease is the human diet. Chapter 9 by Strickland and Groopman focuses on the influence of two specific environmental carcinogens that are a natural component of foods consumed by millions of people. *Aflatoxins* are produced by molds that grow on grain and nuts, and *polycyclic aromatic hydrocarbons* are found in charcoal-broiled foods. These compounds are well-known carcinogens in animals and humans, and the authors discuss how modern molecular techniques are enabling scientists to develop sensitive biological markers to detect exposure to these agents in affected populations.

Environmental agents such as chemicals and radiation may produce a variety of damage to living tissue and cells. In the context of human disease, considerable attention has been paid to the effects of such agents on the genetic material. Damage to DNA, leading to somatic or germ-line mutations of various types, has direct consequences often leading to induction of cancer, birth defects, and other disease. O'Neill et al. discuss in Chapter 10 an important conceptual and technical development with major importance for molecular epidemiology, and the analysis of human exposure and risk from environmental mutagens. These authors describe assays for mutation spectra that give both qualitative and quantitative information regarding genetic damage in circulating lymphocytes.

The measurement of DNA adducts and mutation frequencies discussed in Chapters 9 and 10 are well-documented and established molecular methods for

[1] See *Biomarkers of Environmental Contamination* (J.F. McCarthy and L.R. Shugart, eds.), Lewis Publishers, Boca Raton, FL, 1992.

assessment of environmental toxic exposure and effect. Given the pace of modern research in molecular biological techniques, it is not surprising that new methods are continuously being developed and applied to studies of environmental toxicology and human health. Chapter 11 by Cosma and Garte report on the development and early validation of one such approach — the analysis of gene induction by direct quantitation of messenger RNA of specific inducible genes. It has been known for many decades that a number of genes which code for proteins that protect cells from xenobiotic agents are transcriptionally silent until the cell is challenged by exposure to the particular agent. For example, a wide variety of aromatic organic chemicals may induce transcription of the CYP1A1 (cytochrome P4501A1) gene which codes for the aromatic hydrocarbon hydroxylase enzyme responsible for the first step in detoxification mechanisms. The gene for metallothionein, a protein that binds to toxic heavy metals, is induced by a number of such metals.

The strategy described by Cosma and Garte takes advantage of this phenomenon by using the extent of gene inducibility (using CYP1A1 as an example) as a dosimeter of toxicant exposure. By analyzing the levels of expression of this gene in peripheral lymphocytes, it should be possible to perform field studies and assess exposure and/or response to environmental toxicants in human populations.

Although the contents of this book might appear to be quite eclectic, the unifying theme is the use of the tools of molecular biology in applications related to environmental science. The chapters of this book represent examples of scientists working in several disciplines of biological environmental science who have taken advantage of the wealth of analytical methodology available from the molecular biology "tool kit". There are many other examples that could not be included, and the list is growing. What is most important (and if there is a take-home message of this book, this is it) is that all environmental scientists, whether their specific disciplines be biological, chemical, physical, or geological, become aware that molecular biological approaches to environmental problems have tremendous potential; that they will be used more and more frequently, and that very likely, as has already happened in most areas of biological research, they will become the state-of-the-art methods in environmental sciences.

For those in the environmental community who have had no direct experience with molecular biology or who may even have watched the molecular revolution from afar with some trepidation, this book might be considered a plea for conversion. Our own laboratory is involved in fruitful collaborations with environmental engineers, chemists, and others whose previous knowledge of molecular biology was nil, but who have come to acknowledge the fact that these new methods and ideas can provide the keys to unlock doors we weren't even aware of before, and allow us to move quickly forward in the exciting and challenging study of our natural and human environment.

2

The Use of Molecular Genetic Techniques to Address Conservation Questions

Michael W. Bruford and Robert K. Wayne

Conservation Genetics Group, Institute of Zoology, Regent's Park, London

INTRODUCTION

The past 10 years have seen rapid progress in the development of molecular techniques that can be applied to evolutionary systematics, population genetics, and the genetic characterization of individuals. The recent discovery of techniques such as the polymerase chain reaction[1] and the subsequent development of "universal" primer sequences now allow the rapid characterization of informative DNA sequences in many organisms[2] at both the species[3] and population[4] level, often from very small, degraded, and even ancient samples. Similarly, the discovery of hypervariable DNA repetitive sequences[5,6] in the genomes of most eukaryotes has permitted the unequivocal identification of individuals within populations,[7] allowed the accurate measurement of individual reproductive success,[8] and proved a sensitive indicator of genetic drift[9] and inbreeding.[10] When combined with more established techniques such as allozyme electrophoresis[11] and cytogenetic analysis,[12] we now clearly possess a powerful set of molecular tools with which to approach many important evolutionary questions.

0-87371-631-0/94/$0.00+$.50

Table 1
Levels of Genetic Distinction, Problems, and Appropriate Molecular Tools

Genetic level	Problems	Molecular tools
Species	Phylogenetic distinction Hybridization	Karyology 3 Allozymes (fixed) 2 mtDNA sequence (12s rRNA, cyt. *b*) 1
Subspecies/ metapopulation	Phylogenetic distinction Hybridization Phylogeographic history	Karyology 4 Allozymes (frequency) 3 mtDNA sequence (cyt. *b*, D-loop) 1 Single locus genomic sequence 2
Population	Genetic variability Genetic substructuring	Allozyme (heterozygosity) 1 mtDNA sequence 3 (D-loop) Single locus genomic sequence 2 DNA fingerprinting 3
Individual	Reproductive success Genetic representation	Allozymes 3 Single locus genomic sequence 2 DNA fingerprinting 1

Note: Molecular tools are scored 1, 2, 3, and 4 in descending order of resolution for the level of genetic distinction under examination. Abbreviations: mtDNA 12s rRNA = 12s rRNA gene in the mitochondrial genome; cyt. *b* = cytochrome *b* gene; D-loop = D-loop of the control region in the mitochondrial genome; allozyme (fixed) = analysis of fixed differences among species; allozyme (frequency) = analysis of frequency differences of polymorphic enzymes among populations.

This chapter describes how molecular tools can be applied to fundamental questions in the emerging field of conservation genetics. Our approach will be to define some important conservation questions and the molecular genetic techniques that can be used to address them. We then provide some examples of our own research and that of others which exemplify the application of molecular techniques to conservation problems.

THE QUESTIONS

Many important conservation genetic problems can be approached using currently available molecular techniques. The problems range from the identification of species or subspecies of conservation interest, to the management of genetic variability in small populations. Table 1 matches genetic questions with molecular approaches used to resolve variability at different evolutionary levels. These questions are described in detail below.

Phylogenetic Distinction

The identification of systematic units and the determination of taxonomic uniqueness are crucial information for conservation programs.[13] Given the difficulty of assigning priority to the numerous species that are rare or endangered, molecular techniques can be used to better define the distinctiveness of species and the taxonomic units they contain. Informed decisions can then be made regarding the relative significance of the species and the amount of effort which should be devoted to its conservation. Molecular data can be analyzed using phylogenetic techniques to provide information on the evolutionary heritage of a species and to determine their phylogenetic distinctiveness by ranking them according to the number of close relatives and phylogenetic position.[14] Species which represent monotypic genera may, for example, be regarded as more significant than ones in more specious genera.

Hybridization

Interbreeding among individuals from closely related species, from subspecies, and from different populations of the same species is a common phenomenon in wild populations.[15,16] Hybridization can be a severe problem if it threatens the genetic integrity of endangered taxa. Such introgression often results from habitat alteration or introduction of nonnative species, particularly where the endangered taxon is rare such that hybridization with closely related forms is favored. For example, coyotes and gray wolves in Minnesota, Ontario, and the American South appear to hybridize in disturbed areas where coyotes have recently become abundant.[17,18] Unintentional hybridization can also occur in captive breeding stocks. For example, molecular techniques have shown that Asiatic lions have been found to contain genes from the African subspecies,[19] dramatically altering the Asiatic lion captive breeding program.

Phylogeographic History

Molecular genetic data can be used to reconstruct the phylogenetic relationships of populations within species. Phylogenetic reconstructions can be used to identify isolated and genetically distinct populations, as well as to determine dispersal corridors between populations.[20,21] This level of distinctiveness among populations is of extreme importance for identifying source populations for augmentation or reintroduction programs. Genetic augmentation may be necessary if an endangered population shows evidence of inbreeding depression that significantly increases the probability of extinction and if captive breeding is unlikely to ameliorate inbreeding effects. Given these conditions, augmentation of genetically similar individuals from a similar environment to that of the endangered population is desirable.

A recent example of the use of phylogeographic analysis to address a conservation question concerned relationships of the dusky seaside sparrow.[22] This subspecies became extinct in the wild about 1980, and captive stocks comprised only five male birds. Females from a population near the former range of the subspecies were used to augment the captive population; however, later genetic analysis revealed that these individuals were from a fundamentally different phylogenetic stock and that more appropriate source populations could have been used.

Genetic Variability

Genetic variability within a population has two components: the allelic diversity (or number of alleles at a given locus) and the genetic heterozygosity (or the expected proportion of genes that are heterozygous in the average individual). In outbred populations, typically 10 to 30% of genetically variable structural genes detected by protein electrophoresis have two or more alleles, whereas only 3 to 10% of all genes are actually heterozygous in a typical individual.[23] Analysis of noncoding variable DNA regions, such as mini- and microsatellites reveal far higher values, with greater than 50% of loci being polymorphic, and typical heterozygosity values ranging from 50 to 95%.[24] In small and isolated populations, genetic variation may be drastically reduced, initially by a loss of allelic variability, followed by a decline in heterozygosity.[25] Low levels of heterozygosity, especially when associated with breeding among close relatives, may correspond with decreases in viability and increases in juvenile mortality.[26-29] Genetic variability may be important for the long-term persistence of and adaptive change in populations, and management of captive and wild populations of endangered species are designed to minimize the loss of genetic variation.

Molecular techniques can be used to compare levels of variability among and within populations where subdivision into social groups may result in partitioning of genetic diversity and variation, and to follow the decline of variability in small populations. Examples of this approach have led to the discovery of low levels of genetic variability in, among others, the cheetah,[30] lions of the Ngorongoro crater,[31] Isle Royale gray wolves,[32] and within and among colonies of the highly social naked mole-rat.[33] Captive populations are especially prone to decreased levels of genetic variation, either through poor genetic management or because founder individuals may be related. Relatedness may be a problem if founders are obtained from other inbred captive stocks or from small inbred wild populations.

Reproductive Success and Genetic Representation

In small captive and wild populations, individuals may not reproduce equivalently. Relative to a random breeding population, asymmetry in mating success reduces the effective population size and increases the rate of inbreeding. To minimize the loss of genetic variability in small populations, the genetic relationships of individuals need to be understood so that the number of breeders and their genetic dissimilarity can be maximized. Molecular genetic techniques have been effectively used to deduce parentage in wild and captive populations.[7,8,34]

THE MOLECULAR TOOLS

Karyological Analysis

Chromosome morphology, as revealed by standard metaphase preparations and banding techniques, has been used to distinguish species, subspecies, and populations; delineate hybrid zones; and determine the suitability of individuals for breeding.[35] Although field preparations are possible, chromosomal analyses in mammals and birds commonly involve lymphocyte culture or the establishment of fibroblast cell lines. The application of karyological analysis to conservation questions is limited. Informative differences generally become less discernible as closer taxonomic units are compared and the routine screening of large populations is tedious. Even the phylogenetic analysis of chromosome data is clouded by the arbitrary nature of relative weights applied to chromosome alterations (e.g., see Reference 36). Higher-resolution chromosome analyses may require considerable expertise and expense for the amount of information gained. For conservation purposes, the most appropriate problems to which karyological studies are applied involve phylogenetic questions at the specific/subspecific level and the identification of individuals with rare chromosomal abnormalities.

Allozyme Electrophoresis

The application to systematic and population genetic analyses of methods for separating the protein alleles of a single gene has dramatically changed ideas about the abundance and significance of genetic variation. Hundreds of vertebrate species have been studied over the past 25 years, providing a considerable comparative base.[37] Advances in technology and standardization of protocols has led to allozyme electrophoresis becoming a relatively easy and inexpensive technique. Excellent protocols and technical descriptions are available (see References 11 and 37). Although newer DNA-based methods potentially provide more resolution, allozyme electrophoresis remains an important tool for studying relationships among species and populations.

Allozyme electrophoresis involves identifying protein variants by separation on starch, acrylamide, or cellulose acetate gels in an electric field, followed by histochemical staining for products of a single gene. Usually, 20 to 30 gene products are assayed, although surveys including more loci are always desirable for population genetic purposes, as heterozygosity measures can have a large variance, particularly if only a few loci are polymorphic.

The main difficulties with allozyme electrophoresis include the uncertainty of determining the homology of alleles in some systems, particularly if samples have undergone different degrees of degradation. Additionally, closely related populations, subspecies, and species may differ primarily in allele frequency rather than in the presence or absence of fixed diagnostic alleles, making phylogenetic/phylogeographic inference problematic.[39,40] Moreover, recent research has questioned the implicit assumption of selective neutrality of some allozyme alleles.[41-44]

However, important benefits of the technique are that existing protocols can be applied to a wide variety of taxa to reveal variation at homologous loci, and it requires a minimal amount of capital investment or development.

Mitochondrial DNA Sequence Analysis

The mitochondrion, a cytoplasmic organelle, contains a self-replicating genome which is a closed double-stranded circular molecule, typically between 16,000 and 18,000 base pairs long in vertebrates. The mtDNA sequence evolves five to ten times faster than the average nuclear gene.[45,46] Consequently, it is often possible to discriminate closely related species and populations more effectively than allozyme analysis. Mitochondrial DNA is almost always maternally inherited in a clonal fashion such that, barring mutation, all offspring have a genome identical to their mother.

Owing to the relatively fast rate of mtDNA sequence evolution, the amount of sequence divergence among closely related species may accurately reflect their evolutionary history. When mtDNA genotype data of individuals from different populations is analyzed using phylogenetic algorithms, the structure of evolutionary trees often correlates with the geographic distance between populations or the presence of geographic boundaries.[20]

Mitochondrial DNA sequence differences can either be measured indirectly through restriction site analysis or by sequencing regions of the mitochondrial genome. The former is less expensive and allows rapid screening of many samples, whereas the latter, although laborious, provides data in its most elemental and complete form. In restriction site analysis, mtDNA is digested with a restriction enzyme which cleaves the DNA at specific four, five, or six base sequences. The resulting fragments are separated by electrophoresis. Genotypes are defined as their composite restriction fragment pattern for a panel of usually 10 to 20 restriction enzymes. Many individuals may share a common genotype. The pattern of restriction site gain or loss among genotypes for a panel of enzymes can be used to construct phylogenetic trees of populations or species.[20,47]

Previously requiring considerable effort, the direct sequencing of mtDNA genes is now much easier since the development of the polymerase chain reaction (PCR) and the identification of "universal" primers which enable amplification of several hundred base pair sequences from specific regions of the mitochondrial genome.[1-3] PCR involves the enzymatic amplification, through repeated temperature cycles using a thermostable DNA polymerase, of a segment of DNA bounded by two short primer sequences 10 to 30 base pairs long. Using PCR, sequence can be obtained from very small quantities of DNA, even that isolated from specimens of bone, skin, or hair in museum and archaeological collections.[18,48-50] The sequence data can be used for detailed systematic or phylogeographic analysis; and because mitochondrial genes evolve at different rates, rapidly evolving genes can be analyzed to determine the relationships of recently diverged populations, whereas slowly evolving genes can be used to address systematic questions involving more distantly related taxonomic units.

A difficult problem with phylogenetic analysis of mtDNA sequence data is that repeated substitutions at the same sequence position leads to reversals and parallelisms such that phylogenetically informative changes can be obscured. Additionally, divergence among mtDNA genotypes may precede speciation or population isolation such that the mtDNA tree does not accurately reflect the phylogenetic heritage of species or populations.[51] Moreover, past episodes of hybridization may lead to incorrect phylogenetic inference.[52] Importantly, because mtDNA is maternally inherited, analysis of its sequence provides data on maternal lineages only and may not be precisely associated with nuclear genome evolution. In general, restriction site analysis requires less expertise and expense than direct sequencing, although this discrepancy is becoming less significant.

Hypervariable DNA Analysis

The nuclear genome of most eukaryotes contain blocks of tandemly repeated sequences. The number of repeats in an array may vary dramatically at the same and different loci. The length variation of repeats can be measured by a genomic DNA digest with a restriction enzyme that cleaves in regions flanking the repeat. The restriction fragments are separated by electrophoresis, transferred to nylon membranes, and visualized by hybridization with a radioactively labeled probe based on common sequences in the repeat.[53] The first application of this approach, by Jeffreys et al.,[5,6] assessed variation in tandem arrays of approximately 33 base pairs designated minisatellites, and found fragments ranging from 2000 to over 30,000 base pairs. The multilocus pattern of restriction fragments on autoradiograms, resembling a supermarket barcode, was named a DNA "fingerprint" and found to be unique to each individual.

The presence of a unique combination of diagnostic fragments in parents and offspring has allowed accurate paternity assessment in wild populations,[7,8] and analysis of the overall number of shared fragments between individuals in populations has been used to determine degrees of relatedness and variability.[9,31,33,54] Since Jeffreys' discovery, many other hypervariable repeat sequences have been described.[55,56] Some repeats are much smaller than those detected by the Jeffreys' probes and may even consist of two base (dinucleotide) repeats,[57,58] known as "microsatellites". One of the desirable properties of microsatellites is that the size of a repeat block is generally much smaller (less than 500 bases) than a minisatellite, and therefore can be PCR amplified, even in degraded DNA samples from ancient tissue specimens.[49,50,59,60]

The most apparent problem with DNA fingerprinting as commonly applied is that the multilocus pattern on a fingerprint is, in effect, a phenotype consisting of many linked and allelic bands. Bands are usually assumed to vary independently of each other, thus giving large exclusion probabilities in paternity analyses; however, this assumption is violated for some species.[53,61,62] Additionally, bands in different individuals which appear identical on inspection may actually be slightly different. Nonrandom breeding or past population contractions may also decrease the confidence in paternity estimates.[63,64] However, these problems may be alleviated by the use of single-locus probes and locus-specific primers.

Single-locus analysis presently requires considerable molecular biology expertise and capital investment. Moreover, whereas multilocus probes may be used in most eukaryotic taxa, single-locus probes and primers may need to be developed separately for each species or group of species. However, once developed, they can be routinely used with minimal expertise.

Single Copy DNA Analysis

One approach to generate markers for population genetic and paternity purposes involves the isolation and sequence analysis of single copy DNA of ambiguous function from genomic libraries.[65,66] Such libraries are simply an array of molecular cloning vectors with inserts of DNA fragments of various sizes which together represent the total genomic DNA of an individual. Clones containing single copy DNA are identified and their inserts used directly as radioactive probes or sequenced to develop PCR primers for amplification. Such probes have been used for DNA hybridization to detect restriction site differences among individuals or, if locus-specific PCR primers are developed, sequence difference among individuals can potentially be explored.

Random Amplified Polymorphic DNA (RAPD)

A recently developed, relatively easy, and inexpensive approach involves the use of randomly generated primers to PCR amplify genomic DNA.[67,68] In the Williams et al. RAPD technique,[67] a series of two identical, randomly generated, 10 base pair PCR primers are used. If complementary annealing site sequences to these primers exist that are less than a few thousand bases apart, the PCR reaction will generate the intervening sequence. Usually, several DNA fragments are generated. Mutations at the annealing sites result in the gain or loss of fragments in different individuals and, hence, the generation of markers that may be individual or population specific. Further, because the fragments can be easily generated by PCR and then immediately separated in small agarose or acrylamide gels and visualized by staining with ethidium bromide, data can be obtained in a few hours from a minute sample.

However, the absence of a fragment in two individuals does not indicate allelic identity because a change at any one of the 20 base pairs of the two priming sites may cause a fragment not to be generated. Furthermore, homozygotes may be difficult to distinguish from heterozygotes, as the latter will have one allele with both annealing sites and a second allele that may be one of several null alleles. Thus, while the technique may be useful in paternity analysis or population identification, it cannot easily be used as a population genetic tool. Finally, some difficulty has been reported in reproducing results from this approach. Due to limited published data, RAPD has not been included in the molecular tools matched to questions in Table 1.

EXAMPLES

The examples used in this section are ordered in a phylogenetic hierarchy from species-level systematics to individual characterization (Table 1) and are a partial list chosen to demonstrate a wide range of molecular techniques.

Phylogenetic Distinction

Karyological Analysis

Perhaps one of the most extensive studies of interspecific chromosome differences involved the comparison of diploid number, relative chromosome length, and centromeric indices among 140 species within 12 orders of birds.[69,70] Congeneric species were compared for the presence of unequal translocations, Robertsonian fissions and fusions, centric transpositions, and pericentric inversions. Of the 177 species comparisons, 78 revealed identical karyotypes, suggesting that most *closely related* bird species do not differ in gross chromosome structure. However, some species do have a diagnostic chromosomal morphology,[71] suggesting karyologic analysis may sometimes be important to document phylogenetic distinction.

Allozyme Analysis

A recent example concerns the tuatara (genus *Sphenodon*)[72] which exists on islands off the coast of New Zealand. This group represents the only extant genus of an entire reptilian order and until recently was regarded as one species, *S. punctatus*. Because of its phylogenetic and morphologic uniqueness, conservation of the tuatara is regarded as extremely important. However, the species has a complex taxonomic history, being previously described to include as many as three species, or only one species with three subspecies. Conservation management was based on the premise that there was only one species. In the late 1980s, a survey of allozyme and morphological variation in the tuatara identified three distinct groups. One taxon, which was described in the last century as *S. guntheri*, and exists on one island only, was found to have three fixed allele differences from *S. punctatus* supporting its full species status. A genetically divergent population of *S. punctatus* was also recognized as a separate subspecies. The decisive recognition of these three forms has profound implications for conservation management of the tuatara.

Mitochondrial DNA Analysis

Molecular analyses can also cast doubt on previously recognized species. For example, morphological studies on the red wolf *Canis rufus* had led to its classification as a distinct species. However, the red wolf was known to have hybridized extensively with the coyote before its extinction in the wild about 1980. A captive breeding program was initiated in 1974, and today over 150 red wolves exist in captivity, some having been reintroduced to localities on the southeast coast of the U.S.

MtDNA cytochrome *b* sequence was analyzed from blood samples of gray wolves, coyotes, captive red wolves, and wild red wolves caught immediately before the species vanished. Red wolf cytochrome *b* sequence was found not to be unique and was indistinguishable from coyotes or gray wolves. Cytochrome *b* sequence was also analyzed from DNA isolated from six red wolf museum pelts collected before 1930, a time period when hybridization with coyotes and gray wolves was thought to be less extensive. No red wolf specific genotype was found in the museum specimen DNA, only genotypes that were derived from hybridization with gray wolves and coyotes (Figure 1). These results suggest that the red wolf was entirely a hybrid form or was a distinct taxon that hybridized extensively over its past geographic range.

A similar use of mitochondrial DNA sequence analysis to assess hybridization involved the endangered Florida panther *Felis concolor coryi*.[73] A total of 30 to 50 wild pumas exist in two South Florida populations. Mitochondrial DNA analysis of Florida pumas showed them to be a genetic mix of individuals from the endemic subspecies and individuals originating from a local zoo stock of South American origin. Conservation plans for the Florida puma can therefore focus on saving the remaining pure Florida pumas and risk the increasingly apparent effects of inbreeding, or attempt to save the entire population although it is genetically heterogeneous and may not be protected by the Endangered Species Act.

Phylogeographic History

Mitochondrial DNA Analysis

A recent example of mtDNA being applied to phylogeographic questions concerns the humpback whale *Megaptera novaeangliae*[74] presently existing in two populations in the eastern Pacific and western Atlantic. Individuals in both populations migrate each year between summer feeding grounds in temperate waters and winter breeding grounds in shallow tropical seas. Observations of marked individuals suggest that Atlantic and Pacific populations are divided into distinct subpopulations which are not separated by obvious geographic barriers. A total of 80 individuals darted at feeding and wintering grounds in both oceans were analyzed. Restriction fragment length polymorphisms were identified that distinguished Pacific and Atlantic populations and subpopulations in each ocean. Thus, the mtDNA analysis defined the subpopulations as separate conservation units and provided a forensic tool to trace whale remains to the population of origin.

Genetic Variability

Allozyme Analysis

Perhaps the most well-known example of an allozyme electrophoresis analysis applied to an endangered species concerns the cheetah *Acinonyx jubatus*.[75,76]

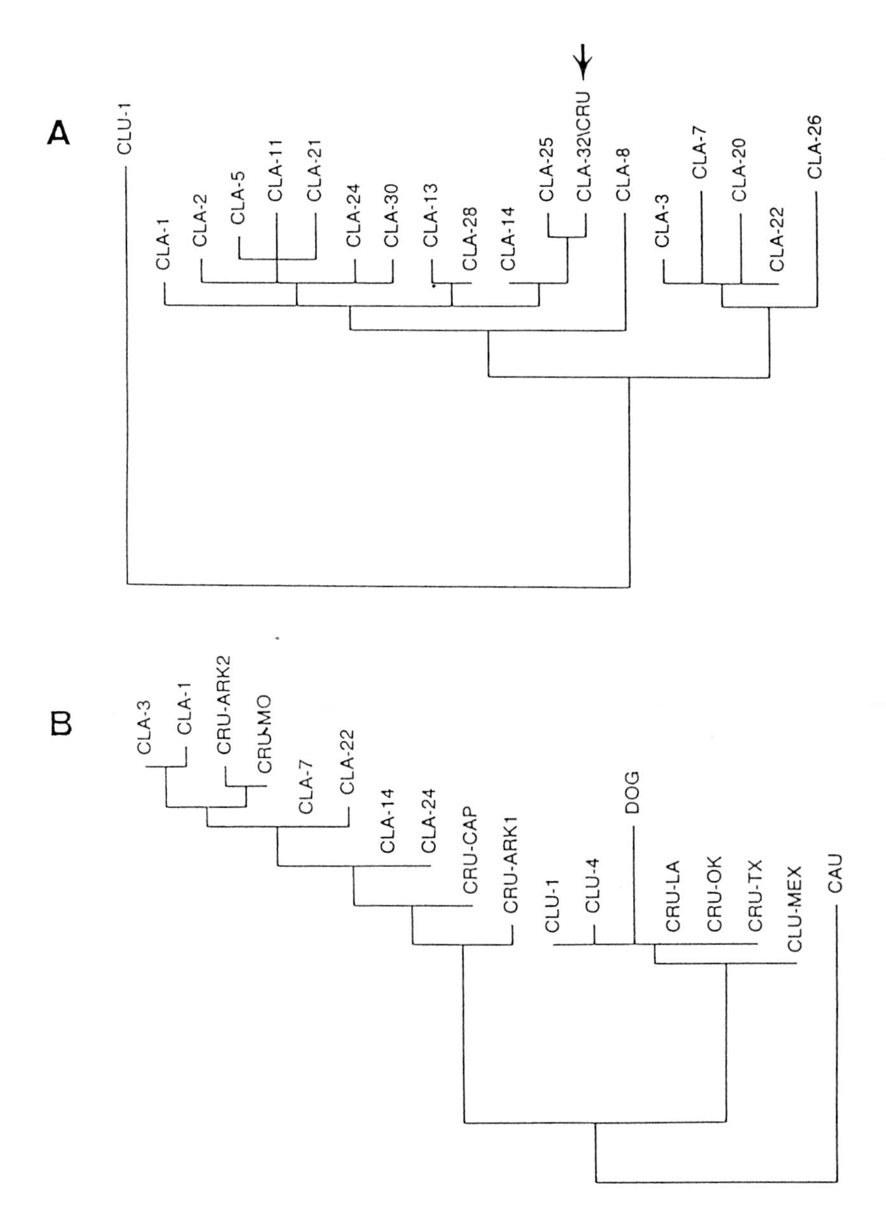

Figure 1. Phylogenetic trees of coyote (CLA), captive red wolf (CRU), grey wolf (CLU-1). Trees were based on sequence differences of 398bp mtDNA cytochrome *b* gene. MO = Missouri, ARK = Arkansas, LA = Louisiana, OK = Oklahoma, and TX = Texas. CLU-1 and CLU-4 are wolves from Minnesota and Alaska, respectively; CLU-MEX = Mexican wolf. The PAUP generated tree *b* was rooted by sequence data from the golden jackal (CAU).

These studies found that a sample of 55 cheetahs from two geographically isolated populations were monomorphic for each of 47 different allozyme loci. No other carnivore has yet been described with such high levels of homozygosity. Moreover, corroborating evidence of low genetic variability was provided by the acceptance of reciprocal skin grafts between unrelated individuals, a result common only among highly inbred animals.[76] The authors suggested that the low genetic variation results from two separate bottlenecks involving East and South African cheetahs.[76] The existence of low genetic variability in the cheetah may have important effects on their ability to survive disease epizootics.[76]

Mitochondrial DNA Sequence Analysis

Mitochondrial DNA variation has recently been characterized in 26 gray wolf populations worldwide.[77] The gray wolf is a highly mobile canid, and high rates of genetic exchange have probably existed in the past even among distantly separated populations. However, recent population declines and habitat fragmentation have isolated previously contiguous populations in the Old World and in western Europe. A single, unique genotype was found in each population. By contrast, in the larger, more continuously distributed populations of the New World, mtDNA genotypes were often shared among different populations. The genetic differences found among present-day Old World gray wolf populations is probably a recent phenomenon reflecting population declines and habitat fragmentation, rather than a long history of genetic isolation.

DNA Fingerprinting

One of the smallest wolf populations in the world inhabits Isle Royale in Lake Superior. This population has been intensively studied over the past 30 years, and recently has declined from 50 to as few as 10 individuals. DNA fingerprinting has indicated that a random sample of individuals from the island population has the same level of band similarity (approximately 70%) as siblings in captive populations and are likely to be inbred.[32] The combination of small population size and isolation from mainland populations has led to inbreeding and loss of genetic variation in the island population, and may be important factors in its recent decline.

Individuality, Reproductive Success, and Genetic Management

Small populations, especially those in captivity, often exhibit low levels of genetic variability. Therefore, genetic techniques that assay variation in highly polymorphic loci are needed for use on captive populations with high degrees of relatedness. DNA fingerprinting is therefore becoming a commonly used method for analyzing small captive populations.

The Mauritius pink pigeon is the worlds' rarest pigeon. Endemic to Mauritius, the wild population probably numbers less than 20. There are two major captive populations at Mauritius and at the Jersey Zoo, both numbering around 50

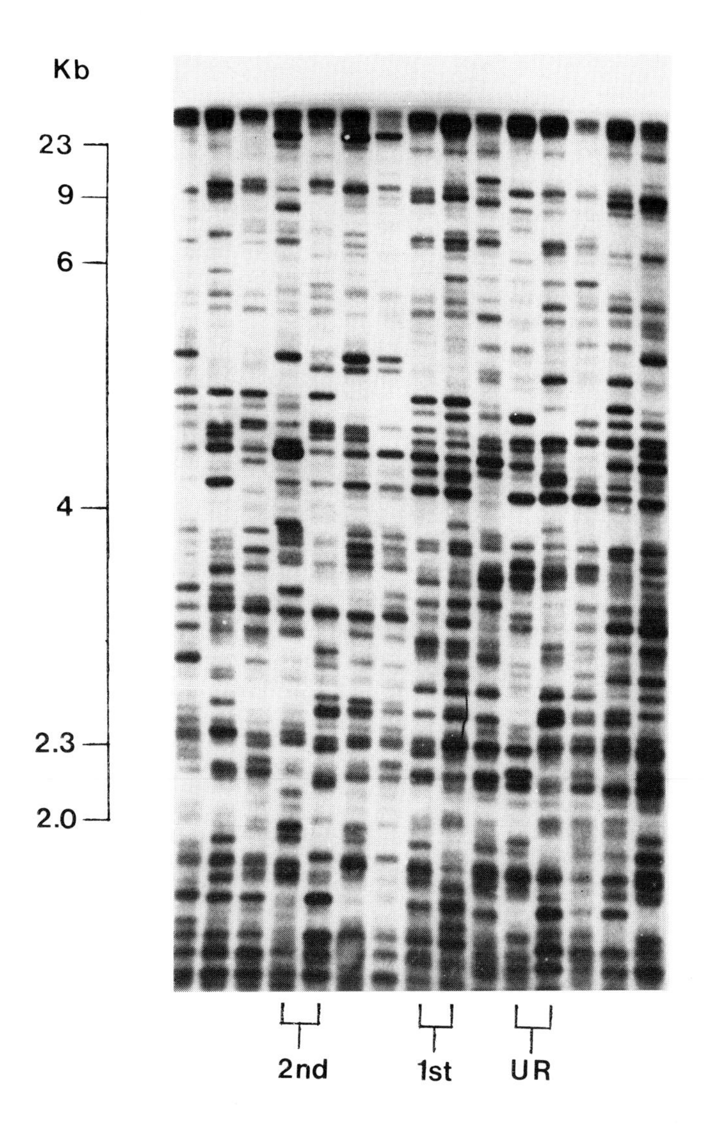

Figure 2. DNA fingerprint of pink pigeon individuals from the captive population at Jersey Zoo. Band sharing coefficients are compared for individuals of different levels of relatedness. 1st = first degree relative; 2nd = second degree relative; and UR = putatively unrelated.

individuals. DNA fingerprinting has been applied to the Jersey population and the level of genetic variation analyzed. The purpose of the analysis is to discriminate different levels of relatedness within the population so that individuals with relatively low levels of genetic similarity can be mated to preserve as much of the extant variation as possible.

Figure 2 shows an example of a fingerprint of individuals in the Jersey population. Band sharing levels can be calculated for first degree, second degree, and unrelated individuals. Any pair of individuals can then be compared and their level of relatedness deduced by their band sharing coefficient. In this example, we found that the band sharing level for each degree of relatedness did not overlap significantly, and therefore breeding among relatives and its related problems can reliably be avoided.

CONCLUSION

The past 10 years of conservation genetic research have shown molecular techniques to be important in the management of endangered species and populations. In the near future, the importance of microsatellite analysis as a population genetic tool is likely to increase. The increased use of PCR amplification will allow an analysis of historic levels of genetic variation in species formerly abundant that are now rare. Such data can be used to assess the effect that habitat fragmentation and reductions in population size have on levels of genetic variation. Analysis of historic material may also provide a reference point for conservation of genetic variation in present-day populations. There is real hope that our increased ability to measure genetic variation could result in both improved preservation of genetic variability and avoidance of inbreeding depression in captive and wild populations of rare and endangered species.

REFERENCES

1. **Saiki, R. K., Gelfand, D. H., Scharf, S., Higuchi, R., Horn, G. T., Mullis, K. B., and Erlich, H. A.,** Primer-directed enzymatic amplification of DNA with a thermostable DNA polymerase, *Science,* 239, 487, 1988.
2. **Kocher, T. D., Thomas, W. K., Meyer, A., Edwards, S.V., Pääbo, S., Villablanca, F. X., and Wilson, A. C.,** Dynamics of mitochondrial DNA evolution in mammals: amplification and sequencing with conserved primers, *Proc. Natl. Acad. Sci. U.S.A.,* 86, 6196, 1989.
3. **Irwin, D. M., Kocher, T. D., and Wilson, A. C.,** Evolution of the cytochrome *b* gene of mammals, *J. Mol. Evol.,* 32, 128, 1991.
4. **Thomas, W. K., Pääbo, S., Villablanca, F. X., and Wilson, A. C.,** Spatial and temporal continuity of kangaroo rat populations shown by sequencing mitochondrial DNA from museum specimens, *J. Mol. Evol.,* 31, 101, 1990.
5. **Jeffreys, A. J., Wilson, V., and Thein, S. L.,** Hypervariable minisatellite regions in human DNA, *Nature,* 314, 67, 1985.
6. **Jeffreys, A. J., Wilson, V., and Thein, S. L.,** Individual specific "fingerprints" of human DNA, *Nature,* 316, 76, 1985.
7. **Burke, T., and Bruford, M. W.,** DNA fingerprinting in birds, *Nature,* 327, 139, 1987.
8. **Burke, T., Davies, N. B., Bruford, M. W., and Hatchwell, B. J.,** Parental care and mating behaviour of polyandrous dunnocks *Prunella modularis* related to paternity by DNA fingerprinting, *Nature,* 338, 249, 1989.

9. **Gilbert, D. A., Lehman, N., O'Brien, S. J., and Wayne, R. K.,** Genetic fingerprinting reflects population differentiation in the Channel Island Fox, *Nature,* 344, 764, 1990.
10. **Kuhnlein, U., Zadworny, D., Dawe, Y., Fairfull, R.W., and Gavora, J.S.,** Assessment of inbreeding by DNA fingerprinting: development of a calibration curve using defined strains of chickens, *Genetics,* 125, 161, 1990.
11. **May, B.,** Starch gel electrophoresis of allozymes, in *Molecular Genetic Analysis of Populations: A Practical Approach,* Hoelzel, A. R., Ed., IRL Press, Oxford, 1992.
12. **Benirschke, K., and Kumamoto, T.,** Mammalian cytogenetics and conservation of species, *J. Heredity,* 82, 187, 1991.
13. **May, R.,** Taxonomy as destiny, *Nature,* 347, 129, 1990.
14. **Vane-Wright, R. I., Humphries, C. J., and Williams, P. H.,** What to protect — systematics and the agony of choice, *Biol. Conserv.,* 55, 235, 1991.
15. **Barton, N. H. and Hewitt, G. M.,** Analysis of hybrid zones, *Annu. Rev. Ecol. Syst.,* 16, 113, 1985.
16. **Barton, N. H., and Hewitt, G. M.,** Adaptation, speciation and hybrid zones, *Nature,* 141, 497, 1989.
17. **Lehman, N., Eisenhawer, A., Hansen, K., Mech, D. L., Peterson, R. O., Gogan, P. J. P., and Wayne, R. K.,** Introgression of coyote mitochondrial DNA into sympatric North American gray wolf populations, *Evolution,* 45, 104, 1991.
18. **Wayne, R. K., and Jenks, S. M.,** Mitochondrial DNA analysis supports a hybrid origin for the endangered red wolf *(Canis rufus), Nature,* 351, 565, 1991.
19. **O'Brien, S., Joslin, P., Smith, G. L., Wolfe, R., Heath, E., Otte-Joslin, J., Rawal, P. P., Bhattacherjee, K. K., and Martenson, J. S.,** Evidence for African origin of founders of the Asiatic lion species survival plan, *Zoo Biol.,* 6, 99, 1987.
20. **Avise, J. C., Arnold, J., Ball, R. M., Bermingham, E., Lamb, T., Neigel, J. E., Reeb, C. A., and Saunders, N. C.,** Intraspecific phylogeography: the mitochondrial DNA bridge between population genetics and systematics, *Annu. Rev. Ecol. Syst.,* 18, 489, 1987.
21. **Avise, J. C., and Ball, R. M.,** Principles of genealogical concordance in species concepts and biological taxonomy, in *Oxford Surveys in Evolutionary Biology,* Vol. 7, Futuyma, D., and Antonovics, J., Eds., Oxford Press, Oxford, U.K., 45, 1990.
22. **Avise, J. C., and Nelson, W. S.,** Molecular genetic relationships of the extinct dusky seaside sparrow, *Science,* 243, 646, 1989.
23. **Nevo, E.,** Genetic variation in natural populations: patterns and theory, *Theor. Popul. Biol.,* 13, 121, 1978.
24. **Burke, T., Hanotte, O., Bruford, M. W., and Cairns, E.,** Multilocus and single locus minisatellite analysis in population biological studies, in *DNA Fingerprinting: Approaches and Applications,* Birkhäuser, Verlag, AG, 154, 1991.
25. **Allendorf, F. W.,** Genetic drift and the loss of alleles versus heterozygosity, *Zoo Biol.,* 5, 181, 1986.
26. **Allendorf, F. W., and Leary, R. F.,** Heterozygosity and fitness in natural populations of animals, in *Conservation Biology: The Science of Scarcity and Diversity,* Soule, M. E., Ed., Sinauer Press, Sunderland, MA, 57, 1986.
27. **Ralls, K., and Ballou, J. D.,** Extinctions: lessons from zoos, in *Genetics and Conservation: A Reference Manual for Managing Wild Animal and Plant Populations,* Schonewald-Cox, C. M., Chambers, S. M., MacBryde, B., and Thomas, L., Eds., Benjamin/Cunnings, Menlo Park, CA, 164, 1983.

28. **Ralls, K., Ballou, J. D., and Templeton, A. R.,** Estimates of lethal equivalents and the costs of inbreeding in mammals, *Conserv. Biol.,* 2, 185, 1988.
29. **Quattro, J. M., and Vrijenhoek, C.,** Fitness differences among remnant populations of endangered sonoran topminnow, *Science,* 245, 976, 1989.
30. **O'Brien, S. J., Wildt, D. E., and Bush, M.,** The African cheetah in genetic peril, *Sci. Am.,* 254, 84, 1986.
31. **Packer, C., Gilbert, D. A., Pussey, A. E., and O'Brien, S. J.,** A molecular genetic analysis of kinship and cooperation in African lions, *Nature,* 351, 562, 1991.
32. **Wayne, R. K., Gilbert, D. A., Lehman, N., Hansen, K., Eisenhawer, A., Girman, D., Peterson, R. O., Mech, L. D., Gogan, P. J. P., Seal, U. S., and Krumenaker, R. J.,** Conservation genetics of the endangered Isle Royale gray wolf, *Conserv. Biol.,* 5, 41, 1991.
33. **Reeve, H. K., Westneat, D. F., Noon, W. A., Sherman, P. W., and Aquadro, C. F.,** DNA fingerprinting reveals high levels of inbreeding in colonies of the eusocial naked mole rat, *Proc. Natl. Acad. Sci. U.S.A.,* 87, 2496, 1990.
34. **Burke, T.,** DNA fingerprinting and other methods for the study of mating success, *Trend. Ecol. Evol.,* 4, 139, 1989.
35. **Sumner, A. T.,** *Chromosome Banding,* Unwin Hyman, London, U.K., 1990.
36. **Modi, W. S.,** Phylogenetic analysis of chromosomal banding patterns among the nearctic Arvicolidae (Mammalia, Rodentia), *Syst. Zool.,* 36, 109, 1987.
37. **Smith, M. W., Aquadro, C. F., Smith, M. H., Chesser, R. K., and Etges, W. J.,** *Bibliography of Electrophoretic Studies of Biochemical Variation in Natural Vertebrate Populations,* Texas Tech. Press, Lubbock, TX, 1982.
38. **Murphy, R. W., Sites, J. W., Buth, D. G., and Haufler, C. H.,** Proteins I: Isozyme Electrophoresis, in *Molecular Systematics,* Hillis, D. M., and Moritz, C., Eds., Sinauer, Sunderland, MA, 45, 1990.
39. **Buth, D. G.,** The application of electrophoretic data in systematic studies, *Annu. Rev. Ecol. Syst.,* 15, 501, 1984.
40. **Swofford, D. L., and Berlocher, S. H.,** Inferring evolutionary trees from gene frequency data under the principle of maximum parsimony, *Syst. Zool.,* 36, 292, 1987.
41. **Allendorf, F. W., Knudson, K. L., and Leary, R. F.,** Adaptive significance of the differences in the tissue-specific expression of a phosphoglucomutase gene in rainbow trout, *Proc. Natl. Acad. Sci. U.S.A.,* 80, 1397, 1983.
42. **Koehn, R. K., Diehl, W. J., and Scott, T. M.,** The differential contribution by individual enzymes of glycolysis and protein catabolism to the relationship between heterozygosity and growth rate in the coot clam, *Mulinia lateralis, Genetics,* 118, 121, 1988.
43. **Pemberton, J. M., Albon, S. D., Guinness, F. E., Clutton-Brock, T. H., and Berry, R. J.,** Genetic variation and juvenile survival in red deer, *Evolution,* 42, 921, 1988.
44. **Mopper, S., Mitton, J. B., Whitham, T. G., Cobb, N. S., and Christensen, K. M.,** Genetic differentiation and heterozygosity in pinyon pine is associated with resistence to herbivory and environmental stress, *Evolution,* 45, 989, 1991.
45. **Brown, W. M.,** The mitochondrial genome of animals, in *Molecular Evolutionary Biology,* MacIntrye, R., Ed., Cornell University Press, New York, 95, 1986.
46. **Shields, G. F., and Wilson, A. C.,** Calibration of mitochondrial DNA evolution in geese, *J. Mol. Evol.,* 24, 212, 1987.

47. **Wilson, A. C., Cann, R. L., Carr, S. M., George, M., Gyllensten, U. B., Helm-Bychowski, K. M., Higuchi, R. G., Palumbi, S. R., Prager, E. M., Sage, R. D., and Stoneking, M.,** Mitochondrial DNA and two perspectives on evolutionary genetics, *Biol. J. Linn. Soc.*, 26, 375, 1985.
48. **Pääbo, S.,** Ancient DNA: Extraction, characterization, molecular cloning, and enzymatic amplification, *Proc. Natl. Acad. Sci. U.S.A.*, 86, 1939, 1989.
49. **Hagelberg, E., Gray, I. C., and Jeffreys, A. J.,** Identification of the skeletal remains of a murder victim by DNA analysis, *Nature*, 352, 427, 1991.
50. **Ellegren, H.,** DNA typing of museum birds, *Nature*, 354, 113, 1991.
51. **Pamilo, P., and Nei, M.,** Relationships between gene trees and species trees, *Mol. Biol. Evol.*, 5, 568, 1988.
52. **Wayne, R. K., Meyer, A., Lehman, N., Van Valkenburgh, B., Kat, P. W., Fuller, T. K., Girman, D., and O'Brien, S. J.,** Large sequence divergence among mitochondrial DNA genotypes within populations of East African black-backed jackals, *Proc. Natl. Acad. Sci. U.S.A.*, 87, 1772, 1990.
53. **Bruford, M. W., Hanotte, O., Brookfield, J. F. Y., and Burke, T.,** Single locus and multi-locus DNA fingerprinting, in *Molecular Genetic Analysis of Populations: A Practical Approach*, Hoelzel, A. R., Ed., IRL Press, Oxford, 1992.
54. **Westneat, D. F.,** Genetic parentage in the indigo bunting: a study using DNA fingerprinting, *Behav. Ecol. Sociobiol.*, 27, 67, 1990.
55. **Nakamura, Y., Leppert, M., O'Connell, P., Wolff, R., Holm, T., Culver, M., Martin, C., Fujimoto, E., Hoff, M., Kumin, E., and White, R.,** Variable number of tandem repeat (VNTR) markers for human gene mapping, *Science*, 235, 1616, 1987.
56. **Vassart, G., Georges, M., Monsieur, R., Brocas, H., Lequarre, A. S., and Christophe, D.,** A sequence in M13 phage detects hypervariable minisatellites in humans and animal DNA, *Science*, 235, 683, 1987.
57. **Tautz, D.,** Hypervariability of simple sequences as a general source for polymorphic DNA, *Nucleic Acids Res.*, 17, 6463, 1989.
58. **Litt, M., and Luty, J. A.,** A hypervariable microsatellite revealed by *in vitro* amplification of dinucleotide repeat within the cardiac muscle actin gene, *Am. J. Hum. Genet.*, 44, 397, 1989.
59. **Rassmann, K., Schlötterer, C., and Tautz, D.,** Isolation of simple-sequence loci for use in polymerase chain reaction-based DNA fingerprinting, *Electrophoresis*, 12, 113, 1991.
60. **Schlötterer, C., Amos, B., and Tautz, D.,** Conservation of polymorphic simple sequence loci in cetacean species, *Nature*, 354, 63, 1991.
61. **Hanotte, O., Bruford, M. W., and Burke, T.,** Multilocus DNA fingerprints in gallinaceous birds: general approach and problems, *Heredity*, 68, 481, 1992.
62. **Brock, M. K., and White, B. N.,** Multifragment alleles in DNA fingerprints of the parrot, *Amazona vantralis*, *J. Hered.*, 82, 209, 1991.
63. **Lynch, M.,** Estimation of relatedness by DNA fingerprinting, *Mol. Biol. Evol.*, 5, 584, 1988.
64. **Lynch, M.,** The similarity index and DNA fingerprinting, *Mol. Biol. Evol.*, 7, 478, 1990.
65. **Quinn, T. W., Quinn, J. S., Cooke, F., and White, B.N.,** DNA marker analysis detects maternity and paternity in single broods of the lesser snow geese, *Nature*, 362, 392, 1987.

66. **Quinn, T. W., and White, B. N.,** Identification of restriction fragment length polymorphisms in genomic DNA of the lesser snow geese *(Anser caerulescens), Mol. Biol. Evol.,* 4, 126, 1987.
67. **Williams, J. G. K., Kubelik, A. R., Livak, K. J., Rafalski, J. A., and Tingey, S. V.,** DNA polymorphisms amplified by arbitrary primers are useful as genetic markers, *Nucleic Acids Res.,* 18, 6531, 1990.
68. **Welsh, J., and McClelland, M.,** Fingerprinting genomes using PCR with arbitrary primers, *Nucleic Acids Res.,* 18, 7213, 1990.
69. **Shields, G. F.,** Comparative avian cytogenetics: a review, *Condor,* 84, 45, 1982.
70. **Shields, G. F.,** Chromosomal Variation, in *Avian Genetics,* Cooke, F., and Buckley, P. A., Eds., Academic Press, London, chap. 3, 1987.
71. **de Boer, L. E. M.,** Karyological problems in breeding owl monkeys *Aotus trivirgatus* in captivity, *Int. Zoo. Yb.,* 22, 119, 1982.
72. **Daugherty, C. H., Cree, A., Hay, J. M., and Thompson, M. B.,** Neglected taxonomy and continuing extinctions of tuatara *(Spenodon), Nature,* 347, 177, 1990.
73. **O'Brien, S. J., Rolke, M. E., Yuhki, N., Richards, K. W., Johnson, W. E., Franklin, W. L., Anderson, A. E., Bass, O. L., Belden, R. C., and Martenson, J. S.,** Genetic introgression within the Florida panther *Felis concolor coryi, Natl. Geo. Res.,* 6, 485, 1990.
74. **Baker, C. S., Palumbi, S. R., Lambertson, R. H., Weinrich, M. T., Calombokodis, J., and O'Brien, S. J.,** Influence of seasonal migration on geographic distribution of mitochondrial DNA haplotypes in humpback whales, *Nature,* 344, 238, 1990.
75. **O'Brien, S. J., Wildt, D. E., Goldman, D., Merril, C. R., and Bush, M.,** The cheetah is depauperate in genetic variation, *Science,* 221, 459, 1983.
76. **O'Brien, S. J., Roelke, M. E., Marker, L., Newman, A., Winkler, C. A., Meltzer, D., Colly, L., Evermann, J. F., Bush, M., and Wildt, D. E.,** Genetic basis for species vulnerability in the cheetah, *Science,* 227, 1428, 1985.
77. **Wayne, R. K., Lehman, N., Allard, M. W., and Honeycutt, R. L.,** Mitochondrial DNA variability of the gray wolf: genetic consequences of population decline and habitat fragmentation, *Conserv. Biol.,* in press.

3

Use of DNA Analyses in the Management of Natural Fish Populations

John R. Waldman[1] and Isaac Wirgin[2]

[1]Hudson River Foundation, New York
and
[2]Nelson Institute of Environmental Medicine, New York University Medical Center, New York

INTRODUCTION

Management strategies for the conservation and enhancement of natural fish populations frequently require the use of markers which can be used to identify the representatives of individual stocks. To distinguish the population of origin of fishes or explore their relationships, diagnostic features must be identified. Any discriminatory feature, be it morphological or genetic, can be used in this context. Ideally, these diagnostic features are unique to individuals of each population; if so, then identification of the population of origin of individuals is unambiguous. In practice, the counts or measurements of features that distinguish populations are rarely statistically discrete, so that the classification of individuals to populations is probabilistic rather than definitive.

0-87371-631-0/94/$0.00+$.50

A complication to probabilistic classifications is that many of the features used to characterize populations for these purposes to date have been ecophenotypic, i.e., the product of interplay between genetically regulated ontogeny and the modifying effects of environment. Phenotypic features such as meristic counts and body or scale morphometry may be influenced by a suite of environmental variables including temperature, salinity, pH, and food availability. Because ecophenotypic variability is introduced on an annual basis with the development of each new cohort, the ability of previously collected data to adequately discriminate among populations may be compromised as new cohorts arise.

DNA-based markers provide two major advantages over phenotypic features as tools with which to discriminate fish populations. One is that DNA sequences are heritable and are free from environmental influences (in the short term) and thus, they offer long-term stability without the need for frequent recalibration. Second, different DNA sequences evolve across a wide range of rates. Therefore, DNA sequence analysis offers the potential to quantify genetic relationships at differing levels of taxonomic divergence, from the individual to the interspecific level.

Initially, genetic studies on fishes focused on the protein level; electrophoretic techniques were used to identify variants (allozymes) at enzyme coding loci (isozymes), and their frequencies were quantified in conspecific and higher taxonomic comparisons. Unfortunately, in many cases this approach suffered from a lack of sufficient levels of detectable protein variability to discriminate taxa due to the inherent limited sensitivity of the technique and its focus on gene products which evolve slowly. Conventional electrophoretic techniques only detect about one third of all amino acid substitutions in protein molecules, and the products of isozyme loci evolve slowly due to their critical functions in cellular activities.

By the 1970s, techniques were developed to isolate purified DNA, visualize selected DNA fragments by radiolabeling or fluorescent techniques, and reproducibly cleave DNA molecules at selected, short, 4 to 6 base pair sequences by the use of restriction endonucleases. Restriction endonuclease digestion of DNA provided a rapid and highly reproducible method to screen any DNAs for differences in their base sequences at randomly distributed recognition sites. The number and size of resulting DNA fragments could be determined by electrophoretic analysis. Probes were developed that allowed for the visualization of distinct DNA segments by their hybridization to complementary sequences in the target DNA. These probes could be used to visualize DNA sequences at coding genes or any region of interest in the nuclear or mitochondrial genomes. The advent of the polymerase chain reaction (PCR) allowed for the direct and rapid determination of sequences of target DNA and placed this approach within the realm of fisheries biology. With these tools, the stage was set to characterize DNA sequences in individuals and compare the frequencies of variant genotypes among individuals, populations, species, and higher taxonomic comparisons.

The range of applications of DNA analysis in fisheries management is extensive and growing. The following is a categorization of fisheries problems that are amenable to DNA-based approaches:

1. Stock Identification: Species are natually divided into populations. For fisheries management purposes, populations are equivalent to unit stocks — a term in fisheries biology that defines a component of a species that can be dealt with as a single management unit. However, a unit stock may incorporate more than one population. Stock identification is the classification of individuals to unit stocks. Individual stocks often exhibit different life history characteristics and natural or fishing mortality rates. As a result of differential biological factors and exploitation rates, individual stocks may require differing levels of conservation.
2. Relative Contribution Estimation: Estimation of the proportions of individuals contributed from discrete stocks to a mixture of stocks. Estimates of the relative contributions of individual stocks to the mixed fisheries may be needed to evaluate the impact of these fisheries on individual stocks.
3. Maximization of Genetic Diversity and Maintenance of Genetic Integrity: For some fish species, hatchery programs have been instituted to augment natural reproduction and restore adult population size. Maintenance of genetic diversity at a level up to that which occurred in undisturbed populations is a goal of fisheries management. However, genetic diversity only describes the range of genetic variation; such diversity should be centered on the presumably best-adapted native genetic composition.
4. Quantification of Levels of Genetic Variability in Stocks: Evaluation of the levels of genetic variability in natural populations can be used to determine the effects of depressed variability on population characteristics.
5. Discrimination of Hatchery-Reared and Wild Fish: Hatchery-reared fish must be identifiable in order to evaluate the relative success of hatchery-reared and wild-produced fish.
6. Identification of Hybrids: Introgression between species may be harmful to the gene pools of populations of one or both species involved. Morphological identification of hybrids is dependent on their demonstrating intermediacy in physical features, which may not always occur, particularly in later generation hybrids.
7. Performance Evaluations: Different populations, subspecies, hybrids, or sexually reciprocal versions of the same cross may exhibit different levels of success in a particular environment. Direct, empirical comparison may be made by co-stocking fish of two or more types and then identifying them at a later time to estimate relative abundance, growth, or other parameters of interest.
8. Gender Identification: The gender of individuals is not always apparent by external examination. Nonfatal determination of gender may be particularly important in the selection of hatchery broodstock and in descriptive studies of life stage abundance in individual unit stocks.
9. Forensic Analysis: Means are sometimes required for the species identification of processed fish specimens, and, less often, for the population identification of unprocessed specimens. This information may be needed to protect consumers or threatened species or populations.
10. Restoration of Extinct Populations: Characterization of genotypes in archived samples representative of extinct populations can permit the selection of genetically similar fish from extant populations to be used in restoration efforts.

MITOCHONDRIAL DNA

The Mitochondrial DNA Molecule

Mitochondrial DNA (mtDNA) is a double-stranded, circular molecule of approximately 17,000 base pairs (range 14,000 to 26,000) found in the cytoplasm of all cells. Mitochondrial DNA is transmitted independently of the DNA found in the nucleus (nDNA) and is believed maternally inherited, although evidence of paternal leakage has been reported in nonfish taxa.[1-3] In the absence of recombination, comparisons of mtDNA genotypes permits the easy and unequivocal identification of the maternal ancestry of individuals.

The major advantage of mtDNA analysis over protein studies is the higher level of resolution it provides in examining relationships among closely related taxa. In mammals, it has been demonstrated that mtDNA evolves approximately an order of magnitude more rapidly than single copy, coding nuclear DNA.[4] However, recent studies suggest that the rate of mtDNA evolution may not be constant across all taxa, and that it may be positively correlated with metabolic rates. It has been shown that mtDNA change in sharks is 5 to 10 times lower than in mammals of similar body size.[5]

Analysis of mtDNA may reveal two forms of interindividual variation: base substitutions (restriction site polymorphisms) and length variants. Restriction site polymorphisms result from mutations in which one base pair of nucleotides is substituted for another within the DNA sequence. Length variants are mtDNA molecules of different total length, these differences resulting from tandem duplications of short 80 to 140 base pair sequences, usually mapped to the D-loop region.[6] Base substitutions predominate across the vertebrates analyzed to date, but certain fish taxa (e.g., American shad *Alosa sapidissima* and striped bass *Morone saxatilis*) display a pronounced tendency toward length polymorphisms. Different mechanisms have been proposed for the generation of length polymorphisms;[6,7] but no matter what the cause, it is clear that in some fishes length polymorphisms are more readily formed than base substitutions and, hence, may be unreliable genetic markers for evolutionary studies. However, on a time scale germane to fisheries — that of decades — length polymorphisms may serve as informative population markers given their demonstrated inheritance, and their temporal stability, as observed in striped bass.[8,9]

Another type of variation in mtDNA that may occur when length polymorphisms exist within a species is heteroplasmy — multiple forms of mtDNA within an individual. Most heteroplasmy found to date in fishes has been length heteroplasmy, but site heteroplasmy has been found at low frequencies in American shad[10] and red drum *Sciaenops ocellatus*.[11] Length heteroplasmy has been observed in a number of invertebrate and vertebrate taxa; fishes showing heteroplasmy bridge much of their phylogenetic spectrum and include white sturgeon *Acipenser transmontanus*,[6,13] bowfin *Amia calva*,[12] American shad,[14] and striped bass.[8] The frequency of heteroplasmic (length variant) individuals in populations that display heteroplasmy can range widely. For example, in two surveys of anadromous fishes, 30 of 244 American shad from 14 rivers along the

Atlantic coast of North America[10] and 55 of 102 white sturgeon from the Fraser River, British Columbia, were found to be heteroplasmic.[13] An extreme case is that of white perch *Morone americana* in Chesapeake Bay. Of 254 individuals analyzed, all were heteroplasmic for at least three, and in some instances 10, different lengthed mtDNA molecules.[15]

The incidence of heteroplasmy may vary significantly among populations[10,13,16] and, therefore, these differences have the potential to serve as population markers. However, in practice, differences in the frequencies of heteroplasmy are not reliable markers because of the high potential for false negatives to occur through uneven visualization of DNA bands among samples. This is because the secondary bands vary in copy number, but to be visualized they must be numerous enough to exceed some threshold of detection, estimated at 5% of the majority fragment.[17] The problem stems from the fact that the threshold of detection is subject to many procedural vagaries, including the purity of the DNA preparation, the ratios of reagents in the restriction endonuclease digestion process, the efficacy of end-labeling or southern blotting, the timespan of autoradiography, and any other aspect of the protocol that affects the visualization of the fragments. As a result of these factors, it is likely that the frequencies of heteroplasmy have been underestimated in many studies conducted to date.

Advantages and Disadvantages of Mitochondrial DNA Analysis

Nucleotide substitution in the mtDNA genome generally occurs at a rate of 5 to 10 times that of the coding nuclear genome. This rapid mutation rate means that mtDNA has the potential to provide considerably more information on genetic divergence over brief evolutionary time scales than alternative approaches. The accelerated mutation rate of mtDNA may be particularly valuable for problems in fishery biology in which historical or ecological factors have not permitted a divergence level that crosses the threshold of generating a significant degree of allozyme variation.[8,18]

Genetic divergence in the mtDNA genome may occur more rapidly than in coding sequences of the nuclear genome because its effective population size is one fourth that of the bisexually inherited nuclear genome and, hence, is more sensitive to erosive population bottlenecks.[19] Mitochondrial DNA lineages are also self-pruning; i.e., lineage extinctions occur in much the same random fashion as male surnames die out when male offspring are not born in paternal surname societies. This stochastic lineage cropping occurs at a counterintuitively rapid rate which continually truncates the spectrum of times to common mtDNA ancestry. For example, Avise and colleagues[20,21] calculated that if adult females within a population produce daughters according to a Poisson distribution and if mothers on average leave one surviving daughter, about 37% of the maternal lineages will by chance go extinct in the first generation. After 100 generations, less than 2% of the original mothers will have contributed mtDNA molecules to the population.

A major drawback to the exclusive use of mtDNA analysis is the fact that the mitochondrial genome represents only a small fraction of the total genetic information within the organism and its transmission may be independent of that of the nuclear genome. For example, studies have demonstrated the introduction of the mtDNA of one species into the nuclear DNA background of another.[22] It is likely that in some fish species, sexual differences in homing fidelity are pronounced. Because of its maternal transmission, analysis of mtDNA provides no information concerning the extent of population mixing resulting exclusively from male vagility.

Methodology Overview

The high copy number of mtDNA makes it relatively easy to purify. We have obtained mtDNA from many fish tissues, including blood, heart, brain, sperm, muscle, liver, and ovary, alone or in combinations, but have achieved the best results from the mitochondria-rich liver and ovaries. However, in a study of orange roughy *Hoplostethus atlanticus*,[23] it was found that autoradiographs were more difficult to interpret if the mtDNA was derived from liver tissue rather than ovarian tissue, which the authors hypothesized was due to degradation of nucleic acids within the metabolically active liver prior to dissection. Taggart and colleagues[24] noted that skeletal muscle gave better results than liver tissue when both were frozen for more than a year.

The isolation and purification of mtDNA has become increasingly rapid as the foundation of the protocols has shifted from physical (ultracentrifugation) to chemical means. Several different methods currently exist to compare mitochondrial DNA genotypes among individuals. These techniques differ in the means by which the base sequences of the mtDNA are determined, the purity of the mtDNA to be analyzed, and in the techniques by which the mtDNA is visualized. Initial studies essentially relied on the protocol outlined by Lansmann et al.,[25] in which densitiy gradient ultracentrifugation is used to isolate highly purified mtDNA. The mtDNA is then subjected individually to digestion with a battery of different restriction endonucleases (restriction frequent length polymorphism [RFLP] analysis), the resulting mtDNA fragments are separated electrophoretically, and composite mtDNA genotypes are compiled for each individual based on the mtDNA fragment pattern for each restriction enzyme. Mitochondrial DNA fragments are visualized by radiolabeling fragments with [^{32}P] or by staining the gels with ethidium bromide. The success of this approach is predicated on the quantity and quality of the mtDNA preparation. Obvious disadvantages of this approach include the cost and time involved in the mtDNA purification, the relatively large amounts of fresh tissue needed to produce sufficient quantities of DNA to execute the analysis, and that only fresh tissue could be used. This had the effect of strongly limiting total sample sizes as the number of fish collected could not exceed the laboratory's capacity to process them *en masse,* inasmuch as additional tissue samples could not be frozen. Consequently, field collectors could not capitalize on concentrations of fish and sample sizes became limited by the vagaries of fish availability. Modifications to this approach have been developed in which sufficient quantities of mtDNA can be purified very rapidly without ultracentrifugations from very small amounts of fresh tissue but which still permit highly sensitive radiolabeling of the digested mtDNA fragments.[26]

An alternative approach has been successfully applied which circumvents the need to produce purified mtDNA and therefore the need for fresh tissue. Southern blot hybridizations have been used to visualize restriction enzyme-digested mtDNA fragments from total DNA preparations. Cloned fish mtDNA[27] or the highly purified mtDNA isolated from a single conspecific fish[8] have been used as probe DNA. The enhanced sensitivity of this approach allows for the detection of very small amounts of mtDNA and therefore permits the analysis of DNA from small fish. Although all Southern blot studies in fishes have used radiolabeling of probes, chemiluminescent protocols are available which permit nonradioactive, but nevertheless sensitive labeling of probe DNAs. The major disadvantage of Southern blot analysis in the context of population discrimination studies is its inherent inability to resolve potentially informative small mtDNA fragments (<500 bp). Thus, the number of mtDNA fragments which can be scored from single restriction enzyme digests is reduced in comparison to end-labeling procedures.

Several rapid preparation methods are currently available and many work well on either fresh[28] or frozen tissue.[26] Mitochondrial DNA, or total DNA suitable for mtDNA analysis, can now be easily prepared within a day and costly ultracentrifuges are not required. Additionally, approaches have been developed that allow for nonlethal sampling of fish using blood[29] or epithelial tissues.[30] It was demonstrated that sufficient mtDNA for use with several informative restriction enzymes could be procured by stripping 20 to 25 mature unfertilized eggs from living brook charr *Salvelinus fontinalis*.[31] We have prepared satisfactory mtDNA from barbels removed from individuals of the Gulf of Mexico form of *Acipenser oxyrhynchus* that were subsequently released. However, prompt preservation of nonlethal tissue samples is even more important than for whole fish because their small volume allows for rapid temperature increase and hastened degradation.

The choice of how many restriction enzymes to use and whether to center on six-base cutter or four-base cutter enzymes depends on the anticipated extent of mtDNA variation present and the level of genetic resolution desired. Assuming no extraneous information exists from which to infer gross genetic diversity, a pilot study of subsamples from all geographic locations of interest may identify informative enzymes and will provide an opportunity to determine which tissue is the most suitable.[32] A means was provided to estimate the number of four-base cutter or six-base cutter restriction enzymes required to reveal at least one variant nucleotide in each genome sampled, given a certain level of sequence divergence.[32] Typically, effort must be applied to finding sufficient informative restriction enzymes, and enzymes with four-nucleotide restriction sites are often the answer; however, if four-base cutter enzymes are used when diversity is high, a counterproductive excess of information may result in which every individual has a different genotype.

A PCR-based method to analyze RFLP patterns in selected mtDNA regions has been proposed.[30] In this case, short stretches of mitochondrial DNA are PCR amplifed through the use of universal PCR primers, and the resulting selected amplified mtDNA is digested with a battery of restriction enzymes. The resulting mtDNA fragments are visualized by ethidium bromide staining and composite genotypes are compiled. This approach provides the ability to analyze mtDNA genotypes from exceedingly small amounts of noninvasively secured tissues such as individual scales. The drawback to this technique is that the relatively short length of the amplified mtDNA will not contain a large number of restriction sites and, therefore, few informative characters are generated. Additionally, since this approach is reliant on the use of highly conserved universal primers, mtDNA sequences which are analyzed may not be rapidly evolving and thus not particulary informative in comparisons among closely related taxa.[33] This approach has been used to determine that armorhead *Pseudopentaceros wheeleri* are not associated as distinct stocks with individual seamounts in the North Pacific Ocean.[34]

An alternative method to study variation is by direct sequencing of mtDNA. Until recently this would have required cloning of the mtDNA sequence of interest, an exceedingly time-consuming endeavor. Today, selected mtDNA sequences of interest, usually the cytochrome *b* (cyt *b*) gene, cytochrome oxidase I, or the control region may be amplified by PCR — thereby providing sufficient number of copies of purified mtDNA to allow for direct sequencing. These three sequences are usually selected for analysis because highly conserved flanking regions that may serve as primers for the amplification procedure have been described across many taxa. Simplified techniques have been developed that allow for the routine direct sequencing of the purified PCR products. Thus, it is possible to rapidly process and analyze a relatively large number of samples using this approach. It should be realized, however, that only intervening mtDNA sequences are analyzed, and thus only a small segment of the mtDNA molecule is screened for polymorphisms as opposed to the entire mtDNA genome via coventional RFLP analysis.

Direct sequencing studies of fish mtDNA have generally detected a slightly greater level of variation than seen in RFLP analysis of the same taxa, but increased sensitivity in discriminating among geographically separate populations has yet to be demonstrated. For example, a moderate level of variation in the cyt *b* sequence was found among Atlantic cod *Gadus morhua* collected from several locations in the western North Atlantic.[35] More than three quarters of the fish exhibited a common genotype, whereas none of the 23 variant genotypes was found in more than 3% of the samples. Similarly, the frequencies of cyt *b* sequence variants were analyzed in both European and North American Atlantic salmon *Salmo salar;* the genetic distinctiveness of these stocks was demonstrated, but these workers were unable to unequivocally determine the continent of origin of all individual fish.[33] Thus, the sensitivity of this approach in mixed stock analysis was no greater than that seen in an mtDNA RFLP study addressing the same question.[36] Finally, it was found that direct sequence analysis of cyt *b* provided less resolution in identifying the genetic structure of armorhead populations in the North Pacific Ocean than RFLP analysis of the control region and rRNA genes in mtDNA.[34]

Given the limited number of studies conducted to date, it is not yet clear if direct sequencing of PCR-amplified mtDNA sequences will prove more sensitive in detecting population-specific variation than observed in conventional RFLP analysis. Given the current state of the technology, the major advantage of sequencing amplified mtDNA regions may be its reliance on PCR, resulting in the ability to analyze very small amounts of partially degraded DNA from archived samples. However, after diagnostic variants have been identified by direct sequencing analysis, it will be possible to use selected restriction enzymes to digest PCR-amplified mtDNA sequences containing these informative restriction sites and determine the frequencies of mtDNA variants without the complexities and expenses of DNA sequencing.

Freshwater Fishes

mtDNA Diversity

Freshwater fishes frequently demonstrate substantial mtDNA divergence among populations, which can often be attributed to mtDNA evolution having occurred in the relative isolation in which many freshwater fish populations persist. However, the inopportunity for mixing and consequent opportunity for genetic divergence which characterize many lacustrine populations has been mediated largely by the historical biogeography of those populations. For example, in North America freshwater fishes have occurred continuously in the rivers of the southeastern coastal plain since at least the late Oligocene (25 mya).[37] As a result, freshwater fishes found there show considerable intraspecific mtDNA diversity (approximately 1.0 to 10% sequence divergence);[19] and may also show more divergent genotypes than seen in northern populations.[38] Mitochondrial DNA diversity of centrarchids in particular has been well surveyed and results have shown extensive mtDNA polymorphism; but even these older populations showed congruent phylogenetic breaks among regions, which reinforces the importance of historical factors in steering how mtDNA diversity becomes subdivided geographically.[39,40]

In contrast, populations of north temperate and boreal freshwater fishes in North America are well fragmented across the many lakes formed as a result of glaciation, but the vast majority of these populations have only occurred for 15,000 years or less following post-Pleistocene deglaciation and subsequent colonization. Consequently, little time has elapsed for mtDNA evolution to occur *in situ* among these populations and much of their current mtDNA diversity (approximately 0.2 to 2.0% sequence divergence[19]) is the result of mtDNA differentiation that occurred before or during the periods in which species were restricted to glacial refugia.[41-43] Secondary contact of populations from two or more refugia may result in increased mtDNA diversity within a drainage basin;[44] Billington and Hebert[45] believed that a clinal change in mtDNA genotype frequencies of walleye *Stizostedion vitreum* from Lake Michigan to Lake Ontario was due to recolonization from both Mississippi and Atlantic refugia. Contributions from three refugia were postulated to account for mtDNA variation in lake trout

Salvelinus namaycush occurring in the Great Lakes.[46] In contrast, seven populations of arctic charr *Salvelinus alpinus* in Scotland were found to show no site substitutions (and only a single length polymorphism in three of the populations) when surveyed with 19 restriction enzymes; this extreme mtDNA homogeneity was attributed to the post-Pleistocene founding of these populations from a single, small population.[47] Likewise, four strongly differentiated morphological variants that may represent incipient species of arctic charr within a volcanic lake in Iceland had a maximum sequence divergence of less than 0.2%.[48]

Applications

One application of mtDNA analysis to freshwater fisheries is the restoration of native stocks by selective breeding in situations where a nonnative stock has been introduced. Striped bass along the Gulf of Mexico are riverine and rarely stray from freshwater. Most Gulf coast populations are believed to have become extinct during mid-century, leaving one remnant population in the Apalachicola River. Considerable numbers of Atlantic coast striped bass were stocked in the Apalachicola River in the 1960s, but from identifications based on differences in lateral-line scale counts, it was concluded that the Atlantic coast fish and their progeny were not as well adapted to the Apalachicola River as were native striped bass.[49]

In order to restore the striped bass population of the Apalachicola River through hatchery augmentation with native fish, Wirgin and colleagues[50] surveyed mtDNA from Apalachicola River striped bass to search for markers that might distinguish the endemic form. Two restriction enzymes revealed base substitutions not seen in extensive surveys of Atlantic coast striped bass; approximately 57% of striped bass from the Apalachicola River displayed an *Xba* I genotype not observed in Atlantic coast fish. These markers, together with nuclear DNA markers, have been used to select native striped bass for hatchery breeding. At this time, all Apalachicola River striped bass broodstock are screened for genotypes at informative mtDNA and nDNA markers prior to release of their progeny in the wild.

The analysis of mitochondrial DNA also has the potential to reveal rare markers that would be useful in identifying hatchery-produced fish. For example, most walleye in the Great Lakes occur as one of two major mitochondrial DNA clones, but many rare genotypes also exist.[45] It would be relatively simple to spawn female broodstock from one of the rare genotypes so that all of their offspring would be genetically marked, thereby allowing straightforward monitoring of their survival and long-term reproductive success. Similar potential was noted for lake trout[46] and brown trout *Salmo trutta*.[51] Such an approach has major advantages over allozyme markers inasmuch as genotypic frequencies return to Hardy-Weinberg equilibrium after a single generation.

RFLP analysis of mtDNA has been demonstrated to be sufficiently sensitive to discriminate among selected hatchery strains and wild brook charr populations in Ontario.[52] These workers found fixed or highly significant frequency differences between two mtDNA clonal lineages of domesticated brook charr from two Ontario hatcheries and wild populations of brook charr in two drainages of Lake Erie. The authors concluded that it should be possible to ascertain the competitive success of hatchery and wild brook charr in this region of Ontario with a high degree of accuracy. It remains to be seen if the same levels of differentiation will be seen among other populations and broodstocks of brook charr, although preliminary evidence suggests a lack of mtDNA divergence among feral populations in Newfoundland.[53] In comparison, a fixed difference in mtDNA genotypes was seen between Ohio River and Atlantic slope drainage-collected brook charr in western Maryland.[31]

Mitochondrial DNA analysis has also been applied to the identification of hybrids; hybridization among fishes is most common in freshwaters, possibly due to the relative recency of many freshwater habitats and the consequent incomplete development of isolating mechanisms.[54] Recently, numerous instances of natural hybridization between brown trout and Atlantic salmon were observed in Canada and in various rivers from Sweden to Spain. In two locations, both involving numerous occurrences,[55,56] protein electrophoresis demonstrated that hybridization occurred, whereas mtDNA analysis indicated the species of the female parents. Concordance was found between allozyme and mtDNA analyses of a hybrid swarm of two strongly differentiated subspecies of cutthroat trout *Onchorynchus clarki* in a Montana lake.[57] Billington et al.[58] found two walleyes that were indistinguishable from typical walleye but which contained a mitochondrial genome identical to that of sauger *S. canadense.*

Mitochondrial DNA genotypes have been used as genetic tags in performance evaluation studies to monitor the relative survivorship, growth rate, and condition factors of strains of fish co-stocked in natural systems. Assuming the neutrality of these base-sequence variants in the mtDNA molecule, these genetic tags offer the advantages of having no effect on the performance of the organism and of offering long-term utility. Mitochondrial DNA variants can be used to distinguish representatives of conspecific populations or separate species. For example, mtDNA variants have been used for several years as genetic markers to distinguish co-stocked striped bass of Gulf of Mexico vs Atlantic descent in Lake Talquin in northwest Florida.[59] This 5-year study was designed to evaluate the relative performance of 5-year classes of these two strains of striped bass in a natural setting.

The hatchery production and release of interspecific hybrids into natural waterways to provide additional recreational possibilities has become more prevalent in recent years. However, which species is male and which species is female in a cross may affect fisheries-related qualities (catchability, growth rate, and migration) in the hybrid offspring, and which cross is preferable for a given situation must be determined empirically. Mitochondrial DNA genotypic differences can easily be used to distinguish between species and thus identify the maternal parentage of interspecific hybrids. This characteristic can be used in performance evaluations of both hybrids released into natural systems. Such an approach is being used to discriminate between co-stocked original (striped bass-mother) and reciprocal (white bass-mother) *Morone* hybrids in Lake Seminole in northwestern Florida.[59]

Another function of mtDNA analysis is the identification of commercially caught fish of uncertain origin, such as where a less valuable species might have been substituted for a more valuable species. For example, RFLP analysis of mtDNA revealed differences between two coregonids of unequal value that correctly identified the species from fresh fillets and fresh or frozen carcasses.[60] Bartlett and Davidson[61] showed that direct sequence analysis of PCR products of mtDNA could be used to determine the species of origin for smoked, pickled, canned, salted, or even partially cooked fish specimens. PCR analysis of mtDNA was also used to discriminate between bluefin tuna (a regulated species) and yellowfin tuna (an unregulated species).[61,62]

Diadromous Fishes

mtDNA Diversity

Although populations of many anadromous species of fishes mix together in the sea, they tend to show high fidelity to their natal streams when spawning, and hence remain sufficiently discrete reproductively to form populations that are genetically differentiated from each other. However, mtDNA variation in anadromous fishes is generally not as pronounced as in freshwater fishes (approximately 0.1 to 1.8% sequence divergence[19]), probably both as a result of some degree of straying among populations (which reduces geographic partitioning of mtDNA diversity) and of the population bottlenecks that are more likely to occur in unstable riverine environments (which reduces total mtDNA diversity). These observations were reinforced by a study which showed that for two forms of Atlantic salmon that were seasonally sympatric within a river system, the nonanadromous form showed a higher nucleon diversity index ($h = 0.52$) than did the anadromous form ($h = 0.37$).[63] Likewise, the nonanadromous form of rainbow trout *Oncorhynchus mykiss* displays greater mtDNA diversity than the anadromous form.[64] As for north temperate freshwater fishes, many anadromous fishes have recolonized waters that were not inhabitable during the Pleistocene, and these recent populations tend to show reduced mtDNA diversity.[14,16,65-67]

Comparatively little analysis of mtDNA in catadromous fishes has occurred, although Avise and colleagues[68,69] estimated sequence divergence between the American and European forms of north Atlantic eels *Anguilla* spp. at 0.3%, a level of genetic differentiation which demonstrates the reproductive isolation of the two taxa. Furthermore, eels from a geographically intermediate location, Iceland, appeared to be the products of hybridization between American and European eels.

Applications

A primary employment of mtDNA analysis of anadromous fishes is the delineation of stock structure. Although most anadromous fish populations appear to show a high degree of homing to their natal rivers, geographically proximate populations may or may not be well differentiated genetically. Even if such stocks are genetically distinguishable, they may for fisheries purposes need to be considered as a single stock if they sustain a common harvest. A prime example is the Chesapeake Bay striped bass stock which is composed of at least 11 tributary-specific populations that experience differential annual recruitment; overall abundance of the Chesapeake Bay stock is the product of recruitment in individual rivers. Considerable effort, without resolution, has been applied using phenotypically based techniques to identify subdivisions of striped bass within Chesapeake Bay on individual or multiriver bases.[70] However, Chapman[71-73] and Wirgin et al.[8,9] have shown indisputable genetic distinctions (frequencies of length variants) among striped bass samples from tributaries of Chesapeake Bay.

Mitochondrial DNA analysis has also been useful in determining the discreteness and probable origins of a small but potentially important striped bass stock in the Delaware River. The Delaware River once supported a moderate-sized migratory striped bass stock, but decades of severe industrial and municipal contamination reduced the population to trace levels. Improved water quality led to a slow but definite increase in the abundance of striped bass in the Delaware River; however, potential management actions could only be authorized if the current population of striped bass in the Delaware River was shown to be discrete, i.e., genetically differentiated from the Chesapeake Bay stock to its south and the Hudson River stock to its north.

Waldman and Wirgin[74] found that the frequencies of major length variants (>100 base pairs) in the Delaware River population differed significantly from those of the Hudson River stock, but not the Chesapeake Bay stock. However, the frequencies of a minor length variant (different electrophoretic mobilities of fragments <500 base pairs) differed significantly between the Delaware River and Chesapeake Bay stocks, but not the Delaware River and Hudson River stocks. They concluded that the Delaware River striped bass stock constituted a genetically unique stock and, together with historical survey information, that the current stock was more likely derived from a relict stock rather than from migrants from other stocks.[74]

Mitochondrial DNA analysis may prove to be a powerful tool in the investigation of fisheries for mixed stocks of anadromous fishes. A qualitative example involves Atlantic salmon, which occur in boreal waters of the north Atlantic. The oceanic fishery is comprised of salmon spawned in North American and European rivers ringing the north Atlantic from Connecticut to France. Adult Atlantic salmon from American and European sources mix off of western Greenland where they are subject to a major fishery. For conservation and economic reasons, it is important to determine the relative contributions of both continents to this fishery on an annual basis.

The relative contributions of the two continents to the intensive West Greenland fishery were previously assessed using scale circuli counts — a phenotypic feature subject to environmental variation that required that discrimination criteria be recalibrated annually. Protein variation was also examined, but proved to lack the required sensitivity. Bermingham et al.[36] addressed the problem by conducting RFLP analysis of mtDNA; 20 restriction endonucleases were used to examine mtDNA variation in 11 hatchery strains of Atlantic salmon from North America and Europe. They found that Atlantic salmon from the two continents differed by a minimum of seven restriction sites — sufficient variability to distinguish their continental origin. Furthermore, only two restriction enzymes (*Dra* I and *Bst* EII) were required to reveal these differences. To test the utility of their approach, hatchery-produced salmon caught in the West Greenland fishery that were originally marked with external tags denoting their source were analyzed with the two restriction enzymes. They correctly identified the continent of origin of 67 of the 68 fish.

Mitochondrial DNA analysis has also been used as the basis for a quantitative analysis of a mixed fishery. The use of genetic data in mixture models was pioneered on west coast U.S. salmonids using protein electrophoretic data.[75-78] Such an approach is also applicable with mtDNA data, provided that the basic caveat to the use of genetic mixture models is fulfilled, i.e., that sufficient genotypic differences exist among stocks.

Almost all striped bass along the northeast U.S. coast are derived from the Hudson River and Chesapeake Bay stocks. Because annual recruitment in these stocks is independent but fluctuates widely, the relative contributions of these stocks to the coastal fishery varies over time. Much effort has been applied to this question but phenotypically based approaches all suffered from environmentally induced biases; unfortunately, the alternative of allozyme studies provided little discriminatory power because protein diversity in striped bass is almost nonexistent.[70] However, a pilot study[9] used the frequencies of major and minor length variants in reference samples of striped bass from the Hudson River and Chesapeake Bay with a constrained estimated generalized least-squares approach to assess the relative contributions of the two stocks to the coastal mixture during 1989; a much larger sample from 1991 is currently being analyzed. We believe that the mixture model approach using mtDNA data will become a major tool in the management of anadromous fishes because of its freedom from the complications of ecophenotypic variance.

Other applications to anadromous fishes include monitoring genetic diversity in artificially propagated populations[79] and defining the genetic structure of natural and hatchery-bred stocks prior to the development of large aquacultural efforts.[80]

Marine Fishes

mtDNA Diversity

Mitochondrial DNA surveys of most marine fishes have shown varying, but generally lesser degrees of diversity than for freshwater and anadromous fishes, and that mtDNA diversity in marine fishes is often not well partitioned geographically. Billington and Hebert[19] found that intraspecific sequence divergence estimates for most marine fishes surveyed were less than 1.0%. The highest levels of diversity have been observed for vagile, planktivorous fishes that have large effective population sizes (e.g., species of Clupeidae[81,82]). Bowen and Avise[83] found an extreme example in two closely related forms of menhaden that have been afforded species status on morphological grounds: the Atlantic form *Brevoortia tyrannus* and the Gulf of Mexico form *B. patronus*. Among 33 assayed specimens of both species, 31 different mtDNA genotypes were observed, yielding a genotypic diversity estimate of 0.996 — the highest value reported for any species and close to the maximum possible value of 1.0. No two individuals within either the Atlantic or Gulf collections shared a mtDNA genotype, but two pairs of the putative sister species were identical at all 49 to 51 restriction sites scored. A similar absence of mtDNA sequence divergence between populations of pelagic predators occurring in the Atlantic and Pacific oceans was found in the skipjack tuna *Katsuwonus pelamis*[84] and albacore tuna *Thunnus alalunga*.[85] They reported a sequence divergence of 0.0% between skipjack tuna from the two oceans, whereas variant mtDNA genotypes were only found in a single albacore tuna.

At the other end of the mtDNA diversity spectrum within marine fishes, species that have localized home ranges and that lack a pelagic larval phase tend to show low levels of mtDNA diversity, but what diversity exists can be structured geographically, at least in accordance with major biotic regions (e.g., toadfishes *Opsanus tau* and *O. beta*[20]). It appears that the relative lack of intraspecific population structure in most marine fishes is related to their generally high dispersal capabilities (in at least some phase of their life cycle, usually the larval phase). The overall low levels of mtDNA sequence divergence among marine fishes is probably the result of bottleneck effects, although Ovenden[32] suggested that marine fishes may have a slower rate of mtDNA evolution and that they may sustain more frequent family-specific mortality. However, Avise[86] noted that mtDNA analysis has generally been applied to those species for which protein electrophoresis yielded insufficient information, so that to date the "test" of marine species has been biased.

Applications

Because of the sheer magnitude and geographical freedom of marine fish populations, marine fisheries are not amenable to the kinds of manipulations that can be performed on freshwater and anadromous fishes. The primary management action concerning marine fisheries is the regulation of harvest; implicit is the notion that harvest be controlled at the level of the unit stock. Thus, the main utility of mtDNA analysis for marine fisheries is the delineation of stock structure. However, the efficacy of mtDNA analysis for marine fisheries has not been as great as for freshwater and anadromous fisheries because of the weak geographic partitioning of mtDNA diversity displayed by most commercially important marine fishes.

North Atlantic cod provided results that were typical of mtDNA surveys of saltwater fish that sustain major fisheries. Although cod range across much of the coastal waters of the north Atlantic, there are numerous geographically isolated populations that show meristic, morphometric, and color differences and for which a high degree of genetic isolation was assumed. However, two mtDNA surveys[7,87] showed very little substructuring to the North Atlantic cod population. In fact, Arnason and Rand[7] estimated that 80% of the diversity encountered was contained within individuals (heteroplasmic length variants), 8% among individuals, and only 12% among localities. Similarily, Atlantic stocks of the capelin *Mallotus villosus* demonstrate radically different spawning modes: one is to spawn on beaches, the other is to spawn on offshore, deepwater banks. Nevertheless, a survey of mtDNA variation among six beach-spawning populations in the Canadian Maritimes and three bottom-spawning populations in Canada, Iceland, and the Barents Sea showed an absence of geographical structured heterogeneity in the frequencies of mtDNA genotypes.[88]

A pattern that has emerged from these studies is that species that were thought to be subdivided into multiple stocks based on studies of phenotypic features such as meristic counts are usually not as genotypically subdivided as the phenotypic studies would suggest. Weakfish *Cynoscion regalis* range along the U.S. mid-Atlantic coast, and a series of phenotypic and behavioral studies inferred that the species is subdivided into two or more stocks. However, an RFLP study of mtDNA found no evidence of geographic heterogeneity; in fact, 91 to 96% of all weakfish in each sample displayed single common genotype.[89] The authors concluded that phenotypic variation in weakfish is ecophenotypic.

Three studies have shown at least some geographic substructuring of mtDNA genotype frequencies in commercially important fishes. Low levels of overall mtDNA diversity but substantial heterogeneity of genotype frequencies were found between samples of orange roughy from the east and west coasts of Tasmania.[22] Mulligan and co-workers[90] found substantial mtDNA variability and evidence of three stocks of walleye pollock *Theragra chalcogramma* in the Bering Sea; these divisions were linked to the effects of prevailing current patterns. These high levels of mtDNA diversity in walleye pollock within a limited geographic range in the North Pacific contrast sharply with the very limited levels of mtDNA variability seen in Atlantic cod populations despite sampling on both sides of the North Atlantic. Historically, both Atlantic cod and Pacific pollock populations should have been of approximately equal age and sufficiently large size to withstand the cropping effects of bottlenecks; thus, expected levels of overall mtDNA diversity and differentiation would be equivalent. Gold and Richardson[91] found that the southeastern Atlantic and the northern Gulf of Mexico populations of red drum were weakly divergent.

Taken as a group, mtDNA studies of marine fishes have shown that frequently there is sufficient genetic exchange among populations to preclude substantial mtDNA differentiation. The degree of interchange required to prohibit or eradicate differentiation is surprisingly small; Slatkin[92] estimated that if selection is not a factor, an average of about one individual exchanged per generation is enough to surmount genetic drift. Of course, when dealing with mtDNA genotypes, only vagility of females impacts on levels of genotypic differentiation between populations. Even low levels of female infidelity will dramatically reduce or preclude mtDNA differentiation. Intuitively, it is not difficult to accept the regular occurrence of at least this low a level of genetic interchange across the violable barriers that separate many marine fish populations.

NUCLEAR DNA

The Structure of Nuclear DNA

In comparison with mtDNA, a far smaller number of studies have directly examined variation in the nuclear DNA of fishes. Three alternative approaches have been used to screen for differences in the nuclear DNA sequences of conspecific or congeneric fish. In all cases, selected regions of nuclear DNA are chosen for visualization by the use of DNA probes. Probes are radio- or chemiluminescent-labeled DNA sequences which bear extensive homology and thus hybridize to regions of fish nuclear DNA. Different probes have homology to different nuclear DNA sequences and thus allow for visualization of different DNA. These approaches differ in the functional or structural characteristics of the probe DNAs used to screen for informative polymorphisms and in the levels of sensitivity they provide in distinguishing closely related taxa. The method employed should reflect the anticipated level of genetic divergence among taxa. In all these approaches, total DNA is isolated from almost any tissue, although fish blood is a particularly easy tissue with which to work. The DNA is digested with restriction enzymes; DNA fragments are electrophoretically separated and analyzed in Southern blot hybridizations. The classes of nuclear DNA probes used include single copy coding sequences, single copy noncoding sequences, and repetitive (minisatellite or microsatellite) sequences. To date, the analysis of nuclear DNA in natural fish populations has been very limited; however, examples of application of all three approaches do exist.

Single Copy, Coding Nuclear DNA Probes

Single copy, coding nuclear DNA probes screen sequences that code for protein products that are under functional constraint and are thus slowly evolving. However, in RFLP analysis there is a strong likelihood that flanking regions and intronic sequences will also hybridize to these probes and be visualized. Given the relaxation of functional constraints on these classes of DNA, it is probable that this direct examination of coding nuclear DNA sequences will provide higher levels of resolution than encountered in the analysis of the protein products of these loci, yet lower levels than seen with other nuclear DNA sequences. Despite the hypervariability likely to be encountered in noncoding regions, heterologous probes will still hybridize to these sequences due to the extreme conservation of the core coding sequences.

Because of their anticipated slow rate of change, single copy coding genes should only be informative in delineating genetic relationships among somewhat divergent taxa; i.e., interspecific or intergeneric comparisons. At this level of organization, management of fish populations often focuses on discrimination of first generation (F_1) and later generation hybrids from native parental species. If hybrids are identified in systems, the impact of these interbreeds on parental gene pools may be of concern and require an evaluation. In this regard, analysis of morphological characters, either meristic or morphometric measurements, have been used to distinguish F_1 hybrids from parental species based on the intermediacy of these characters in first generation hybrids. However, this approach is of limited utility in distinguishing F_1 from later generation hybrids. Electrophoretic analysis of proteins was an advancement because it potentially allowed for discrimination of later generation hybrids; however, in many interspecific comparisons, an insufficient number of polymorphic loci were identified to allow for unequivocal identification of lineage of hybrids beyond the first generation.

Reports of natural hybridization between North American species in the genus *Morone* have increased in the recent past.[93] Additionally, hatchery-reared hybrids between white bass *Morone chrysops* and striped bass have been released into tributaries of the Chesapeake Bay and in major reservoirs in the southeastern U.S. because of their favorable growth rates, survivorship, and attractiveness as a sport fish. However, concern was voiced over potential introgression of white bass genes into native striped bass gene pools in these systems. The potential severity of this problem would be greatest in Chesapeake Bay striped bass because introgression would combine genes from the nonmigratory white bass with those of a highly migratory striped bass stock. Due to the release of F_1 *Morone* in nearby drainages, concern over this problem in the Chesapeake Bay has been voiced. To address this issue, Wirgin and coworkers[94] used RFLP analysis of a highly conserved set of nuclear DNA genes, cellular oncogenes, and found fixed differences among all four North American *Morone* species at a minimum of nine independently segregating genetic loci. Inheritance of these diagnostic nuclear DNA fragments was evaluated in the F_1 hybrids of controlled hatchery crosses of white bass × striped bass and striped bass × white perch. In both cases, all DNA fragments observed in the F_1 hybrids were also detected in the two parental species, thus confirming the Mendelian inheritance of these characters. Furthermore, analysis of nuclear DNA genotypes in the progeny of hatchery-controlled backcrosses between striped bass × white bass F_1 hybrids × striped bass demonstrated the independent assortment of these DNA fragments and also the utility of these markers in unequivocally identifying all these backcrosses as later generation hybrids. In summary, the results from these controlled crosses demonstrated that this approach has a degree of certainty exceeding 97% in distinguishing among parental species, F_1 hybrids, and later generation hybrids. This approach was then applied to distinguishing the lineage of *Morone* hybrids of unknown descent collected in Smith Lake, Alabama, which is utilized as a refuge for striped bass broodstock of Gulf of Mexico origin. Nuclear DNA was also used to determine the ancestry of suspected *Morone* hybrids collected from a tributary of the Potomac River which feeds into the Chesapeake Bay. In both cases, analyses demonstrated with a high degree of certainty that all fish were F_1 hybrids.

Single copy, coding genes have also been examined for identification of salmonid stocks. While these genes are single copy in the sense that they reside at a single genetic locus, internal regions may be repeated multiple times and thus provide diagnostic polymorphisms. The nuclear ribosomal RNA genes (rDNA) contain three regions, each of which evolves at different rates. The coding regions for 5.8S, 18S, and most of the 28S rRNA evolve slowly, the internal and external transcribed spacer regions (ITS-1, ITS-2, 5′ ETS) as well as certain variable portions of the 28S rRNA evolve more rapidly, and the intergenic spacer region (IGS) evolves most rapidly.[95,96] Variation in the IGS was examined in several species, including Arctic charr. In Arctic charr, fixed differences were seen among different subspecies and some stock differences were seen in all species, probably resulting from variability in the number of small repeats. Fixed differences in the 5′ ETS were found between allopatric populations of Arctic charr,[97] but intraspecific variation has not yet been examined in other species. Similarly, a polymorphism was found in the spacer region of the rRNA genes that distinguishes between Atlantic salmon from North America and Europe.[98]

Single Copy, Noncoding Nuclear DNA

Alternatively, single copy, noncoding nuclear DNA sequences may be screened for informative polymorphisms. This approach is founded on the premise that the vast majority of the nuclear genome is noncoding and therefore most randomly selected nDNA sequences will probably be nonfunctional and thus rapidly evolving. A simple strategy exists to select for single copy nDNA and exclude repetitive nDNA sequences from analysis. Focusing on single copy nDNA sequences offers two major practical advantages. First, gel banding patterns will be rather simple and allow for easy interpretation. Given the species specificity of these probes, this approach also guarantees that the hybridization signal will be strong and easy to read, with little interference from background signal. Second, analysis of single copy nDNA sequences allows for the application of classical Hardy-Weinberg analysis of population equilibrium and the use of perturbations from equilibrium for mixed stock estimates. The major drawback to the utilization of this approach is the need to develop species-specific nDNA probes. Development of nDNA probes requires the establishment of a nuclear DNA library in plasmid or viral vectors for the species under investigation, selection of recombinant clones representative of nonrepetitive nDNA, and empirical testing of each clone DNA in conjunction with various restriction enzymes to determine which combinations reveal informative polymorphisms in regards to the problems under investigation. An almost infinite number of clones can be generated, each representative of different nDNA sequences, and each potentially informative in population studies. Unfortunately, there is no *a priori* way of determining which clones will be informative, and therefore each must be tested on individuals representative of the populations to be investigated. In fact, it is likely that within a species' distribution, different clone-restriction enzyme combinations will be effective in addressing different population level comparisons.

Initial studies involving single copy, noncoding nDNA in snow geese,[99] honeybees,[100] and humans[101] in all cases revealed high levels of genetic diversity; frequencies of genotypes often differed significantly among geographically separate populations. Almost every clone tested among conspecific individuals revealed variability.

To date, this approach has witnessed limited application in fisheries. Wirgin et al.[102,103] tested the utililty of this methodology in identifying polymorphisms among populations of striped bass. Of 20 different clones tested, only seven revealed polymorphisms among striped bass samples representative of most of the species' North American distribution. This level of variability was much lower than seen in other non-fish species,[99-101] but is consistent with the limited genetic diversity detected in striped bass by analyses of isozymes, mtDNA, and repetitive nDNA. Polymorphisms were detected that were informative in discriminating between Gulf of Mexico and Atlantic coast striped bass populations,[102] between Hudson River and Chesapeake Bay populations, and in estimating the relative contributions of these two Atlantic coast populations to the mixed ocean striped bass fishery.[103] For example, using one probe, 43% of Gulf of Mexico striped bass exhibited a genotype which was absent from all Atlantic samples. Similarly, the use of a second probe revealed that 93% of Gulf striped bass shared a nDNA genotype which was only present in 56% of Atlantic fish. As a result of these initial findings, candidate striped bass broodstock used to restore Gulf populations are being genotyped at these informative single copy loci prior to the use of their progeny in this program. Additionally, informative single copy loci nDNA genotypes are being used in combination with mtDNA genotypes to provide more discriminatory markers in identifying ancestry of striped bass in a long-term performance evaluation of co-stocked Gulf vs Atlantic fish in a Gulf coast reservoir. Given the demonstrated Mendelian inheritance of these polymorphic DNA sequences, we also anticipate the use of these probes in evaluating the effects of decreased genetic variability on characteristics such as growth rates, bilateral asymmetry, and the incidence of hermaphroditism in introduced striped bass populations.

Single locus probes have also been developed from and applied to salmonid populations.[104,105] Stevens and co-workers[106] tested the utility of a single locus Atlantic salmon probe, 3.15.34, developed by Taggart and Ferguson[105] in determining the relatedness of full sibs, half sibs, and unrelated progeny of controlled hatchery crosses of chinook salmon *Oncorhynchus tshawytscha.* This approach could then be scaled up and applied to management of broodstock in hatchery-based enhancement programs and in aquaculture. The 3.15.34 probe revealed nine DNA fragments in seven candidate broodstock fish and pedigree analysis confirmed that this probe detected both alleles of a single polymorphic locus. However, the level of information provided by this single locus probe in determining relatedness among juvenile chinook salmon was low compared to that revealed with the multilocus B2-2 probe.

A modification of this procedure was used by Karl and Avise[107] to screen for RFLPs in single copy nDNA sequences of American oysters. Sets of 24 to 25 mer

PCR primers were designed to amplify polymorphic intervening sequences, the products were digested with restriction enzymes, separated electrophoretically, and visualized by ethidium bromide staining. Thus, unlike the RAPD methodology, variant DNA banding patterns represent restriction site polymorphisms within the amplified intervening DNA sequences. This approach revealed a sharp genetic discontinuity between Atlantic coast and Gulf of Mexico oyster populations which supported results from an earlier mtDNA study on the same populations. Genotype frequencies in these oyster populations conformed to Hardy-Weinberg expectations, confirming the Mendelian inheritance of these DNA markers.

A final application of single copy nDNA probes to fisheries is the development of a Y-chromosome probe capable of determining gender in chinook salmon.[108] The male-specific DNA sequence on the Y chromosome was isolated by its inabililty to hybridize to enriched female genomic DNA. A sex-specific DNA pattern was observed in individuals representative of five separate chinook salmon stocks and further studies demonstrated that this pattern segregates from father to son. Long exposures of autoradiographs suggest that closely related DNA sequences may also be found in females. A PCR-based technique to rapidly screen individuals has already been developed for this Y-chromosome-specific locus in chinook salmon and it can be routinely used to identify the sex of hatchery broodstock.

Unlike the use of repetitive DNA probes, analysis of DNA fragments produced by single copy probes generally offers clear and easy-to-interpret banding patterns. The demonstration of Mendelian inheritance of alternative alleles at these loci will permit the application of conventional Hardy-Weinberg analysis of genotype frequencies which will aid in understanding factors impacting on population structure and the use of diagnostic markers in mixed stock estimates. The immense size of the nuclear genome offers the promise of an almost unlimited supply of potentially informative sequences to be surveyed. Thus, in many cases, the level of sensitivity obtained may reflect the resolve of the investigator.

Repetitive Nuclear DNA Probes

Repetitive nuclear DNA sequences have not received significant usage in fish studies; however, given their potential applicability in the fields of aquaculture and animal behavior, their use is likely to increase rapidly. Minisatellite and microsatellilte sequences are short DNA sequences which are dispersed throughout the nuclear genome. Probe DNAs contain either short (10 to 20 bp) or very short (4 to 6 bp) core sequences. Due to an apparent lack of functional contraints, the number or positions of these repetitive DNA sequences may vary greatly among individuals within a species. This absence of known function and selective pressures suggests that the patterns of DNA polymorphisms within and among populations strictly reflect the biogeographical histories of these taxa. The number of fragments seen in DNA fingerprints is large and, when considered in total, each individual may exhibit a unique profile. However, all the DNA fragments need not be considered in the analysis, and it is likely that specific fragments may be diagnostic at the kinship or even population level. Thus, it is probable that analyses of repetitive DNA can provide informative markers at the individual, family, and population levels.

Analyses of DNA fingerprints in fishes have frequently involved the use of heterologous probes isolated and characterized in higher vertebrates; thus, this approach offers the advantage of commercial availability of probe DNAs. Most probes used in familial, forensic, or behavioral studies in mammals and birds have sufficient homology to repetitive fish sequences to provide strong hybridization signals if hybridization conditions are relaxed. Heterologous probes successfully tested on fish DNA include the viral genome M13,[109-111] mouse sequences closely related to the *Drosophila* Per gene,[110-112] Jeffrey's[114] human probes,[104,105,110,115] BKm minisatellite sequences,[116] and human alpha globin 3′ HVR sequences.[110]

To date, DNA fingerprinting probes have been applied to three types of problems in fisheries: discrimination of natural populations which heretofore had displayed depauperate levels of genetic variability using less sensitive approaches; determination of the extent of relatedness among progeny of controlled hatchery crosses; and assessing the mating behavior of nest-building fishes.

Turner and colleagues[112,113] applied simple-sequence DNA fingerprinting probes, such as CAC-5 and GACA-4, to the selfing hermaphrodite *Rivulus marmoratus*. Even in this species in which nDNA genotypes are inherited clonally, substantial levels of intraindividual variability were observed, and attributed to the generation of *de novo* mutations. Simple sequence probes were also used to determine if genetic variability could be detected in striped bass, a species in which allozyme and mtDNA analyses had revealed remarkably low levels of genetic polymorphisms.[113] By considering all of the DNA fragments, it was found that most individuals from even hatchery stocks of striped bass exhibited individual-specific nDNA genotypes. Additionally, Jeffrey's human probes[114] were used to distinguish among three morphotypes of brown trout in a lake in Scotland.[115] While no diagnostic DNA fragments were detected which could be used to unequivocally distinguish any of the populations, a comparison of the degree of band sharing within and between populations confirmed the genetic distinctiveness of these populations.

We used a slight modification of conventional DNA fingerprinting analysis to screen for stock-specific DNA fragments in striped bass. Electrophoretic, hybridization, and autoradiographic exposure conditions were optimized to visualize DNA fragments which in initial surveys proved informative in discriminating striped bass among different populations. To illustrate this point, Wirgin and coworkers[111] used DNA fingerprinting analysis with M-13 and Per probes to detect single DNA fragments that were diagnostic in distinguishing striped bass of Gulf of Mexico vs Atlantic descent in the Apalachicola-Chattohoochee-Flint river system in northwestern Florida. Population-level diagnostic DNA fingerprinting markers were also detected among selected Atlantic coast striped bass stocks.

Controlled hatchery crosses of striped bass were conducted in which the heritability of these nDNA fragments were confirmed in the progeny of several matings.

Stevens et al.[106] used a multilocus probe (B2-2), developed from a chinook salmon genomic DNA library, to determine if DNA fragment patterns could be used to distinguish full sibs, half sibs, and unrelated progeny from controlled hatchery crosses. The level of band sharing among unrelated parents was low (0.18) and pedigree analyses demonstrated that DNA fragments were inherited in a Mendelian fashion at a minimum of 10 unlinked, autosomal, polymorphic loci. Furthermore, it was demonstrated that levels of band sharing among individual fish could be used to determine their relatedness even in the absence of parental genotypic data.

The B2-2 probe, which gives multilocus DNA fingerprinting hybridization patterns on *Hae* III (four-base cutter) digested chinook salmon DNA, can also provide stock-specific fragment patterns when hybridized to *Bam*H I (six-base cutter) digested salmon DNA. In this case, the probe recognizes highly repetitive, possibly telomeric, sequences in the chinook salmon genome. It produces hybridization patterns in which DNA fragment number, size, and optical density vary among stocks.[117]

Quantification of band sharing among related progeny detected by DNA fingerprinting has also been used to characterize mating behavior in a nesting species of fish. In the three-spined stickleback *Gasterosteus aculeatus,* males guard the nests. By comparing the DNA fingerprinting patterns of the known male parent to the developing eggs within the nest, Rico et al.[118] were able to identify maternally inherited DNA fragments using a fingerprinting probe isolated from human DNA, pYNZ132. The extent of band sharing of these maternally inherited DNA fragments was determined among the progeny within a single nest. These interindividual band sharing indices were then used to determine how many different females were likely parents of the developing embryos.

A modification of the DNA fingerprinting approach was used by Turner et al.[113] who developed species-specific probes from tandem repeat nDNA sequences located at single genomic locations which were isolated from the sheepshead minnow *Cyprinodon variegatus*. Probes were prepared from the digested nDNA of a single fish, resulting DNA fragments were electrophoretically separated, the gel was stained with ethidium bromide, and within the expected nDNA smear, characteristic discrete DNA fragments were evident. These DNA fragments were isolated, cloned, and sequenced. Analysis of the core sequence among conspecifics revealed that it was homogenous within individuals and within populations; however, every population sample yielded a different characteristic sequence. In total, seven different natural populations were investigated and no intraindividual or interindividual variation was found. These results support the assertion that "concerted evolution" may be operative even at the population level resulting in DNA sequence homogeneity of these repetitive sequences within populations, but heterogeneity in their sequences between populations.[119] If that hypothesis is validated in many species and among their different populations, it would provide an excellent *a priori* strategy to develop population-specific probes that would be effective in distinguishing individuals representative of populations of interest. Once these populations have been characterized for sequence variation, one can envision the development of a PCR-based technique to easily screen large numbers of samples for diagnostic polymorphisms in these repeats.

RAPD Analysis

To date, there are no publications addressing the use of the random amplified polymorphic DNA technique (RAPD) on populations of fish; however, it has seen wide application in domesticated plants.[120,121] In most cases, the RAPD approach screens for variability in the nuclear genome. The technique is based on the premise that nDNA contains multiple copies of short, inverted DNA sequences close to one another that can serve as primers to PCR amplify intervening sequences. Unlike conventional PCR amplification of known target genes or DNA sequences, only a single short primer of random sequence is used, and the chromosomal location or identity of amplified sequences is unknown. PCR products are separated electrophoretically, and the number and size of bands is compared among individuals. The expectation is that the number of copies of these exact sequences may differ among individuals and thus the number of PCR products. Given the sensitivity of the amplification process to primer mismatch, a single nucleotide change in the primer sequence will alter the number of PCR products generated. Also, the length of the intervention sequence may vary and thus the molecular size of PCR products.

The appeal of this procedure lies in its simplicity. No *a priori* knowledge of target sequences is required and a large number of commercially available PCR primers can be tested. Species specificity in primer complementarity is not an issue, and small amounts and partially degraded DNA samples can be analyzed. Given the vast preponderance of noncoding sequences in the nuclear genome, the probability of focusing on rapidly evolving sequences is great. For example, studies on conifers[120] and tomatoes[121] have revealed substantial levels of genetic variability in taxa, whereas previous studies detected only low levels of genetic polymorphism. Additionally, the heritability of diagnostic fragments was demonstrated.[120] However, given the vagaries of the PCR protocol, the reproducibility of this approach when screening large numbers of outbred individuals needs to be demonstrated.

RETROSPECTIVE DNA ANALYSES

One area of research that is certain to increase in the future is the analysis of DNA sequences from archived samples. These may include tissue samples maintained in liquid preservatives such as formalin or alcohols, uncleaned scales, or museum or personal mounts of whole fish. For example, DNA sequence analyses have been routinely performed on formalin fixed-parafin embedded human tissue sections. These studies are reliant on the use of PCR to provide sufficient copies of target sequences to allow for their analysis. The main limitation to these retrospective studies is the maximum size of intact DNA that can be extracted from these specimens. Provided that a small number of intact DNA fragments can be isolated that span the intervening sequences between the two PCR primers, amplification should be possible. Immediate application of PCR amplification to archived samples will likely focus on analysis of selected mtDNA regions by the use of universal primers. However, we can envision the appeal of this approach extending to other diagnostic mtDNA or nDNA sequences. Once informative polymorphisms have been identified, flanking regions will be sequenced, thus permitting the development of species or locus-specific PCR primers.

We are currently evaluating the use of archived samples for the management of two striped bass populations. Although striped bass were once abundant in the St. Lawrence River estuary, they have not been reported for more than two decades. Efforts to restore this population are now being considered; however, the genetic relatedness of the extinct St. Lawrence stock to other populations is unknown. To address this question, we are characterizing mtDNA length variants from scales taken from St. Lawrence River striped bass collected in the 1940s and 1950s. This should permit the identification of an extant striped bass population which is most genetically similar to the extinct St. Lawrence stock. Additionally, efforts are underway to restore striped bass populations along their entire endemic distribution along the Gulf of Mexico by using hatchery-reared progeny of broodstock collected annually from the naturally reproducing Apalachicola-Chattohoochee-Flint (A-C-F) river population. However, given the introduction of striped bass of Atlantic descent into the A-C-F system in the 1960s and 1970s and their potential natural reproduction, it is important to be able to identify and select broodstock that most closely resembles that of "pure" Gulf heritage. In this regard, we are in the process of characterizing mtDNA from archived samples collected in the A-C-F system prior to the introduction of Atlantic fish. In this case, the diagnostic mtDNA polymorphism is the *Xba* I restriction site variant found in a high percentage of A-C-F fish, but absent from all Atlantic populations. Characterization of this mtDNA marker in archived samples requires the development of striped bass-specific PCR primers that immediately flank the polymorphic *Xba* I restriction site.

ANALYSIS OF DNA DATA FOR FISHERIES PURPOSES

There are two general approaches to analyzing DNA data for fisheries purposes. One is the simple testing of whether genotype frequency differences among populations or stocks are statistically significant. The other approach is the ordering of relationships among populations or stocks by estimating the genetic distances among the groups of interest. Clustering techniques may be used to create dendrograms that illustrate these relationships. Basic forms of statistical genetic analysis are well summarized by Hillis and Moritz.[122]

Two recent advances have occurred that increase the efficacy of DNA data for fisheries applications. Differences in genotype frequencies among populations or stocks are normally tested using the Chi-square test. However, genotypes often occur in several predominant forms and many additional rare types. If the expected values within cells are very small, the calculated Chi-square may be inflated upward, so that tabulated Chi-square values cannot be used to evaluate the significance of the observed value. A widely followed set of guidelines was provided by Cochran,[123] who stated that no expected frequency should be <1.0 and that <20% of the expected frequencies should be <5.0. Typically, when data don't fit these criteria, the rare genotypes are grouped, which sacrifices information.[124,125]

Roff and Bentzen[126] overcame the problem of rare genotypes by developing a Monte Carlo-based approach in which the marginal row and column totals of the contingency table are preserved, but the interior matrix is randomized. The exact probability of obtaining a Chi-square value as great as that observed can be determined directly by considering all possible arrangements of the data set. Although this is impractical except for very small data sets, it can be approximated with a Monte Carlo approach — the greater the number of trials, the closer the approximation. The Roff and Bentzen[126] approach has been incorporated into a software package called REAP that is designed to manipulate DNA data.

Another statistical advance is the development of AMOVA, an analogue of analysis of variance (ANOVA) suitable for mtDNA data.[127] AMOVA is an improvement over earlier approaches to partitioning genetic variation because it does not require a particular assumption about the evolution of the genetic system; i.e., the distance metric may be computed using (1) a standard Euclidean metric counting the differences among genotypes, (2) an equidistant metric based on the idea that genotypes are merely distinguishable, (3) a distance metric measured along a mutation network, and (4) a matrix allowing for nonlinearity of changes along the network. Two or more of these alternative distance matrices, one for each particular set of assumptions, may be input and their influence on the outcome evaluated. This approach allows for quantitative estimates of the hierarchical components to genetic variability within a species, i.e., the among-regions, the among-populations within regions, and the within-populations variance components.

CONCLUSIONS

Much of the motivation for the intensive application of molecular appoaches to fisheries problems has been due to the unambiguous nature of the results; classical approaches to the differentiation of fishes have been confounded by phenotypic variation. The obstacle of phenotypic variation was eased with the advent of protein electrophoresis; however, the inflexibility of the approach and the lack of sufficient protein variation in many taxa left protein electrophoresis only a partial solution. The analysis of DNA sequences first began to be routinely applied to fisheries problems in the mid-1980s and some of these early studies showed considerable DNA sequence diversity in species that exhibited little protein variation. In summary, these studies have demonstrated that sufficient levels of DNA sequence diversity exist to aid in population or familial studies even among species previously characterized as genetically monomorphic. In less than a decade, molecular analysis has emerged as one of the most powerful tools available to fisheries managers.

Indeed, enthusiasm for molecular approaches to fisheries problems is growing as a result of the great methodological refinements and expansions that have taken place. During its brief history, many new advances occurred that simplified and speeded the protocols. Faster processing and relaxed constraints on the preservation of samples have helped obviate an early debility of many molecular studies of fishes — the combination of small sample sizes and an absence of fixed genotypic differences. Additionally, while the amount of tissue from individual fish needed to conduct DNA analysis at one time was large, today many forms of analyses can be performed on the amount of DNA contained in a single cell. This will allow for investigations on the smallest of life stages, which are often the most informative in population studies due to their reduced ability to emigrate from natal sites. The same relaxation of demands on tissue size also applies to preservation conditions of the samples. Whereas, most early DNA studies utitlized tissues from freshly killed organisms, today all forms of DNA analyses can deal with frozen samples and many studies have been undertaken with archived specimens.

With no *a priori* information as to the genomic locations of informative polymorphisms, early studies were burdened with screening vast stretches of sequences for useful markers. Many informative polymorphisms have since been found allowing for subsequent efforts to concentrate on quantifying their comparative frequencies in taxa. Sample sizes in published studies appear to be becoming progressively larger, thereby permitting the detection of statistically significant frequency differences among taxa with modest levels of sequence diversity. Additionally, as studies could be conducted more efficiently and as more laboratories became involved, a better understanding was gained as to how DNA sequences varied in fishes and how that variation might be tapped to yield relevant information. The development of new probes also allowed further exploration of the nuclear genome — in particular, rapidly evolving sequences. PCR analysis of mtDNA using universal primers and direct sequencing permitted fine scale screening of a few selected regions of the mitochondrial genome. The use of PCR analysis will expand beyond its current limitations. As more diagnostic sequences are identified in other regions of mtDNA and nDNA, new PCR primers will be developed which will allow for the amplification and the rapid quantification of variant genotypes.

Today, the challenge of DNA sequence analysis may be in selecting the right approach to the problem at hand. No single technique is best suited for all applications and careful thought should go into the selection process. Characteristics of the taxa under investigation should guide this decision. For example, low levels of mtDNA divergence among stocks is expected when female vagility is substantial. Similarily, populations which undergo bottlenecks, paticularly those in the pool of spawning females, will exhibit only low levels of mtDNA variation.

Although mtDNA analysis is by far the most frequently used of the DNA-based approaches currently applied in fisheries, it may not offer the levels of genetic variation or differentiation needed to distinguish closely related taxa. Investigation of single copy coding nDNA sequences provides more resolution than has been utilized because of the high levels of polymorphism in their introns and flanking sequences. Analysis of single copy anonymous nDNA loci probably affords an even higher level of sensitivity, an infinite number of potentially informative loci, and Mendelian inheritance of diagnostic fragments. Repetitive sequences in nDNA are best studied to determine the degree of relatedness of individuals within populations and to quantify levels of their overall genetic diversity. However, in unusual cases where levels of conspecific genetic diversity are low, repetitive sequences may even be used to differentiate populations.

The breadth of information that can be gained with the suite of DNA-based techniques now available may be best exemplified by studies of striped bass *Morone saxatilis* — a species of high recreational and commercial value that receives intensive management across its considerable range. Several populations of migratory striped bass occur along the U.S. Atlantic coast, but attempts to morphologically differentiate these populations provided inconsistent results and protein surveys showed virtually no useful variation. Although, RFLP analysis of mtDNA did not reveal informative restriction site substitutions among these populations, significant and consistent differences in length genotype frequencies were detected.[8,9] Additional diagnostic markers for this purpose were then found by cloning single copy noncoding nDNA to develop informative species-specific probes.[102,103]

When markers were required to distinguish native Gulf of Mexico drainage striped bass from Atlantic coast fish, RFLP analysis of mtDNA did uncover useful site polymorphisms.[50] These markers were later augmented by nuclear DNA fragment differences revealed by two DNA fingerprinting probes.[111] Finally, when markers were needed to distinguish among striped bass and F_1 and later generation hybrids, RFLP analysis of oncogenes — highly conserved, single copy coding genes — provided a powerful tool.[94] Together, these studies demonstrate how different portions of single species genomes can be assayed to yield information vital to a series of diverse questions concerning that species.

Despite the growing popularity of molecular genetic approaches in fisheries research, we feel that some of the studies conducted to date have failings which should be addressed. First, many publications fail to display photographs of the most informative polymorphisms supportive of the main conclusions of the study, or alternatively, they provide only line diagrams of gel banding patterns. If management decisions are to be based on these results, it is not unreasonable to expect one photograph of suitable quality for publication to illustrate the most important data from the study. This would preclude misinterpretations of banding patterns that may arise, such as incomplete digestion of DNA molecules. Second, premature conclusions concerning a lack of variability among taxa are often reached, when in actuality an insufficient number of restriction enzymes or enzymes that only recognize a limited number of recognition sites have been used. Third, to distinguish among taxa or to estimate the relative contributions of individual populations to mixed assemblages, it is necessary that discriminatory mtDNA genotypes be in moderate frequencies within populations. In this regard, composite genotypes are often generated based on data from a large number of restriction enzymes, each of which detects independent polymorphisms. Alternatively, the data from a single or a subset of the restriction enzymes used can prove diagnostic in establishing relationships among all or a subset of taxa investigated.

We anticipate that the application of molecular approaches to fisheries problems will become more and more routine. As the number of diagnostic markers that are revealed continues to grow, the need to regularly screen large numbers of fish will also increase. However, it is not clear how this will be best accomplished. Future efforts would be enhanced by a better partnership between fisheries managers and researchers. It is our impression that some fisheries managers still view DNA-based research as an expensive luxury and, consequently, they don't make the effort to become conversant about it. We also believe that the academic community that performs most molecular research has not gone far enough to explain the results and potential of these approaches. Part of the problem also stems from the lack of concordance in the apportionment of managers and researchers. Whereas managers are employed by government agencies and every government jurisdiction has a management staff, molecular biologists with an interest in fisheries questions are distributed haphazardly. In the U.S., DNA-based fisheries research might be facilitated by the establishment of regional centers specializing in molecular approaches to fisheries problems.

REFERENCES

1. **Gyllensten, U., Wharton, D., Josefsson, A., and Wilson, A. C.,** Paternal inheritance of mitochondrial DNA in mice, *Nature,* 352, 255, 1991.
2. **Hoeh, W. R., Blakely, K. H., and Brown, W. M.,** Heteroplasmy suggests limited biparental inheritance of *Mytilus* mitochondrial DNA, *Science*, 251, 1488, 1991.
3. **Zouros, E., Freeman, K. R., Ball, A. O., and Pogson, G. H.,** Direct evidence for extensive paternal mitochondrial DNA inheritance in the marine mussel *Mytilus, Nature,* 359, 412, 1992.
4. **Brown, W. M., George, M., Jr., and Wilson, A. C.,** Rapid evolution of animal mitochondrial DNA, *Proc. Natl. Acad. Sci. U.S.A.,* 76, 1967, 1979.
5. **Martin, A. P., Naylor, G. J. P., and Palumbi, S. R.,** Rates of mitochondrial DNA evolution in sharks are slow compared with mammals, *Nature*, 357, 153, 1992.
6. **Buroker, N. E., Brown, J. R., Gilbert, T. A., O'Hara, P. J., Beckenbach, A. T., Thomas, W. K., and Smith, M. J.,** Length heteroplasmy of sturgeon mitochondrial DNA: an illegitimate elongation model, *Genetics,* 124, 157, 1990.
7. **Arnason, E., and Rand, D. M.,** Heteroplasmy of short tandem repeats in mitochondrial DNA of Atlantic cod, *Gadus morhua, Genetics,* 132, 211, 1992.

8. **Wirgin, I. I., Silverstein, P., and Grossfield, J.,** Restriction endonuclease analysis of striped bass mitochondrial DNA: the Atlantic coastal migratory stock, *Am. Fish Soc. Symp.*, 7, 475, 1990.
9. **Wirgin, I., Maceda, L., Waldman, J. R., and Crittenden, R. N.,** Use of mitochondrial DNA polymorphisms to estimate the relative contributions of the Hudson River and Chesapeake Bay striped bass stocks to the mixed Atlantic coastal fishery, *Trans. Am. Fish. Soc.*, 122, 1993.
10. **Bentzen, P., Leggett, W. C., and Brown, G. G.,** Length and restriction site heteroplasmy in the mitochondrial DNA of American shad *(Alosa sapidissima), Genetics,* 118, 509, 1988.
11. **Gold, J. R., and Richardson, L. R.,** Restriction site heteroplasmy in the mitochondrial DNA of the marine fish *Sciaenops ocellatus* (L.), *Anim. Genet.*, 21, 313, 1990.
12. **Bermingham, E., Lamb, T., and Avise, J. C.,** Size polymorphism and heteroplasmy in the mitochondrial DNA of lower vertebrates, *J. Hered.*, 77, 249, 1986.
13. **Brown, J. R., Beckenbach, A. T., and Smith, M. J.,** Mitochondrial DNA length variation and heteroplasmy in populations of white sturgeon *(Acipenser transmontanus), Genetics,* 132, 221, 1992.
14. **Nolan, K., Grossfield, J., and Wirgin, I.,** Discrimination among Atlantic coast populations of American shad *(Alosa sapidissima)* using mitochondrial DNA, *Can. J. Fish. Aquat. Sci.*, 48, 1724, 1991.
15. **Mulligan, T. J., and Chapman, R. W.,** Mitochondrial DNA analysis of Chesapeake Bay white perch, *Morone americana, Copeia,* 679, 1989.
16. **Wirgin, I. I., Waldman, J. R., Maceda, L., Moore, D. W., and Courtenay, S.,** Mitochondrial DNA variation in Canadian striped bass, *Can. J. Fish. Aquat. Sci.*, 50, 80, 1993.
17. **Avise, J. C., and Lansman, R. A.,** Polymorphism of mitochondrial DNA in populations of higher animals, in *Evolution of Genes and Proteins,* Nei, M., and Koehn, R. K., Eds., Sinnauer, Sunderland, MA, 1983, chap. 8.
18. **Ward, R. D., Billington, N., and Hebert, P. D. N.,** Allozyme and mitochondrial DNA variation in populations of walleye, *Stizostedion vitreum, Can. J. Fish. Aquat. Sci.*, 47, 2074, 1989.
19. **Billington, N., and Hebert, P. D. N.,** Mitochondrial DNA diversity in fishes and its implications for introductions, *Can. J. Fish. Aquat. Sci.*, 48 (Suppl. 1), 80, 1991.
20. **Avise, J. C., Arnold, J., Ball, R. M., Bermingham, E., Lamb, T., Neigel, J. E., Reeb, C. A., and Saunders, N. C.,** Intraspecific phylogeography: the mitochondrial DNA bridge between population genetics and systematics, *Annu. Rev. Ecol. Syst.*, 18, 489, 1987.
21. **Avise, J. C., Neigel, J. E., and Arnold, J.,** Demographic influences on mitochondrial DNA lineage survivorship in animal populations, *J. Mol. Evol.*, 20, 99, 1984.
22. **Ferris, S. D., Sage, R. D., Huang, C. M., Nielsen, J. N., Ritte, U., and Wilson, A. C.,** Flow of mitochondrial DNA across a species boundary, *Proc. Natl. Acad. Sci. U.S.A.*, 80, 2290, 1983.
23. **Ovenden, J. R., Smolenski, A. J., and White, R. W. G.,** Mitochondrial DNA restriction site variation in Tasmanian populations of orange roughy *(Hoplostethus atlanticus),* a deep-water marine teleost, *Aust. J. Mar. Freshwater Res.*, 40, 1, 1989.
24. **Taggart, J. B., Hynes, R. A., Prodohl, P. A., and Ferguson, A.,** A simplified protocol for routine total DNA isolation from salmonid fishes, *J. Fish Biol.*, 40, 963, 1992.

25. **Lansman, R. A, Shade, R. O., Shapira, J. F., and Avise, J. C.,** The use of restriction endonucleases to measure mitochondrial DNA sequence relatedness in natural populations. III. Techniques and potential applications, *J. Mol. Evol.,* 17, 214, 1981.
26. **Tamura, K., and Aotsuka, T.,** Rapid isolation method of animal mitochondrial DNA by the alkaline lysis procedure, *Biochem. Genet.,* 26, 815, 1988.
27. **Gonzalez-Villasenor, L. I., Burkhoff, A. M., Corces, V., and Powers, D. A.,** Characterization of cloned mitochondrial DNA from the teleost *Fundulus heteroclitus* and its usefulness as an interspecies hybridization probe, *Can. J. Fish. Aquat. Sci.,* 43, 1866, 1986.
28. **Chapman, R. W., and Powers, D. A.,** A method for the rapid isolation of mtDNA from fishes, University of Maryland Sea Grant Program, Technical Report UM-SG-TS-84-05, College Park, MD, 1984.
29. **Billington, N., and Hebert, P. D. N.,** Technique for determining mitochondrial DNA markers in blood samples from walleyes, *Am. Fish. Soc. Symp.,* 7, 492, 1990.
30. **Whitmore, D. H., Thai, T. H., and Craft, C. M.,** Gene amplification permits minimally invasive analysis of fish mitochondrial DNA, *Trans. Am. Fish. Soc.,* 121, 170, 1992.
31. **Quattro, J. M., Morgan, R. P., II, and Chapman, R. W.,** Mitochondrial DNA variability in brook trout populations from western Maryland, *Am. Fish. Soc. Symp.,* 7, 470, 1990.
32. **Ovenden, J. R.,** Mitochondrial DNA and marine stock assessment: a review, *Aust. J. Mar. Freshwater Res.,* 41, 835, 1990.
33. **McVeigh, H. P., Bartlett, S. E., and Davidson, W. S.,** Polymerase chain reaction/direct sequence analysis of the cytochrome *b* gene in *Salmo salar, Aquaculture,* 954, 225, 1991.
34. **Martin, A. P., Naylor, G. J. P., and Palumbi, S. R.,** Rates of mitochondrial DNA evolution in sharks are slow compared with mammals, *Nature,* 357, 153, 1992.
35. **Carr, S. M., and Marshall, H. D.,** A direct approach to the measurement of genetic variation in fish populations: applications of the polymerase chain reaction to studies of Atlantic cod, *Gadus morhua* L., *J. Fish Biol.,* 39 (Suppl. A), 101, 1991.
36. **Bermingham, E., Forbes, S. H., Friedland, K., and Pla, C.,** Discrimination between Atlantic salmon *(Salmo salar)* of North American and European origin using restriction analyses of mitochondrial DNA, *Can. J. Fish. Aquat. Sci.,* 48, 884, 1991.
37. **Swift, C. C., Gilbert, C. R., Bortone, S. A., Burgess, G. H., and Yerger, R. W.,** Zoogeography of the freshwater fishes of the southeastern United States: Savannah River to Lake Ponchartrain, in *The Zoogeography of North American Freshwater Fishes,* Hocutt, C. H., and Wiley, E. O., Eds., John Wiley & Sons, New York, 1986, chap. 7.
38. **Billington, N., Barrette, R. J., and Hebert, P. D. N.,** Management implications of mitochondrial DNA variation in walleye stocks, *N. Am. J. Fish. Manage.,* 12, 276, 1992.
39. **Avise, J. C., and Saunders, N. C.,** Hybridization and introgression among species of sunfish *(Lepomis):* analysis by mitochondrial DNA and allozyme markers, *Genetics,* 108, 237, 1984.
40. **Bermingham, E., and Avise, J. C.,** Molecular zoogeography of freshwater fishes in the southeastern United States, *Genetics,* 113, 939, 1986.
41. **Bailey, R. M., and Smith, G. R.,** Origin and geography of the fish fauna of the Laurentian Great Lakes basin, *Can. J. Fish. Aquat. Sci.,* 38, 1539, 1961.

42. **Crossman, E. J., and McAllister, D. E.,** Zoogeography of freshwater fishes of the Hudson Bay drainage, Ungava Bay and the arctic archipelago, in *The Zoogeography of North American Freshwater Fishes,* Hocutt, C. H., and Wiley, E. O., Eds., John Wiley & Sons, New York, 1986, chap. 3.
43. **Schmidt, R. E.,** Zoogeography of the northern Appalachians, in *The Zoogeography of North American Freshwater Fishes,* Hocutt, C.H., and Wiley, E. O., Eds., John Wiley & Sons, New York, 1986, chap. 5.
44. **Bernatchez, L., and Dodson, J. J.,** Allopatric origin of sympatric populations of lake whitefish *(Coregonus clupeaformis)* as revealed by mitochondrial DNA restriction analysis, *Evolution,* 44, 1263, 1990.
45. **Billington, N., and Hebert, P. D. N.,** Mitochondrial DNA variation in Great Lakes walleye *(Stizostedion vitreum)* populations, *Can. J. Fish. Aquat. Sci.,* 45, 643, 1988.
46. **Grewe, P. M., and Hebert, P. D. N.,** Mitochondrial DNA diversity among broodstocks of the lake trout, *Salvelinus namaycush, Can. J. Fish. Aquat. Sci.,* 45, 2114, 1988.
47. **Hartley, S. E., Bartlett, S. E., and Davidson, W. S.,** Mitochondrial DNA analysis of Scottish populations of arctic charr, *Salvelinus alpinus, J. Fish. Biol.,* 40, 219, 1992.
48. **Danzmann, R. G., Ferguson, M. M., Skulason, S., Snorrason, S. S., and Noakes, D. L.,** Mitochondrial DNA diversity among four sympatric morphs of Arctic charr, *Salvelinus alpinus* L., from Thingvallavatn, Iceland, *J. Fish. Biol.,* 39, 649, 1991.
49. **Wooley, C. M., and Crateau, E. J.,** Biology, population estimates, and movement of native and introduced striped bass, Apalachicola River, Florida, *N. Am. J. Fish. Manage.,* 3, 383, 1983.
50. **Wirgin, I. I., Proenca, R., and Grossfield, J.,** Mitochondrial DNA diversity among populations of striped bass in the southeastern United States, *Can. J. Zool.,* 67, 891, 1989.
51. **Hynes, R. A., Duke, E. J., and Joyce, P.,** Mitochondrial DNA as a genetic marker for brown trout, *Salmo trutta* L., populations, *J. Fish. Biol.,* 35, 687, 1989.
52. **Danzmann, R. G., Ihssen, P. E., and Hebert, P. D. N.,** Genetic discrimination of wild and hatchery populations of brook charr, *Salvelinus fontinalis* (Mitchill), in Ontario using mitochondrial DNA analysis, *J. Fish Biol.,* 39 (Suppl. A), 69, 1991.
53. **Ferguson, M. M., Danzmann, R. G., and Hutchings, J. A.,** Incongruent estimates of population differentiation among brook charr, *Salvelinus fontinalis,* from Cape Race, Newfoundland, Canada, based upon allozyme and mitochondrial DNA variation, *J. Fish Biol.,* 39 (Suppl. A), 79, 1991.
54. **Hubbs, C. L.,** Hybridization between fish species in nature, *Syst. Zool.,* 4, 1, 1955.
55. **McGowan, C., and Davidson, W. S.,** Unidirectional natural hybridization between brown trout *(Salmo trutta)* and Atlantic salmon *(S. salar)* in Newfoundland, *Can. J. Fish. Aquat. Sci.,* 49, 1953, 1992.
56. **Youngson, A. F., Knox, D., and Johnstone, R.,** Wild adult hybrids of *Salmo salar* L. and *Salmo trutta* L., *J. Fish. Biol.,* 40, 817, 1992.
57. **Gyllensten, U., Leary, R. F., Allendorf, F. W., and Wilson, A. C.,** Introgression between two cutthroat trout subspecies with substantial karyotypic, nuclear and mitochondrial genomic divergence, *Genetics,* 11, 905, 1985.
58. **Billington, N., Hebert, P. D. N., and Ward, R. D.,** Evidence of introgressive hybridization in the genus *Stizostedion:* interspecific transfer of mitochondrial DNA between sauger and walleye, *Can. J. Fish. Aquat. Sci.,* 45, 2035, 1988.
59. **Mesing, C.,** unpublished data.

60. **Shields, B. A., Kapuscinski, A. R., and Guise, K. S.,** Mitochondrial DNA variation in four Minnesota populations of lake whitefish: utility as species and population markers, *Trans. Am. Fish. Soc.*, 121, 21, 1992.
61. **Bartlett, S. E., and Davidson, W. S.,** FINS (forensically informative nucleotide sequencing): a procedure for identifying the animal origin of biological specimens, *BioTechniques*, 12, 408, 1992.
62. **Barlett, S. E., and Davidson, W. S.,** Identification of *Thunnus* tuna species by the polymerase chain reaction and direct sequence analysis of their mitochondrial cytochrome *b* genes, *Can. J. Fish. Aquat. Sci.*, 48, 309, 1991.
63. **Birt, T. P., Green, J. M., and Davidson, W. S.,** Mitochondrial DNA variation reveals genetically distinct sympatric populations of anadromous and nonanadromous Atlantic salmon, *Salmo salar, Can. J. Fish. Aquat. Sci.*, 48, 577, 1991.
64. **Wilson, G. M., Thomas, W. K., and Beckenbach, A. T.,** Intra- and inter-specific mitochondrial DNA sequence divergence in *Salmo:* rainbow, steelhead, and cutthroat trouts, *Can. J. Fish. Aquat. Sci.*, 63, 2088, 1985.
65. **Bernatchez, L., Dodson, J. J., and Boivin, S.,** Population bottlenecks: influence on mitochondrial DNA diversity and its effect in coregonine stock discrimination, *J. Fish. Biol.*, 35 (Suppl. A), 233, 1989.
66. **Brown, J. R., Beckenbach, A. T., and Smith, M. J.,** Influence of Pleistocene glaciations and human intervention upon mitochondrial DNA diversity in white sturgeon *(Acipenser transmontanus), Can. J. Fish. Aquat. Sci.*, 49, 358, 1992.
67. **Wilson, G. M., Thomas, W. K., and Beckenbach, A. T.,** Mitochondrial DNA analysis of Pacific northwest populations of *Oncorhynchus tshawytscha, Can. J. Fish. Aquat. Sci.*, 44, 1301, 1987.
68. **Avise, J. C., Helfman, G. S., Saunders, N. C., and Hales, L. S.,** Mitochondrial DNA differentiation in North Atlantic eels: population genetic consequences of an unusual life history pattern, *Proc. Natl. Acad. Sci. U.S.A.*, 83, 4350, 1986.
69. **Avise, J. C., Nelson, W. S., Arnold, J., Koehn, R. K., Williams, G. C., and Thorsteinsson, V.,** The evolutionary genetic status of Icelandic eels, *Evolution*, 44, 1254, 1990.
70. **Waldman, J. R., Grossfield, J., and Wirgin, I.,** Review of stock discrimination techniques for striped bass, *N. Am. J. Fish. Manage.*, 8, 410, 1988.
71. **Chapman, R. W.,** Changes in the population structure of male striped bass, *Morone saxatilis*, spawning in three areas of the Chesapeake Bay from 1984 to 1986, *U.S. Nat. Mar. Fish. Serv. Fish. Bull.*, 85, 167, 1987.
72. **Chapman, R. W.,** Spatial and temporal variation of mitochondrial DNA haplotype frequencies in the striped bass *(Morone saxatilis)* 1982 year class, *Copeia*, 344, 1989.
73. **Chapman, R. W.,** Mitochondrial DNA analysis of striped bass populations in Chesapeake Bay, *Copeia*, 355, 1990.
74. **Waldman, J. R., and Wirgin, I. I.,** Origin of the current Delaware River striped bass population based on analysis of mitochondrial DNA, *Trans. Am. Fish. Soc.*, in press.
75. **Milner, G. B., Teel, D. J., Utter, F. M., and Winans, G. A.,** A genetic method of stock identification in mixed populations of Pacific salmon, *Oncorhynchus* spp., *U.S. Natl. Mar. Fish. Serv. Mar. Fish Rev.*, 47, 1, 1985.
76. **Beacham, T., Gould, A., Withler, R., Murray, C. B., and Barner, L. W.,** Biochemical genetic survey and stock identification of chum salmon *(Oncorhynchus keta)* in British Columbia, *Can. J. Fish. Aquat. Sci.*, 44, 1702, 1987.

77. **Shaklee, J. B., and Phelps, S. R.,** Operation of a large-scale, multiagency program for genetic stock identification, *Am. Fish. Soc. Symp.,* 7, 817, 1990.
78. **Smouse, P. E., Waples, R. S., and Tworek, J. A.,** A genetic mixture analysis for use with incomplete source population data, *Can. J. Fish. Aquat. Sci.,* 47, 620, 1990.
79. **Ginatulina, L. K.,** Genetic differentiation among chum salmon, *Onchorynchus keta* (Walbaum), from Primorye and Sakhalin., *J. Fish. Biol.,* 40, 33, 1992.
80. **Hovey, S. J., King, D. P. F., Thompson, D., and Scott, A.,** Mitochondrial DNA and allozyme analysis of Atlantic salmon, *Salmo salar* L., in England and Wales, *J. Fish. Biol.,* 35 (Suppl. A), 253, 1989.
81. **Kornfield, I., and Bogdanowicz, S. M.,** Differentiation of mitochondrial DNA in Atlantic herring, *Clupea harengus, Fish. Bull.,* 85, 561, 1987.
82. **Schweigert, J. F., and Withler R. E.,** Genetic differentiation of Pacific herring based on enzyme electrophoresis and mitochondrial DNA analysis, *Am. Fish. Soc. Symp.,* 7, 459, 1990.
83. **Bowen, B. W., and Avise, J. C.,** The genetic structure of Atlantic and Gulf of Mexico populations of sea bass, menhaden, and sturgeon: the influence of zoogeographic factors and life history patterns, *Mar. Biol.,* 107, 371, 1990.
84. **Graves, J. E., and Dizon, A. E.,** Mitochondrial DNA sequence similarity of Atlantic and Pacific albacore tuna *(Thunnus alalunga), Can. J. Fish. Aquat. Sci.,* 46, 870, 1989.
85. **Graves, J. E., Ferris, S. D., and Dizon, A. E.,** Close genetic similarity of Atlantic and Pacific skipjack tuna *(Katsuwonus pelamis)* demonstrated with restriction endonuclease analysis of mitochondrial DNA, *Mar. Biol.,* 79, 315, 1984.
86. **Avise, J. C.,** Identification and interpretation of mitochondrial DNA stocks in marine species, in *Proceedings of the Stock Identification Workshop*, Kumpf, H. E., Vaught, R. N., Grimes, B., Johnson, A. G., and Nakamura, E.L., Eds., NOAA Tech. Mem. NMFS SEFC-199, Panama City, FL, 1987, 105.
87. **Smith, P. J., Birley, A. J., Jamieson, A., and Bishop, C. A.,** Mitochondrial DNA in Atlantic cod, *Gadus morhua:* lack of genetic divergence between eastern and western populations, *J. Fish. Biol.,* 34, 369, 1989.
88. **Dodson, J. J., Carscadden, J. E., Bernatchez, L., and Colombani, F.,** Relationship between spawning mode and phylogeographic structure in mitochondrial DNA of North Atlantic capelin *Mallotus villosus, Mar. Ecol. Prog. Ser.,* 76, 103,1991.
89. **Graves, J. E., McDowell, J. R., and Jones, M. L.,** A genetic analysis of weakfish *Cynoscion regalis* stock structure along the mid-Atlantic coast, *Fish. Bull.,* 90, 469, 1992.
90. **Mulligan, T. J., Chapman, R. W., and Brown, B. L.,** Mitochondrial DNA analysis of walleye pollock, *Theragra chalcogramma,* from the eastern Bering Sea and Shelikof Strait, Gulf of Alaska, *Can. J. Fish. Aquat. Sci.,* 49, 319, 1992.
91. **Gold, J. R., and Richardson, L. R.,** Genetic studies in marine fishes. IV. An analysis of population structure in the red drum *(Sciaenops ocellatus)* using mitochondrial DNA, *Fish. Res.,* 12, 213, 1991.
92. **Slatkin, M.,** Rare alleles as indicators of gene flow, *Evolution,* 39, 53, 1985.
93. **Waldman, J. R., and Bailey, R. M.,** Early occurrence of natural hybridization within *Morone* (Perciformes), *Copeia*, 553, 1992.
94. **Wirgin, I. I., Maceda, L., and Mesing, C.,** Use of cellular oncogene probes to identify *Morone* hybrids, *J. Hered.,* 83, 375, 1992.
95. **Phillips, R. B., and Pleyte, K. A.,** Nuclear DNA and salmonid phylogenetics, *J. Fish Biol.,* 39 (Suppl. A), 259, 1991.

96. **Phillips, R. B., Pleyte, K. A., and Brown, M. R.,** Salmonid phylogeny inferred from ribosomal DNA restriction maps, *Can. J. Fish. Aquat. Sci.,* 49, 2345, 1992.
97. **Zhuo, L.,** Cytogenetic and Molecular Analysis of the Ribosomal RNA Genes in Lake Trout, Doctoral dissertation, University of Wisconsin, Milwaukee, WI, 1991.
98. **Cutler, M. G., Bartlett, S. E., Hartley, S. E., and Davidson, W. S.,** A polymorphism in the ribosomal RNA genes distinguishes Atlantic salmon *(Salmo salar)* from North America and Europe, *Can. J. Fish. Aquat. Sci.,* 48, 1655, 1991.
99. **Quinn, T. W., and White, B. N.,** Identification of restriction fragment-length polymorphisms in genomic DNA of the lesser snow goose *(Anser caerulescens caerulescens), Mol. Biol. Evol.,* 4, 126, 1987.
100. **Hall, H. G.,** Parental analysis of introgressive hybridization between African and European honeybees using nuclear DNA RFLPs, *Genetics,* 125, 611, 1990.
101. **Barker, D., Schafer, M., and White, R.,** Restriction sites containing CpG show a higher frequency of polymorphism in human DNA, *Cell,* 36, 131, 1984.
102. **Wirgin, I. I., and Maceda, L.,** Development and use of striped bass-specific RFLP probes, *J. Fish Biol.,* 39 (Suppl. A), 159, 1991.
103. **Wirgin, I. I.,** unpublished data.
104. **Taggart, J. B., and Ferguson, A.,** Hypervariable minisatellite DNA single locus probes for the Atlantic salmon, *Salmo salar* L., *J. Fish Biol.,* 37, 991, 1990.
105. **Taggart, J. B., and Ferguson, A.,** Minisatellite DNA fingerprints of salmonid fishes, *Anim. Genet.,* 21, 377, 1990.
106. **Stevens, T. A., Withler, R. E., and Beacham, T. D.,** A new multi-locus probe for DNA fingerprinting in chinook salmon *(Oncorhynchus tshawytscha),* and comparisons with a single locus probe, *Can. J. Fish. Aquat. Sci.,* in press.
107. **Karl, S. A., and Avise, J. C.,** Balancing selection at allozyme loci in oysters: implications from nuclear RFLPs, *Science,* 256, 100, 1992.
108. **Devlin, R. H., McNeil, B. K., Groves, T. D. D., and Donaldson, E. M.**, Isolation of a Y-chromosome DNA probe capable of determining genetic sex in chinook salmon *(Oncorhynchus tshwaytscha), Can. J. Fish. Aquat. Sci.,* 48, 1606, 1991.
109. **Fields, R. D., Johnson, K. R., and Thorgaard, G. H.,** DNA fingerprints in rainbow trout detected by hybridization with DNA of bacteriophage M13, *Trans. Am. Fish. Soc.,* 118, 78, 1989.
110. **Castelli, M., Philippart, J.-C., Vassart, G., and Georges, M.,** DNA fingerprinting in fish: a new generation of genetic markers, *Am. Fish. Soc. Symp.,* 7, 514, 1990.
111. **Wirgin, I. I., Grunwald, C., Garte, S. J., and Mesing, C.,** Use of DNA fingerprinting in the identification and management of a striped bass population in the southeastern United States, *Trans. Am. Fish. Soc.,* 120, 273, 1991.
112. **Turner, B. J., Elder, J. F., Jr., Laughlin, T. F., and Davis, W. P.,** Genetic variation in clonal vertebrates detected by simple-sequence DNA fingerprinting, *Proc. Natl. Acad. Sci. U.S.A.,* 87, 5653, 1990.
113. **Turner, B. J., Elder, J. F., Jr., and Laughlin, T. F.,** Repetitive DNA sequences and the divergence of fish populations: some hopeful beginnings, *J. Fish Biol.,* 39 (Suppl. A), 131, 1991.
114. **Jeffreys, A. J., Wilson, V., and Thein, S. L.,** Hypervariable 'minisatellite' regions in human DNA, *Nature,* 314, 67, 1985.
115. **Prodohl, P. A., Taggart, J. B., and Ferguson, A.,** Genetic variability within and among sympatric brown trout *(Salmo trutta)* populations: multi-locus DNA fingerprint analysis, *Hereditas,* 117, 45, 1992.

116. **Lloyd, M. A., Fields, M. J., and Thorgaard, G. H.,** BKm minisatellite sequences are not sex associated but reveal DNA fingerprint polymorphisms in rainbow trout, *Genome,* 32, 865, 1989.
117. **Withler, R.,** unpublished data.
118. **Rico, C., Kuhnlein, U., and Fitzgerald, G. J.,** Spawning patterns in the three-spined stickleback *(Gasterosteus aculeatus):* an evaluation by DNA fingerprinting, *J. Fish Biol.,* 39 (Suppl. A), 151, 1991.
119. **Arnheim, N.,** Concerted evolution of multigene families, in *Evolution of Genes and Proteins,* Nei, J., and Koehn, R. K., Eds., Sinauer, Sunderland, MA, 1983, chap. 3.
120. **Carlson, J. E., Tulsieram, L. K., Glaubitz, J. C., Luk, V. W., Kauffeldt, C., and Rutledge, R.,** Segregation of random amplified DNA markers in F_1 progeny of conifers, *Theor. Appl. Genet.,* 83, 194, 1991.
121. **Klein-Lankhorst, R. M., Vermunt, A., Weide, R., Liharska, T., and Zabel, P.,** Isolation of molecular markers for tomato *(L. esculentum)* using random amplified polymorphic DNA (RAPD), *Theor. Appl. Genet.,* 83, 108, 1991.
122. **Hillis, D. M., and Moritz, C.,** *Molecular Systematics,* Sinauer Associates, Sunderland, MA., 1990.
123. **Cochran, W. G.,** Some methods for strengthening the common χ^2 tests, *Biometrics,* 10, 417, 1954.
124. **Bentzen, P., Brown, G. G., and Leggett, W. C.,** Mitochondrial DNA polymorphisms, population structure, and life history variation in American shad *(Alosa sapidissima), Can J. Fish. Aquat. Sci.,* 46, 1446, 1989.
125. **Snyder, T. P., Larsen, R. D., and Bowen, S. H.,** Mitochondrial DNA diversity among Lake Superior and inland lake ciscoes *(Coregonus artedi* and *C. hoyi), Can. J. Fish. Aquat. Sci.,* 49, 1902, 1992.
126. **Roff, D. A., and Bentzen, P.,** The statistical analysis of mitochondrial DNA polymorphisms: χ^2 and the problem of small samples, *Mol. Biol. Evol.,* 6, 539, 1989.
127. **Excoffier, L., Smouse, P. E., and Quattro, J. M.,** Analysis of molecular variance inferred from metric distances among DNA haplotypes: application to human mitochondrial DNA restriction data, *Genetics,* 131, 479, 1992.

4

Genetic Aspects of PCB Biodegradation

James R. Yates[1] and Frank J. Mondello[2]

[1]Department of Biology, University of South Carolina at Aiken
and
[2]Bioremediation Laboratory, General Electric Co., Schenectady, NY

INTRODUCTION

Bacteria are remarkable in their ability to enzymatically transform a tremendous variety of organic compounds. Considering the relatively small number of environmental bacteria which have been examined, it is certain that their actual abilities far exceed those already identified. The introduction of organic pollutants into the environment is a nationwide problem for which bioremediation (biological destruction of pollutants) is an attractive solution. Environmental contaminants differ in the ease with which they can be biologically degraded. One of the most difficult groups of these compounds are the polychlorinated biphenyls (PCBs), due to their inherent chemical stability, low water solubility, and the fact that they were used and discarded as complex mixtures such as the Aroclors. They consist of between 60 and 80 different PCB molecules (congeners), and therefore require organisms with broad substrate specificity for effective remediation. Many bacteria have been isolated that are capable of utilizing the parent compound biphenyl (BP) as a sole carbon and energy source, and some of them are also capable of degrading PCBs.

0-87371-631-0/94/$0.00+$.50

Catabolism of BP involves a relatively well-defined biochemical pathway. The initial step involves the incorporation of both atoms of molecular oxygen (O_2) at the 2 and 3 positions of one ring. This reaction is catalyzed by the enzyme biphenyl dioxygenase. The product of this first reaction is a 2,3-*cis*-dihydrodiol, which is the substrate for the second enzyme dihydrodiol dehydrogenase. The third intermediate in the pathway is 2,3-dihydroxybiphenyl. Conversion of 2,3-dihydroxybiphenyl to the yellow ring-cleavage (or "*meta*-cleavage") compound is catalyzed by the enzyme 2,3-dihydroxybiphenyl dioxygenase. Assays for this enzyme are extremely useful because the *meta*-cleavage product (2-hydroxy-6-oxo-6-phenylhexa-2,4-dienoic acid) is intensely colored. Application of an ether spray containing 2,3-dihydroxybiphenyl to colonies expressing 2,3-dihydroxybiphenyl dioxygenase results in the appearance of a yellow compound in several minutes. The *meta*-cleavage product is then acted upon by the fourth enzyme (2-hydroxy-6-oxo-6-phenylhexa-2,4-dienoic acid hydrase) to give benzoic acid and the degraded ring. Degradation of PCBs is thought to follow this pathway, and several intermediates have been isolated and characterized.

The genetics of BP catabolism and PCB degradation are not fully understood, but recent investigations have shed some light on the problem. It is obvious that a diverse group of genes are involved, but the number of types is not known. DNA sequence analysis has been performed for some of these genes, but in no case have all of the genes in a given organism been sequenced. Finally, the mechanisms for regulating gene expression are not known and future efforts must be directed to this area.

PSEUDOMONAS SP. LB400

Investigators at the General Electric Co. (GE) Research and Development Center have identified and characterized a bacterium capable of growth on BP and degradation of a wide variety of PCB congeners. This Gram-negative strain was isolated from soil taken from a disposal site that contained various pollutants, including BP and PCBs, via enrichment on a minimal salts medium with BP.[1] Attempts to identify strain LB400 using standard biochemical techniques resulted in this organism being classified as a species of Pseudomonad. Attempts at more specific identification using techniques such as comparing its fatty acid profile and metabolic profile to those of other organisms were inconclusive. For example, when the metabolic profile of LB400 was determined using the Biolog GN microplate™ system, and compared to the entries in the GN database, the best match was to *Pseudomonas solanacearum* A. The SIM value between these two strains (0.293) was not sufficient to be considered a match, but the utilization patterns for 84 of the 95 carbon sources were identical (88%) (J. Yates, unpublished results). The inability to determine the phylogeny of LB400 is not too surprising since this is an environmental isolate, and although this database contains 569 entries, it is too small to identify most bacteria isolated from the environment.

LB400 has been shown to degrade an impressive array of PCB congeners, including some containing five or six chlorines. Interestingly, both the number and position of the chlorines on the BP nucleus affect the ability of LB400 to degrade PCB congeners. For example, in resting cell assays, LB400 degrades more of the pentachlorobiphenyls 2,3,4,2′,5′ and 2,4,5,2′,5′ than di- and trichlorobiphenyls containing chlorines in both *para* positions (such as 4,4′ and 2,4,4′). There are several possible explanations for this specificity of degradation. For example, there are clear differences in the physical characteristics (e.g., solubility in water) of the congeners and this could affect degradation rates. Also, substrate binding or catalysis of one or more enzymes may be affected by the structures of the different congeners.

Identification and Characterization of the *bph* Genes

Analysis of the LB400 genes involved in BP metabolism and PCB degradation (designated the *bph* genes) was facilitated when these genes were isolated and cloned into *Escherichia coli*.[2] The method used to clone these genes involved partial digestion of total LB400 DNA with *Sau*3A and subsequent ligation into the broad host-range cosmid pMMB34. A novel selection procedure was employed to detect the presence of one of the enzymes involved in BP catabolism. Recombinant colonies were tested for the presence of 2,3-dihydroxybiphenyl dioxygenase by addition of 2,3-dihydroxybiphenyl and selecting those that produced a yellow compound, 2-hydroxy-6-oxo-6-phenylhexa-2,4-dienoic acid (called the *meta*-cleavage product for convenience). Although the initial selection procedure required the presence of only one of the *bph* genes, two of the five cosmids encoding 2,3-dihydroxybiphenyl dioxygenase had all of the *bph* genes needed to convert PCBs to chlorobenzoic acids. In fact, recombinant *E. coli* with each of the remaining three cosmids were shown to have at least two of the enzymatic activities required for this conversion. Subsequent analysis has shown that the *bph* genes are closely linked on the LB400 chromosome.

One of the original five cosmids (pGEM410) was chosen for further analysis since it conferred all of the enzymatic activities involved in the catabolism of BP to benzoic acid and was stable in *E. coli*. Subcloning of the 26.2 kb insert of pGEM410 was employed to localize the *bph* genes and to determine the gene order.[2] These observations and more recent investigations[3] have shown that the gene order is *bphA, E, F, G, B, C, D* and they are contained within a 12.4 kb region. There are four genes (*bphAEFG*) that encode subunits of biphenyl dioxygenase, and the remainder of these genes (i.e., *bphB*, *bphC*, and *bphD* encode dihydrodiol dehydrogenase, 2,3-dihydroxybiphenyl dioxygenase, and 2-hydroxy-6-oxo-6-phenylhexa-2,4-dienoic acid hydrase, respectively). There is some evidence that all of the *bph* genes are not organized as an operon. *E. coli* containing recombinant plasmids with portions of the pGEM410 insert had only dihydrodiol dehydrogenase and 2,3-dihydroxybiphenyl dioxygenase activity. Another recombinant *E. coli* had only 2-hydroxy-6-oxo-6-phenylhexa-2,4-dienoic acid hydrase activity. The expression of *bph* genes from isolated fragments of the pGEM410 insert argues strongly that there are promoters for at least *bphAEFG* and *bphBC*. Sequence analysis of a DNA fragment containing *bphAEFG*, in addition to primer extension experiments, indicates that there are multiple promoters for these genes.[3]

Comparison of *bph* Genes

A number of bacteria capable of degrading PCBs have been isolated by GE scientists. This group of bacteria includes a variety of different species with a range of degradative capabilities. The exact relationships between the genes involved in PCB degradation present in these isolates is not known. However, the cloned *bph* genes of LB400 have been used as a probe for related sequences in some of these bacteria.[4]

Total DNA preparations from five strains — including *P. cepacia*, *P. testosteroni*, *Alcaligenes eutrophus*, *A. fecalis*, and *Corynebacterium* sp. — were examined for the presence of sequences related to the LB400 *bph* genes, via Southern hybridization experiments. In these experiments, pGEM410 was used as a probe, and under moderately stringent hybridization conditions (T_m-25°C), binding was detected only to DNA from *A. eutrophus* (strain H850). The *bph* genes of LB400 and H850 were compared in a separate experiment. Genomic DNAs from both organisms were digested with restriction enzymes (*Eco*R1, *Pst*I, and *Pvu*II) and probed with pGEM410.[4] The results of this experiment indicate that the *bph* genes of LB400 and H850 are very similar. A total of 16 restriction sites located within, or immediately adjacent to, the *bph* genes were conserved. However, restriction sites outside this region are not conserved, and regions of pGEM410 >7 kb from the *bph* genes did not exhibit significant hybridization to H850 DNA. It is of interest to note that the region conserved in LB400 and H850 seems to extend past the *bph* genes (3 to 7 kb). This striking similarity between the *bph* genes of LB400 and H850 provides an explanation for the observation[5] that these organisms have similar PCB degradative capabilities.

One possible explanation for the presence of closely related *bph* genes in LB400 and H850 is that these two organisms are closely related. Although tests performed by the American Type Culture Collection show numerous biochemical differences between LB400 and H850, another Southern hybridization experiment was used to compare non-*bph* DNA from these strains.[4] A probe was chosen that contained ~25 kb of LB400 sequences that did not cross-hybridize with pGEM410. Under hybridization conditions identical to those described above, there was no significant binding of this probe to H850 DNA. Clearly, there are large regions (>25 kb) of the LB400 and H850 genomes that exhibit much less sequence similarity than their respective *bph* genes. Therefore, it was concluded that LB400 and H850 are not closely related bacteria. The presence of closely related *bph* genes in these two dissimilar bacteria is a strong indication that these genes may be mobile. It is possible that these genes may be on a mobile genetic element, as has been shown for the *xyl* genes of the plasmid pWWO.[6] Usually, gene transfers between bacteria are plasmid mediated, but no evidence of plasmid-borne *bph* genes has been found in LB400 or H850.

Table 1
Stability of *bph* Expression in LB400

Experiment	No. of colonies	No. Bph-[a]	No. stable[b]	% loss[c]
1	452	9	8	1.8
2	310	10	10	3.2
3	591	6	6	1.0
4	1067	5	6	0.6
5	1515	22	21	1.4
6	1287	11	11	0.9
		Average = 1.5%		

[a] The number of colonies unable to convert 2,3-dihydroxybiphenyl to the yellow *meta*-cleavage product.
[b] These mutants reverted to the Bph^+ phenotype at a frequency of less than 1 in 10^{10}.
[c] Percent loss determined by number of stable mutants per number of colonies examined.

Instability of the *bph* Genes

It has been observed that long-term growth of LB400 under conditions that do not require the expression of the *bph* genes results in the loss of PCB degradative capabilities. Although no attempts have been made to quantify this phenomenon, it has been observed repeatedly. While investigating several cosmids containing the *bph* genes, we have noticed that cells with some cosmids lose the enzymatic activities associated with these genes (J. Yates and F. Mondello, unpublished results). Both of these observations could be due to instability of the *bph* genes in LB400 and recombinant *E. coli*.

Experiments were designed to test the stability of the *bph* genes and the results indicate that they are lost at a high frequency. These experiments involved growth of LB400 under nonselective conditions, and subsequent testing for colonies unable to express one or more *bph* genes. Colonies of LB400 were grown on agar plates containing a minimal salts medium (PAS) with BP supplied as the sole carbon source.[2] Several colonies were removed from these plates and added to BP-free broth cultures (PAS + 0.5% succinate). When each of the broth cultures had reached stationary phase, an aliquot was removed and added to fresh PAS + 0.5% succinate. This process was repeated until ~100 doublings had occurred. At this time, aliquots of the final broth culture were removed, serially diluted, and plated onto PAS + succinate agar plates. Three different procedures were used to calculate the percentage of Bph^- colonies present. Some of the plates were replica plated onto PAS + BP medium, and the number of colonies unable to grow were scored. Other plates were treated with a solution of 2,3-dihydroxybiphenyl and the number of colonies unable to produce the yellow ring-fission product were scored. The last procedure involved application of dilutions of the final broth culture onto PAS + BP + 0.02% succinate agar plates. Bph^+ cells produce normal sized colonies, while Bph^- cells produce very small colonies on this type of plate. These experiments were repeated six times and the results are shown in Table 1. The number of cells screened by the procedures described above varied from 310 to 1515. In all cases, Bph^- cells were detected, and

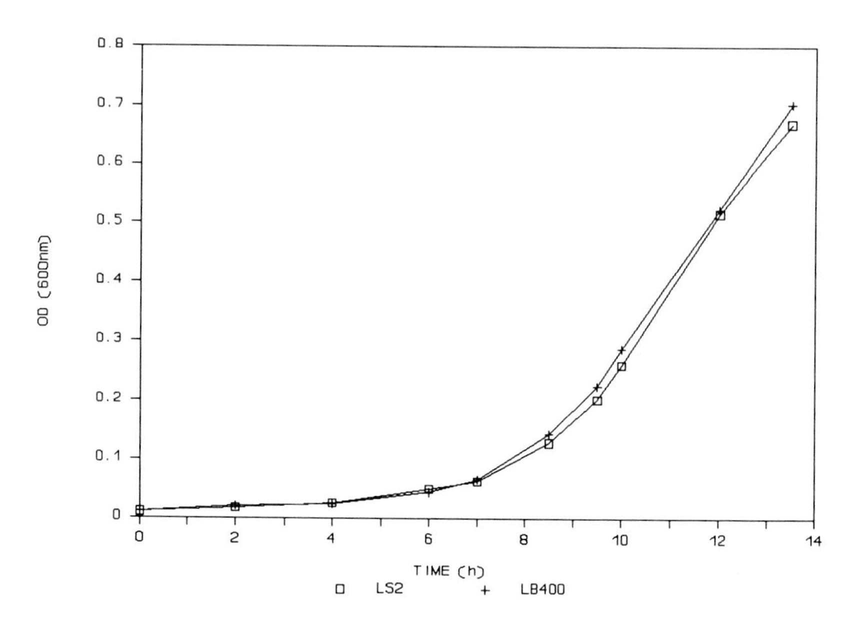

Figure 1. Comparison of the LB400 and LS2 growth curves. Aliquots of LB400 and LS2 broth cultures were diluted (1:100) into PAS + 0.5% succinate and grown at 30°C with shaking (200 RPM). Cell growth was estimated by measuring the OD_{600} periodically. The curve labeled with (+) corresponds to LB400, and the one with (□) corresponds to LS2.

64 mutants were isolated over the 6 trials. A total of 62 of these mutants were stable and reverted to Bph^+ at a frequency of <1 in 10^{10}. For each experiment, the percent loss was calculated by dividing the number of stable mutants by the total number of colonies tested. The percent loss ranged from 0.6 to 3.2%, and the mean for the six experiments was 1.5%. None of the 62 stable mutants was able to utilize BP as a growth substrate. They were tested for biphenyl dioxygenase and dihydrodiol dehydrase activities and all were negative (data not shown). Mutants isolated in these and similar experiments were designated LS (LB400 Spontaneous Bph^-).

Calculations of percent loss would clearly be affected by differences in the growth characteristics of LB400 and LS mutants. For example, if the LS mutants had shorter generation times than LB400 on PAS + succinate, then the number of Bph^- colonies present at the end of the experiment would appear to be too high. This would result in an overestimation of the percent loss. To test this possibility, the growth of one LS mutant (designated LS2) was compared to wild-type LB400. Typical growth curves for LB400 and LS2 are shown in Figure 1. As can easily be seen, the growth curves are essentially identical. The lag and exponential phases are the same for both cell types. These data can also be used to calculate the generation time for LB400 and LS2, and in both cases a value of ~2.1 hr was obtained. There appears to be no difference in the growth characteristics of LB400 and LS2 on PAS + succinate and therefore, the percent loss is a reasonable estimate of the proportion of cells in a population that have been converted to the Bph⁻ phenotype.

The data presented in Table 1 cannot be used to calculate the rate at which LS mutants arise in a population of LB400 cells. Further experiments will be required to determine this rate. However, it is interesting to note that cells derived from a colony growing on BP (presumably, wild-type LB400) give rise to a mixed population (~1.5% Bph⁻) after a relatively short period of time (equivalent to 100 doublings).

It is somewhat surprising that stable LS mutants lose all of the enzymatic activities associated with BP catabolism. One explanation for this observation involves deletion of the *bph* genes from the LB400 chromosome. A similar situation has been shown to occur with the *xyl* genes on pWWO.[7] This possibility was examined by testing for the presence of *bph* genes in LS mutants using Southern blots. Genomic DNA was prepared from five different LS mutants (LS1-5), digested with *Eco*R1, and probed with pGEM410. LS1 is an unstable mutant while LS2-5 are stable. The autoradiogram of this blot is shown in Figure 2. The data for LS2-5 are straightforward and easy to interpret (see Figure 2, lanes 3 to 6). Several bands (corresponding to the 6.7, 2.3, 2.0, and one of the 2.9 kb fragments) present when this probe is hybridized to wild-type LB400 genomic DNA are absent (indicated with arrowheads). It should be noted that pGEM410 contains two 2.9 kb *Eco*R1 fragments (see Figure 2, lane 1); and when LB400 DNA is probed with pGEM410, the band corresponding to these fragments is significantly darker than the 2.3 kb band. Only one of these 2.9 kb fragments is deleted from the LS2-5 genomes. The fragments that are absent correspond to those previously shown to contain the *bph* genes.[2] However, regions near the *bph* genes are present (i.e., the 6.1, 0.8, and the other 2.9 kb fragments). In addition, two fragments hybridizing to the ends of the pGEM410 insert are present (bands indicated with stars). The simplest interpretation of these data is that a deletion of ~12 kb has occurred in the LS2-5 genomes, and the *bph* genes are no longer present. The deleted region corresponds to the region of the H850 genome that is similar to the *bph* genes of LB400.[4]

The LS1 mutant (see Figure 2) has six of the fragments known to be present in wild-type LB400 and two extra bands (located above the 6.1 kb band). Inspection of lane 2 shows that a fragment slightly larger than the 6.1 kb fragment produces a relatively weak band. This blot was reprobed with a subclone of pGEM410 (called pGEM456) that contains the 6.7 and 2.9 kb fragments. This probe hybridizes to two fragments of the LS1 genome (i.e., the 2.9 kb and the largest fragment in lane 2). It appears likely that a segment of DNA (~3.9 kb) has inserted into the 6.7 kb fragment to create a 10.6 kb fragment. The weak band slightly above the 6.1 kb band does not hybridize to this probe. At this time, nothing is known about the fragment producing the weak band or how to explain its presence in LS1. The presence of an insertion into a fragment containing *bph* genes correlates well with the observation that LS1 is unstable and reverts to Bph⁺ at a much higher frequency than LS2-5.

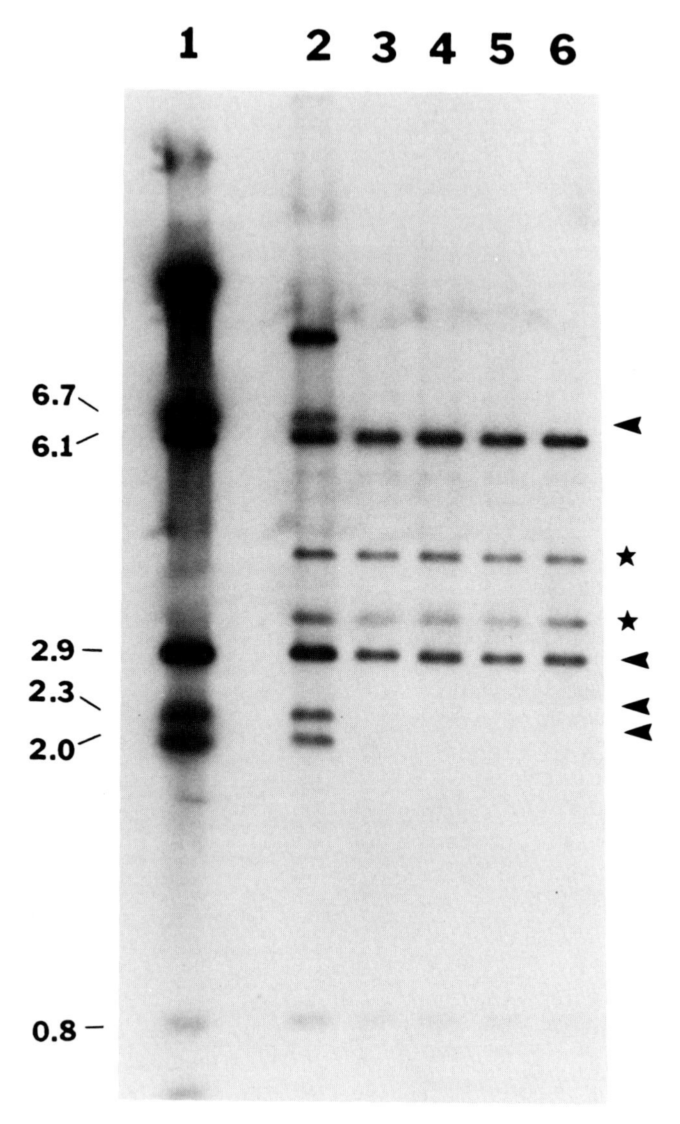

Figure 2. Southern blot of several LS mutants. Approximately 2 μg of genomic DNA from each of the LS mutants, and ~0.1 μg of pGEM410 were digested with *Eco*R1; the resultant fragments were separated by electrophoresis in a 1% agarose gel. The DNA was denatured and transferred to a nylon membrane. The membrane was prehybridized and hybridized according to the manufacturer's specifications (42°C in 50% formamide). The probe was pGEM410 (labeled via nick translation using [^{32}P]dCTP). After removal of the nonspecifically bound probe, autoradiography was performed. Lane 1, pGEM410; lane 2, LS1; lane 3, LS2; lane 4, LS3; lane 5, LS4; lane 6, LS5. The bands that are absent in LS2-5 are indicated with arrowheads. The bands that correspond to fragments hybridizing to the ends of the pGEM410 insert are indicated with stars.

PSEUDOMONAS PSEUDOALCALIGENES KF707

A PCB-degrading strain of *P. pseudoalcaligenes* was isolated from soil taken from a site near a BP processing facility, via an enrichment procedure using BP as a sole carbon and energy source.[8] The PCB degradative capabilities of this strain (designated KF707) were tested using a resting-cell type assay. In these assays, one monochloro (4-Cl-BP), three dichloro (2,3-, 3,4-, and 2,4′-diCl-BP), and 2,4,5-triCl-BP were catabolized to chlorinated benzoic acids. However, degradation of 2,4,4′-triCl-BP was incomplete and the corresponding *meta*-cleavage product was detected.

Identification and Characterization of the *bph* Genes

The procedure used to clone the genes involved in BP catabolism and PCB degradation was slightly different than that described for LB400. Genomic DNA was prepared from KF707 cells and digested with *Xho*I. A genomic library was created by inserting these fragments into the broad host-range plasmid pKF330 and transformation of the chimeric plasmids into *P. aeruginosa*.[8] The selection procedure involved screening of transformants for 2,3-dihydroxybiphenyl dioxygenase activity. Out of the 8000 colonies that were tested for the ability to convert 2,3-dihydroxybiphenyl to the yellow *meta*-cleavage product, only one tested positive. The recombinant plasmid (designated pMFB1) conferring this activity was purified and analyzed. The plasmid contains a 7.9 kb insert; Southern hybridization experiments were used to demonstrate that the *bph* genes originated from the KF707 chromosome, and not from an indigenous plasmid.

Recombinant *P. aeruginosa* bearing pMFB1 were capable of converting BP to the *meta*-cleavage product.[8] From these observations, it was concluded that the first three enzymes involved in the catabolism of BP were encoded on this plasmid. These genes are also involved in PCB degradation, and this was demonstrated by incubation of the recombinant bacterium with 4-Cl-BP; 3,4-diCl-BP; 2,4′-diCl-BP; and 2,4,5-triCl-BP.[8] In each case, a yellow compound (assumed to be the *meta*-cleavage product) was produced. Two metabolites from 2,4,5-triCl-BP were analyzed via gas chromatography-mass spectrometry (GC-MS). A dihydroxylated intermediate and a ring-opened intermediate (i.e., the *meta*-cleavage product) were identified. These results show clearly that the *bphABC* genes of KF707 are involved in PCB degradation as well as BP metabolism. The genes encoding biphenyl dioxygenase, dihydrodiol dehydrogenase, and 2,3-dihydroxybiphenyl dioxygenase were designated *bphA*, *B*, and *C*, respectively. Accumulation of the *meta*-cleavage product indicates that pMFB1 probably does not contain a functional *bphD* gene (encoding 2-hydroxy-6-oxo-6-phenylhexa-2,4-dienoic acid hydrase).

The levels of *bphC* and *bphD* activity in KF707 and recombinant *P. aeruginosa* were compared in cell extracts.[8] KF707 grown on succinate had lower levels of *bphC* and *D* expression than those grown on BP. It was suggested that this was due to induction of *bph* gene expression by BP. Cell extracts from *P. aeruginosa* with pMFB1 grown on either BP or succinate had approximately twice the 2,3-dihydroxybiphenyl dioxygenase activity as KF707 grown on BP. Expression of *bphD* in these recombinant cells was not detected. This experiment provided direct evidence that *bphD* is either incomplete, absent, or not expressed in pMFB1. Construction of pMFB1 involved insertion of the KF707 DNA insert immediately downstream of the Km^R promoter of pKF330. The suggestion has been made[8] that the failure of BP to induce *bphABC* gene expression in recombinant bacteria is due to transcription of these genes from the Km^R promoter. However, subsequent subcloning experiments indicated that this was unlikely.

Several subclones of the pMFB1 insert were constructed and used to examine *bph* gene positions and expression. Digestion of the pMFB1 insert with *Xho*I results in the formation of a 0.8 kb fragment and a ~7.1 kb fragment. Two subclones were created by inserting the larger fragment into pKF330 in both orientations. Cells bearing a subclone containing the 7.1 kb fragment in the same orientation as in pMFB1 converted BP to the *meta*-cleavage product. Measurements of 2,3-dihydroxybiphenyl dioxygenase activity indicated that the levels were higher (53%) than in the original construct. The subclone with the 7.1 kb fragment inserted in the opposite orientation had levels of 2,3-dihydroxybiphenyl dioxygenase activity similar to the original construct. It appears unlikely that expression of *bphC* was due to the Km^R promoter of pKF330 in this case. Digestion of the pMFB1 insert with *Sma*I results in two fragments (2.5 and 5.4 kb). A subclone containing the 2.5 kb fragment expressed *bphC*, and a subclone containing the 5.4 kb fragment expressed *bphA*. Expression of *bphB* was not detected in either subclone, suggesting that the internal *Sma*I site probably is located within *bphB*.

DNA Sequence Analysis of *bphC*

A plasmid containing the *bphC* gene was obtained by inserting a 2.1 kb *Pst*I fragment of the pMFB1 insert into pUC8 and selecting for recombinant *E. coli* that were capable of converting 2,3-dihydroxybiphenyl to the yellow *meta*-cleavage product.[9] DNA sequence analysis of this 2040 bp insert showed it to contain one complete open reading frame (ORF) consisting of 912 bp (encoding 303 aa). The amino acid sequence predicted from this ORF indicates that a protein with a molecular weight of 33,074 would be produced and would be very hydrophobic. The predicted aa sequence of this ORF was compared to the N-terminal sequence of purified 2,3-dihydroxybiphenyl dioxygenase. Assuming that the N-terminal methionine is removed posttranslationally, the predicted and observed amino acid sequences were identical, demonstrating that this ORF did indeed encode the *bphC* gene. The *bphC* coding sequence was preceded by a potential ribosome binding site (RBS) located 5 bp upstream from the initiator ATG. No sequences similar to either the consensus promotor sequence of *E. coli* or the promotor sequences of *Pseudomonas* were detected upstream of the ORF. The ORF is

59.8% G+C, which is slightly lower than the overall value for *P. pseudoalcaligenes*. This difference is probably not significant.

The fragment containing the *bphC* gene also had two incomplete ORFs that were presumably truncated during the subcloning.[9] An incomplete ORF was detected upstream of *bphC*, which may be the C-terminal portion of *bphB*. Also, an incomplete ORF was detected downstream of *bphC*, which was presumed to be the N-terminal portion of *bphD*. However, a later report demonstrated that *bphD* is actually located ~3.3 kb away from *bphC* (see Reference 12).

Comparison of the KF707 and Q1 *bphC* Genes

A strain of *P. paucimobilis*, designated Q1, was isolated from soil via enrichment on BP. Since this strain is capable of utilizing BP as a growth substrate, and catabolism proceeds via the usual 2,3-dioxygenase pathway, the presence of a gene analogous to the KF707 *bphC* gene was postulated.[10] Obviously, a comparison of these two genes would shed some light on the diversity of the enzymes (and genes) involved in the metabolism of 2,3-dihydroxybiphenyl. Cloning of the Q1 *bphC* gene was relatively straightforward. Chromosomal DNA was digested with *Xho*I, inserted into pKF330, and transformed into *P. aeruginosa*. Transformants expressing *bphC* were selected using 2,3-dihydroxybiphenyl as previously described. A clone that tested positive was shown to have a 2.6 kb insert derived from the chromosome of Q1. The nucleotide sequence of a portion of this insert was determined. The sequence contained one ORF of 900 bp (encoding a protein of 299 aa) that was preceded by a potential RBS, located 3 bp upstream from the ATG start codon. Inspection of the adjacent sequences revealed no evidence of a promoter. The ORF was shown to be 62% G+C and this is similar to the KF707 *bphC* gene (59.8%). The sequence of amino acids 2-16 predicted from the DNA sequence was identical to the N-terminal portion of the purified protein. The assumption was made that the N-terminal methionine was removed posttranslationally, as seen in the KF707 enzyme.

A comparison of the nucleotide sequences of *bphC* from KF707 and Q1 revealed that they exhibited 60% similarity when optimally aligned. There were regions with extensive similarities and also very dissimilar regions. One small region was detected where 20 out of 21 bases were identical. Codon usage in both genes was similar, although there were slight differences.

A comparison of the primary structures of the two proteins was presented[10] and identical amino acids were detected in 38% of the positions, but the percentage of positions with similar amino acids was not given. The differences between these two proteins were reflected in the fact that antisera raised against the Q1 enzyme did not react with the KF707 enzyme, and vice versa. The differences in the structures of these two enzymes is not surprising, but they do have several common features: number of subunits (8), subunit molecular weight (~33,000), native molecular weight (~260,000), and cofactor (Fe^{+2}).

Other *bph* Genes Related to Those in KF707

The cloned *bphABC* genes of KF707 were used to detect related sequences in 15 strains of bacteria.[11] All were capable of growth on BP and benzoic acid, and five of the strains were reported to grow on 4-Cl-BP. Genomic DNA was prepared from each strain and Southern hybridization experiments were used to detect strains with *bph* genes similar to KF707. When the *bphA* gene was used as a probe, six strains had banding patterns identical to that of KF707. Therefore, restriction sites are likely to be conserved in these organisms. DNA from three other strains hybridized to this probe, but showed restriction length polymorphisms (RFLPs). The remaining six strains, one of which was Q1, did not bind the probe under the hybridization conditions used.

In another experiment, 13 of these strains were examined for sequences related to the KF707 *bphC* gene.[11] Two strains (Q1 and another that did not bind the *bphA* probe in the previous experiment) were omitted from this experiment. Of the strains tested, the six that showed identical banding patterns to KF707 with the *bphA* probe again showed identical banding patterns with the *bphC* probe. RFLPs were detected with the *bphC* probe in the same three strains that exhibited RFLPs with the *bphA* probe. DNA from the four remaining strains did not bind the *bphC* probe. These data lead to the conclusion that there are at least two types of *bph* genes related to those in KF707. One type is very similar to the KF707 *bphA* and *C* genes (exhibiting no RFLPs), and the other is somewhat divergent, but still related. As has been shown with LB400 and H850,[4] the conserved regions containing *bph* genes are surrounded by very divergent sequences, suggesting that the genes were obtained via some type of lateral transfer mechanism. At this time, no direct comparisons of the *bph* genes of KF707 and LB400/H850 have been reported.

Surprisingly, two strains that exhibited RFLPs with either *bphA* or *C* had four sites in common with KF707 when probed with a larger probe containing *bphABC*. The *bph* genes of one of these strains (KF715) were compared to those of KF707 in a later report.[12] The third strain had no restriction sites in common with KF707. However, it should be noted that quite strong hybridization was detected to all three strains. The degree of sequence divergence between KF707 and these three strains cannot be estimated from these data.

A probe containing the *bphC* gene of Q1 was used to identify genes related to the 15 strains tested above. None of these strains, including the five that did not hybridize to any of the KF707 probes, exhibited detectable hybridization to the Q1 gene.

Comparisons of the *bph* Genes in KF707 and KF715

A strain of *P. putida* (designated KF715) has been shown to posses a *bph* region similar, but not identical, to KF707. The KF715 *bph* genes were cloned and analyzed in order to compare the sequence and organization of these related genes.[12] Previously,[11] it had been shown that the *bphABC* genes of KF715 were present on a 9.4 kb *Xho*I fragment. Deletion analysis of a plasmid containing this fragment that expressed 2,3-dihydroxybiphenyl dioxygenase revealed that the gene order was *bphA, B, C, D* and the four genes spanned a region of ~7.8 kb.

Several of the restriction sites present in the *bph* region of KF715 were mapped and compared to those in KF707.[12] The *bphA* genes in both strains are relatively large (~4 kb), and appear to be quite similar within the first 2.7 kb because all of the sites mapped in this region (a total of nine) were identical. However, no common sites were found in the remainder of the region containing *bphA, B, C,* or *D* (15 nonconserved sites were mapped). The absence of DNA sequence information makes it difficult to estimate the degree of similarity between the *bphA* genes of KF707 and 715; however, such information is available for the *bphC* genes of these strains. The DNA sequences of these two genes were 92.4% identical, and derived amino acid sequences were 91.4% identical. The only major difference is an 18-bp deletion at the 3′ end of the KF715 gene.

DNA sequence comparisons of the partial ORFs immediately upstream of *bphC* in KF707 and KF715 showed that these regions were ~95% identical; thus, the putative *bphB* genes of these two strains are closely related. In KF707, there appears to be an extra 3.3 kb of DNA between the *bphC* and *D* genes that is not present in KF715. This could represent sequences that have been lost by KF715. This 3.3 kb region has not been shown to be involved in BP catabolism (or to have any other function).

Transposon Mutagenesis of the *bph* Genes

A recent report[13] describes the creation of a series of mutations in the cloned *bph* genes of KF707 by random insertion of a Tn*5* derivative, and the exchange of these mutated genes for the wild-type chromosomal genes. This procedure was also used to introduce insertional mutations into the chromosome of three strains previously shown to have *bph* genes similar to KF707.[11] Presumably, these mutants will be useful for determining the effect of *bph* mutations on BP and PCB degradation in KF707 and strains with related genes.

PSEUDOMONAS SP. KKS102

This unspeciated Pseudomonad (designated KKS102) was isolated via enrichment on BP and PCBs from soil near an oil refinery. Mixtures of PCBs containing tetrachlorobiphenyls were efficiently degraded by a mixed culture of KKS102 and a strain of *P. fluorescens*. Degradation of PCBs was shown to be dependent on the presence of KKS102. Clearly, the genes in this bacterium should be compared to other *bph* genes.

Gene Cloning and Analysis

Using a two-step cloning procedure,[14] a 29 kb insert that expressed 2,3-dihydroxybiphenyl dioxygenase in *P. putida*, was subcloned to identify a 3.2 kb DNA fragment that expressed this enzyme in *E. coli*. The approximate position of the gene encoding 2,3-dihydroxybiphenyl dioxygenase was determined using transposon mutagenesis.

DNA sequence analysis was performed in order to identify and characterize the genes present in this 3.2 kb insert.[14] Approximately 2.1 kb of the insert was sequenced and two ORFs were detected. Both were oriented in the same direction and were designated ORFI and ORFII. ORFI was 882 bp long and encoded 2,3-dihydroxybiphenyl dioxygenase. This ORF is preceded by an RBS and a sequence similar to the consensus sequence proposed for *P. putida* promoters. The nucleotide sequence of ORFI was compared to the KF707 *bphC* gene and exhibited 68% identity. Alignment of these two sequences shows that there are regions of high similarity interspersed with less similar ones. In addition, the KKS102 ORFI has a deletion of 30 bp at the 3′ end (similar to the 18-bp deletion seen in KF715). The amino acid sequences predicted from ORFI and KF707 *bphC* were 66% identical. It therefore appears that these two genes are related, although ORFI is not as closely related to the KF707 *bphC* gene as is the gene in KF715.

ORFII is 834 bp long and probably encodes the gene for 2-hydroxy-6-oxo-6-phenylhexa-2,4-dienoic acid hydrolase. This ORF is preceded by a potential RBS 23 bp downstream from the ORFI termination codon. In a later report,[12] the amino acid sequences of ORFII and KF715 *bphD* were compared. Identical amino acid residues were found in 79.5% of the positions. At this time, information on other genes involved in BP and PCB degradation in KKS102 is not available, and therefore it is not possible to compare them to the KF707 *bphA* and *B* genes. However, it does appear that the KKS102 genes are organized more like those in KF715; i.e., the *bphC* and *D* genes are adjacent to each other and not separated by 3.3 kb as is seen in KF707.

PSEUDOMONAS PUTIDA OU83

This strain is capable of utilizing BP and 4-Cl-BP as sole carbon and energy sources, and analysis of the genes involved has been initiated. Some of the intermediates in the degradation pathway have been identified and are consistent with those produced by other bacteria. For instance, oxidation of BP[15] and 4-Cl-BP[16] to the corresponding 2,3-dihydroxybiphenyls has been confirmed using GC-MS. In addition, unlike other known PCB-degrading bacteria, strain OU83 can convert 4-Cl-BP to both 4-chlorobenzoic acid and benzoic acid.

Identification and Characterization of the *cbp* Genes

A two-step procedure was employed to clone the genes involved in BP catabolism and 4-Cl-BP degradation into *E. coli*.[15,16] First, a genomic library was prepared by inserting large DNA fragments into the broad host-range cosmid pCP13 and infection of *E. coli* with the packaged cosmids. Cosmids were transferred from recombinant *E. coli* to *P. putida* (AC812) via conjugation. This step was thought to be necessary because expression of the *cbp* genes might be too low to detect in a heterologous system (i.e., *E. coli*). Colonies of recombinant *P. putida* were assayed for 2,3-dihydroxybiphenyl dioxygenase activity via the standard procedure. Recombinant plasmids were prepared from colonies that tested positive, and when introduced into *E. coli* via transformation, expression of 2,3-dihydroxybiphenyl dioxygenase was detected. Therefore, it appears that expression of OU83 genes in *E. coli* is not severely impaired.

Analysis of recombinant cosmids from colonies that tested positive for 2,3-dihydroxybiphenyl dioxygenase revealed the presence of common restriction fragments.[16] One of these cosmids (designated pOH88) contained a 23.7 kb insert and was chosen for further analysis. Fragments of this insert were cloned into pUC18 and tested for 2,3-dihydroxybiphenyl dioxygenase activity. The gene encoding this enzyme (designated *cbpC*) was found in a recombinant plasmid (designated pAW6194) containing a 9.8 kb *Eco*R1 fragment. The *cbpC* gene of OU83 appears to be analogous to the *bphC* genes of LB400/H850 and KF707/KF715. Recombinant *E. coli* with pAW6194 expressed 2,3-dihydroxybiphenyl dioxygenase at levels that were approximately three times higher than recombinant *P. putida* with pOH88. SDS polyacrylamide gel electrophoresis was used to identify two protein bands (~55 and ~22 kd) that were expressed from the insert of pAW6194 in recombinant *E. coli*.

The growth of several recombinant *P. putida* on BP and 4-Cl-BP was compared to the wild-type OU83 strain.[16] Four of the recombinants grew better than OU83 on BP, and two of these recombinants also grew better than the parental strain on 4-Cl-BP. Degradation of 4-Cl-BP by both the recombinants and OU83 resulted in the accumulation of 4-chlorobenzoic acid and benzoic acid. Therefore, it appears that all of the genes necessary for the initial degradation of 4-Cl-BP are present on pOH88. The production of benzoic acid from 4-Cl-BP can be explained by (1) attack on the chlorinated ring, or (2) dechlorination of the ring followed by degradation of either ring.

Further analysis of the OU83 *cbp* genes has been described in a recent report.[17] Recombinant *E. coli* with pAW6194 were capable of converting 4-Cl-BP, 3-Cl-BP, 2,4-DiCl-BP, and 2,4,5-TriCl-BP to their corresponding *meta*-cleavage products. Analysis of the degradation products of 4-Cl-BP showed that this compound was converted to 4-chlorobenzoic acid. Clearly, all of the required genes must be expressed in *E. coli*. Several restriction sites in the pAW6194 insert have been mapped. Transposon mutagenesis and subcloning were used to determine the location and organization of the *cbp* genes in pAW6194. The gene order was determined to be *cbpADCB*. The *cbpA* gene was localized to a 2.8 kb *Sal*I fragment and is separated from the *DCB* genes by ~3 kb. The presence of a promoter located upstream of these genes was detected by subcloning an 8.0 kb fragment of pAW6194 into pKK232-8 (a promoter selection plasmid). Expression of chloramphenicol acetyltransferase was detected with the insert in only one orientation.

DNA sequence data on the OU83 *cbp* genes is not available at this time and, therefore, it is not possible to compare them to other sequences. However, if the gene order is correct, then the *cbp* genes may be very different from all of those previously described. Also, the suggestion has been made[16] that OU83 has two types of 2,3-dihydroxybiphenyl dioxygenase with different substrate specificities. Further investigations into the *cbp* genes of OU83 are clearly indicated.

PSEUDOMONAS TESTOSTERONI B-356

This organism, which was isolated from activated sludge, is capable of growth on BP and 4-Cl-BP although the doubling times are rather long (7 and 26 hr, respectively). In addition, B-356 has been shown to degrade two other monochloro (2-Cl and 3-Cl), four dichloro (2,3-diCl; 2,4-diCl; 3,4-diCl; 2,4′-diCl; and 4,4′-diCl), and two trichloro (2,4,5-triCl and 2,4,4′-triCl) biphenyls.[18] The major metabolite that accumulates when B-356 is grown on 4-Cl-BP is 4-chlorobenzoic acid. It has been proposed that degradation of BP and 4-Cl-BP proceeds via the usual 2,3-dioxygenase pathway.[18]

Identification and Characterization of the *bph* Genes

The presence of *bph* genes analogous to those seen in other bacteria was inferred from the catabolic pathway deduced from the 4-Cl-BP metabolites. The genes encoding the enzymes of this pathway were cloned by inserting fragments of B-356 DNA into the broad host-range cosmid pPSA842.[18] Recombinant *P. putida* were screened by spraying a film of 4-Cl-BP onto the plates and selecting those that produced a clear ring around the colony. Two cosmids were isolated from colonies selected by this procedure. One of them (designated pDA1) had a 21.6 kb insert, and the other (designated pDA2) had a 24.9 kb insert. Partial restriction site maps of inserts in pDA1 and pDA2 showed a 9.1 kb region common to both.

Although recombinant *P. putida* with either pDA1 or pDA2 were unable to grow on BP or 4-Cl-BP, metabolism of both compounds was detected. In resting cell assays, *P. putida* with pDA1 converted 4-Cl-BP to 4-chlorobenzoic acid; with pDA2, the *meta*-cleavage product was detected. The presence of 4-Cl-BP during the growth of these recombinants did not affect the expression of the *bph* genes, but increased expression was detected in B-356. This observation led to the suggestion[18] that the *bph* genes are inducible in B-356, but other interpretations are possible. Recombinant *P. putida* with either cosmid were capable of degrading the same mono, di- and tri-chlorobiphenyls that were attacked by B-356.

The insert of pDA2 was subcloned in an attempt to locate the *bph* genes.[18] The results of these experiments indicated that the *bphABC* genes are present within a 9.1 kb region also found in pDA1. The location of *bphD* was not identified, but the insert of pDA2 is truncated immediately after the region shown to have the *bphA*, *B*, and *C* genes. It is unfortunate that pDA1 was not analyzed in the same manner. This cosmid probably has a functional *bphD* gene because recombinant *P. putida* with this cosmid are able to convert 4-Cl-BP to 4-chlorobenzoic acid. The gene order was not determined.

An attempt was made to examine other PCB-degrading bacteria for sequences similar to those present on pDA1 and pDA2. A dot-blot experiment was performed in which DNA from 12 PCB-degrading bacteria, and two plasmids (pKF1 and pSS50) that have been implicated in PCB degradation, were probed with a mixture of pDA1 and pDA2. Genomic DNAs from six strains bound these probes, but strong hybridization was not detected to any of them. It was somewhat surprising that these probes hybridized weakly to LB400 DNA, but no hybridization was detected to DNA from H850. As stated previously, the *bph* genes in these two strains appear to be very similar. Possibly, the observed hybridization is due to sequences other than the *bph* genes on pDA1 and pDA2. Southern blots with the 9.1 kb fragment containing *bphABC* would unambiguously show the presence of related sequences in the strains tested.

REFERENCES

1. **Bopp, L. H.,** Degradation of highly chlorinated PCBs by *Pseudomonas* strain LB400, *J. Ind. Microbiol.*, 1, 23, 1986.
2. **Mondello, F. J.,** Cloning and expression in *Escherichia coli* of *Pseudomonas* strain LB400 genes encoding polychlorinated biphenyl degradation, *J. Bacteriol.*, 171, 1725, 1989.
3. **Erickson, B. D., and Mondello, F. J.,** accepted for publication.
4. **Yates, J. R., and Mondello, F. J.,** Sequence similarities in the genes encoding polychlorinated biphenyl degradation by *Pseudomonas* strain LB400 and *Alcaligenes eutrophus* H850, *J. Bacteriol.*, 171, 1733, 1989.
5. **Bedard, D. L., Unterman, R., Bopp, L. H., Brennan, M. J., Haberl, M. L., and Johnson, C.,** Rapid assay for screening and characterizing microorganisms for the ability to degrade polychlorinated biphenyls, *Appl. Environ. Microbiol.*, 51, 761, 1986.
6. **Tsuda, M., and Iino, T.,** Genetic analysis of a transposon carrying toluene degrading genes on a TOL plasmid pWWO, *Mol. Gen. Genet.*, 210, 270, 1987.
7. **Meulien, P., Downing, R. G., and Broda, P.,** Excision of the 40 kb segment of the TOL plasmid from *Pseudomonas putida* mt-2 involves direct repeats, *Mol. Gen. Genet.*, 184, 97, 1981.
8. **Furukawa, K., and Miyazaki, T.,** Cloning of a gene cluster encoding biphenyl and chlorobiphenyl degradation in *Pseudomonas pseudoalcaligenes*, *J. Bacteriol.*, 166, 392, 1986.
9. **Furukawa, K., Arimura, N., and Miyazaki, T.,** Nucleotide sequence of the 2,3-dihydroxybiphenyl dioxygenase gene of *Pseudomonas pseudoalcaligenes*, *J. Bacteriol.*, 169, 427, 1987.
10. **Taira, K., Hayase, N., Arimura, N., Yamashita, S., Miyazaki, T., and Furukawa, K.,** Cloning and nucleotide sequence of the 2,3-dihydroxybiphenyl dioxygenase gene from the PCB-degrading strain of *Pseudomonas paucimobilis* Q1, *Biochemistry*, 27, 3990, 1988.
11. **Furukawa, K., Hayase, N., Taira, K., and Tomizuka, N.,** Molecular relationship of chromosomal genes encoding biphenyl/polychlorinated biphenyl catabolism: some soil bacteria possess a highly conserved *bph* operon, *J. Bacteriol.*, 171, 5467, 1989.

12. **Hayase, N., Taira, K., and Furukawa, K.,** *Pseudomonas putida* KF715 *bphABCD* operon encoding biphenyl and polychlorinated biphenyl degradation: cloning, analysis and expression in soil bacteria, *J. Bacteriol.*, 172, 1160, 1990.
13. **Furukawa, K., Hayashida, S., and Taira, K.,** Gene-specific transposon mutagenesis of the biphenyl/polychlorinated biphenyl-degradation-controlling *bph* operon in soil bacteria, *Gene*, 98, 21, 1991.
14. **Kimbara, K., Hashimoto, T., Fukuda, M., Koana, T., Takagi, M., Oishi, M., and Yano, K.,** Cloning and sequencing of two tandem genes involved in degradation of 2,3-dihydroxybiphenyl to benzoic acid in the polychlorinated biphenyl-degrading soil bacterium *Pseudomonas* sp. KKS102, *J. Bacteriol.*, 171, 2740, 1989.
15. **Khan, A., Tewari, R., and Walia, S.,** Molecular cloning of 3-phenylcatechol dioxygenase involved in the catabolic pathway of chlorinated biphenyl from *Pseudomonas putida* and its expression in *Escherichia coli*, *Appl. Environ., Microbiol.*, 54, 2664, 1988.
16. **Khan, A., and Walia, S.,** Cloning of bacterial genes specifying degradation of 4-chlorobiphenyl from *Pseudomonas putida* OU83, *Appl. Environ. Microbiol.*, 55, 798, 1989.
17. **Khan, A. A., and Walia, S. K.,** Expression, localization and functional analysis of polychlorinated biphenyl degradation genes *cbpABCD* of *Pseudomonas putida*, *Appl. Environ., Microbiol.*, 57, 1325, 1991.
18. **Ahmad, D., Masse, R., and Sylvestre, M.,** Cloning and expression of genes involved in 4-chlorobiphenyl transformation by *Pseudomonas testosteroni*: homology to polychlorobiphenyl-degrading genes in other bacteria, *Gene*, 86, 53, 1990.

5

Molecular Analysis of Aromatic Hydrocarbon Degradation

Gerben J. Zylstra

Center for Agricultural Molecular Biology, Cook College, Rutgers University

INTRODUCTION

Aromatic hydrocarbons are found ubiquitously in nature. It is generally accepted that the majority of these compounds are formed through the pyrolysis of organic matter. Unsubstituted polycyclic aromatic hydrocarbons are generally formed at high temperatures of pyrolysis, while alkyl-substituted aromatic hydrocarbons are generally formed at low temperatures of pyrolysis.[1] Crude petroleum is formed under the latter conditions, and microorganisms that have the ability to utilize aromatic hydrocarbons as carbon sources can readily be isolated from locations where there is a history of petroleum seepage to the surface.[2] Many of these aromatic hydrocarbons are suspected carcinogens. The amount of aromatic hydrocarbons found in air and soil samples has increased over the last century.[3,4] This is mainly due to the increased use of petrochemicals by modern society. Environmental contamination by aromatic hydrocarbons stems from the isolation, processing, combustion, and disposal of fossil fuels. In addition, the production of value-added petroleum products such as polymers, plastics, pesticides, solvents,

0-87371-631-0/94/$0.00+$.50

explosives, and even pharmaceuticals can lead to release of aromatic hydrocarbons into the environment.

Given the ubiquitous and historical nature of aromatic hydrocarbons found in the environment, it is not surprising that microorganisms have evolved that are capable of exploiting this catabolic niche. It is relatively easy to isolate microorganisms from the environment that have the ability to utilize simple aromatic hydrocarbons as carbon sources. In addition, with the right environmental source material, patience, and time, one can isolate — through the enrichment culture technique — microorganisms that have the ability to degrade more complex aromatic hydrocarbons. During the past two decades, much research has centered on determining not only the biochemical and genetic mechanisms through which such biodegradation occurs, but also on determining the diversity of mechanisms by which aromatic hydrocarbons are biodegraded. Much activity is focused on applying the knowledge that is learned on improving the ability of microorganisms to biodegrade aromatic hydrocarbons in the environment, waste treatment plants, and industrial effluents.

CLONING AND ANALYSIS OF THE GENES FOR AROMATIC HYDROCARBON DEGRADATION

Molecular Approaches to Toluene Degradation

Toluene Catabolic Pathways

The degradation of toluene by microorganisms serves as the paradigm for aromatic hydrocarbon degradation. Multiple catabolic pathways encoded by different microorganisms are known for toluene degradation, exhibiting the multiple ways by which aromatic compounds could potentially be degraded. Five aerobic catabolic pathways have been elucidated for toluene degradation by different microorganisms. The catabolic pathways for toluene degradation under anaerobic conditions have only recently been described.[5,6] Aerobically, catabolism of toluene proceeds through oxygenated intermediates to eventually produce a 1,2-dihydroxylated aromatic compound. The latter compound is subject to ring cleavage, either between the hydroxyl groups (*ortho*-cleavage) or adjacent to the hydroxyl groups (*meta*-cleavage). Further metabolic reactions lead to products that enter the tricarboxylic acid cycle. These multiple aerobic pathways are shown in Figure 1. The best studied toluene catabolic pathway is that encoded by the TOL plasmid (pathway A in Figure 1). The TOL catabolic pathway[7,8] initially involves the oxidation of the methyl substituent group of toluene successively to benzyl alcohol, benzaldehyde, and then to benzoic acid. Benzoate is metabolized through a *cis*-dihydrodiol intermediate to form catechol. Catechol is then the substrate for *meta*-cleavage of the aromatic ring to form 2-hydroxymuconic semialdehyde. Subsequent reactions lead to acetaldehyde and pyruvate.

In addition to initiating degradation of toluene by attack on the methyl substituent, catabolic pathways are known that initiate degradation through oxidation of

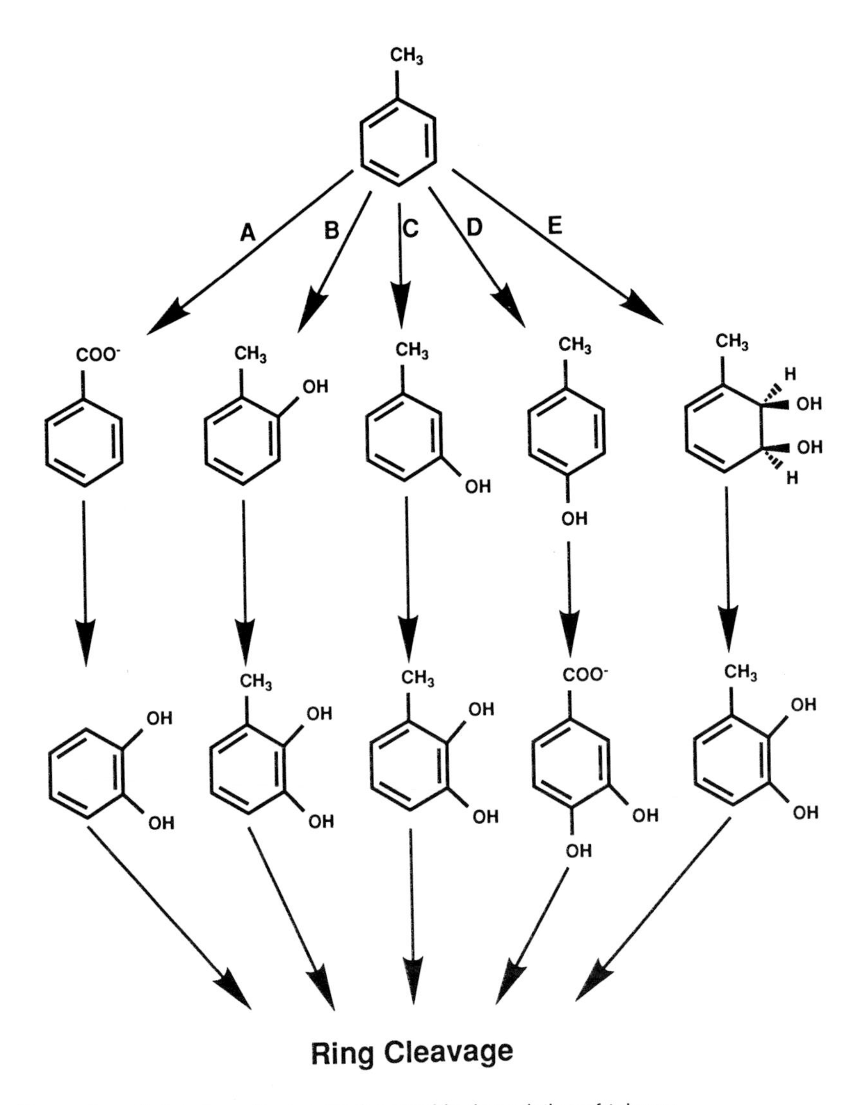

Figure 1. Catabolic pathways for the aerobic degradation of toluene.

the aromatic ring through oxygenase attack. For instance, *Pseudomonas cepacia* strain G4 initiates catabolism of toluene by hydroxylation to form *o*-cresol (pathway B in Figure 1).[9] Subsequent metabolic reactions involve a second hydroxylation to form 3-methylcatechol, which is the substrate for *meta*-cleavage of the aromatic ring. Subsequent reactions lead to TCA cycle intermediates. Analogously, *P. picketii* strain PKO1 initiates catabolism of toluene by hydroxylation to form *m*-cresol (pathway C in Figure 1).[10] A subsequent hydroxylation results in the formation of 3-methylcatechol which is then cleaved at the *meta* position to form 2-hydroxy-6-oxo-2,4-heptadienoate. Subsequent metabolic reactions lead to TCA cycle intermediates. Degradation of toluene by *P. mendocina* strain KR combines elements of both the aromatic ring hydroxylating catabolic pathways and the TOL catabolic pathway (pathway D in Figure 1).[11,12] Initially, toluene is hydroxylated to form *p*-cresol by toluene 4-monooxygenase. The methyl sustituent is then successively oxidized to a carboxylic acid moiety (to form *p*-hydroxybenzoate) by metabolic conversions similar to that seen in the initial steps of toluene oxidation by TOL plasmid strains. Subsequent hydroxylation of *p*-hydroxybenzoate yields protocatechuate which is the substrate for *ortho*-cleavage of the aromatic ring. Further metabolism proceeds by the β-ketoadipate pathway to tricarboxylic acid cycle intermediates.

The best-studied toluene degradative system in terms of the enzymology involved is that of *P. putida* strain F1.[13] This strain initiates catabolism of toluene through dioxygenase attack on the aromatic nucleus to incorporate both atoms of diatomic oxygen and two electrons from NADH into the resulting (+)-*cis*-1(S),2(R)-dihydroxy-3-methylcyclohexa-3,5-diene (*cis*-toluene dihydrodiol).[14-16] This reaction is catalyzed through the action of toluene dioxygenase, a multicomponent enzyme system[17] consisting of a flavoprotein reductase, a 2Fe·2S ferredoxin, and a terminal iron sulfur protein. The *cis*-toluene dihydrodiol is further dehydrogenated in an NAD^+-dependent reaction to form 3-methylcatechol. The latter compound is then the substrate for *meta*-cleavage of the aromatic nucleus to form 2-hydroxy-6-oxo-2,4-heptadienoate. Further metabolism eventually leads to TCA cycle intermediates.

Molecular Analysis of the TOL Plasmid Pathway

The TOL (pWW0) plasmid was originally described by Williams and co-workers[7,8] as obtained from *P. putida* strain mt-2 (or as originally characterized, *P. arvilla* strain mt-2). Variants of the TOL plasmid or of the TOL plasmid pathway have been reported by other researchers. For instance, Cruden et al.[18] described a *P. putida* strain that tolerates high concentrations (up to 50% toluene) of aromatic compounds in the growth medium. This strain, *P. putida* Idaho, utilizes a TOL pathway for the catabolism of toluene in two-phase culture systems. In addition to toluene, the TOL catabolic pathway is capable of metabolizing the related aromatic compounds *m*-xylene, *p*-xylene, 1,2,4-trimethylbenzene (pseudocumene), and 3-ethyltoluene.[19] The ability to degrade these other methyl-substituted aromatic compounds stems from the enzymology of the TOL catabolic pathway. The metabolic pathway involves oxidation and eventual removal of a methyl group from the aromatic nucleus. Therefore, a methyl group is an absolute requirement since it is the initial starting point for degradation. Since many of the enzymes in the catabolic pathway have a relaxed specificity, it is not surprising that a number of substituted methylbenzenes can also be metabolized by the TOL pathway. The catabolic pathway encoded by the TOL plasmid is shown in detail in Figure 2. As mentioned previously, a methyl group directly attached to the aromatic ring is oxidized to a carboxylic acid through the successive actions of xylene oxygenase, benzyl alcohol dehydrogenase, and benzaldehyde dehydrogenase. The resulting benzoic acid is

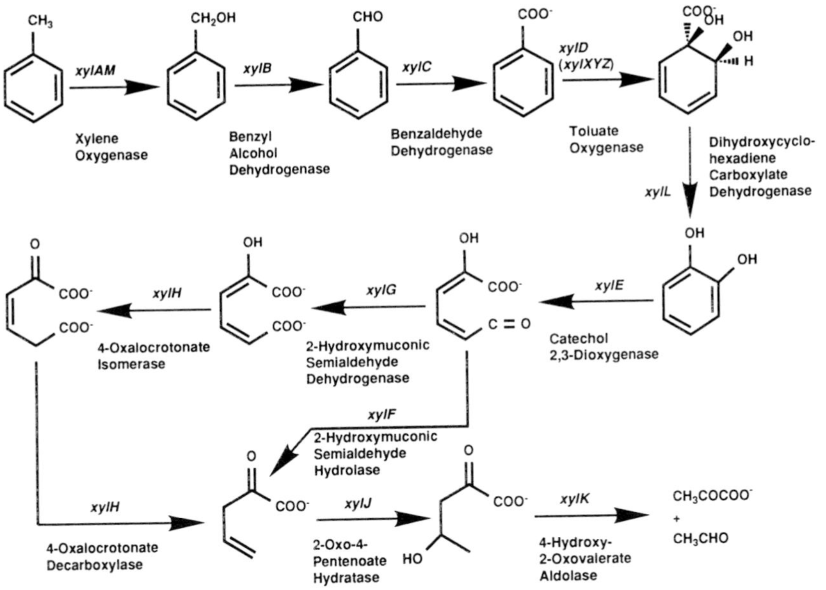

Figure 2. TOL plasmid catabolic pathway for toluene degradation. Gene designations for each enzyme are indicated.

oxidized to 1,2-dihydroxycyclo-3,5-hexadiene carboxylic acid (*cis*-benzoate dihydrodiol) through the action of the multicomponent enzyme system, toluate 1,2-dioxygenase. The action of a dehydrogenase on the *cis*-benzoate dihydrodiol catalyzes the re-aromatization of the ring and the concomitant decarboxylation to form catechol. Catechol then undergoes *meta*-ring-cleavage through the action of catechol 2,3-dioxygenase to form 2-hydroxymuconic semialdehyde. The TOL catabolic pathway then branches into two separate possible sets of metabolic reactions. The 2-hydroxymuconic semialdehyde may be converted directly to 2-oxo-4-pentenoate through the action of a hydrolase. Conversely, 2-hydroxymuconic semialdehyde may first be oxidized to a diacid (2-hydroxy-2,4-hexadiene-1,6-dioate), isomerized to 2-oxo-3-hexene-1,6-dioate, and then decarboxylated to form 2-oxo-4-pentenoate. It has been shown that *m*-toluate is degraded through the hydrolytic route, while benzoate and *p*-toluate are degraded through the longer route.[20] This is due to the specificities and capabilities of each of the two branches of the pathway. The subsequent actions of a hydratase and an aldolase result in the formation of the central metabolites pyruvate and acetaldehyde.

The genetic organization of the TOL plasmid is quite straightforward. Less than half of the approximately 115 kb pair plasmid is responsible for the degradation of toluene. The remainder of the plasmid encodes for enzymes involved in replication and conjugal transfer, although additional functions, currently unknown, may also be encoded by this region. The region encoding the genes is readily lost under selective pressure and this is attributed to the fact that the entire region involved in toluene degradation constitutes a transposon.[21,22] The region

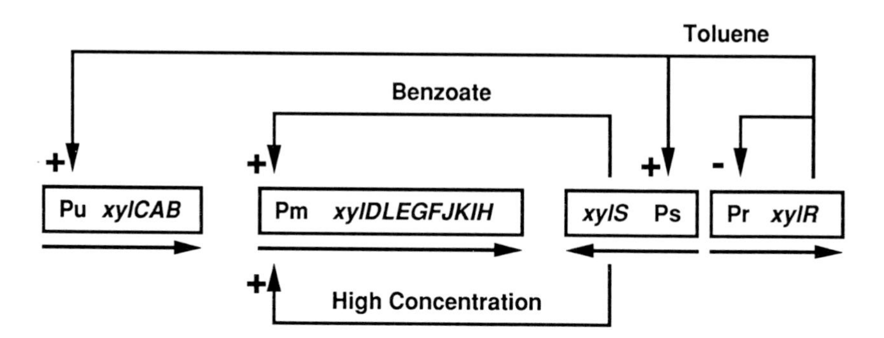

Figure 3. Structural organization and regulation of the TOL plasmid toluene catabolic pathway. Each operon is enclosed in a box. Pu, Pm, Ps, and Pr indicate the promoter for each operon. Arrows immediately below the boxes indicate the direction of transcription. Arrows with either a plus or minus sign indicate regulatory controls.

containing the toluene degrading genes may undergo reciprocal recombination between the 1.4 kb direct repeats flanking the region, resulting in complete loss of degradative ability. Alternatively, the entire region may transpose into the chromosome or into another replicon. The TOL plasmid genes encoding for toluene degradation are organized into two operons, with approximately 14 kb pairs of DNA separating them. This organization is shown diagramatically in Figure 3. Essentially, the genes coding for the enzymes responsible for the oxidation of the methyl group to a carboxylic acid (designated *xylABC*) are located in one operon, while the genes responsible for the conversion of the aromatic acid to pyruvate and acetaldehyde (designated *xylDEFGHIJKL*) are located in a second operon. The catabolic steps encoded by these two operons have often been referred to as the "upper" and "lower" pathway, respectively. The order of the genes[23,24] in the upper operon is *xylC*, *xylA*, and *xylB*, while the order in the lower operon is *xylD*, *xylL*, *xylE*, *xylG*, *xylF*, *xylJ*, *xylK*, *xylI*, and *xylH*. All of these genes have been cloned.[25-28] The nucleotide sequence is known for *xylE*, the gene coding for catechol 2,3-dioxygenase;[29,30] for the *xylXYZ* genes, coding for the three components of toluate 1,2-dioxygenase[31] (*xylD* is the designation for the entire locus); for the *xylL* gene, coding for the *cis*-dihydrodiol dehydrogenase;[32] for the *xylGFJ* genes, coding enzymes in the *meta*-cleavage pathway;[33] and for the *xylAM* genes, coding for the two subunits of xylene monooxygenase.[34] In addition, the sequences for the regulatory genes *xylR*[35] and *xylS*[36] are known.

The TOL plasmid has proved to be a model system to elucidate the various regulatory control mechanisms that are involved in catabolic gene expression in *Pseudomonas*.[37] The regulatory system can be divided into two parts, just as the *xyl* structural genes were divided into two operons (Figure 3). Two regulatory proteins are involved, encoded by the *xylR* and *xylS* genes. The protein products of both of these genes act as positive regulatory elements. The XylR activator directly recognizes toluene and other aromatic compounds such as *m*-xylene and *p*-xylene and acts to promote transcription at the promoter for the *xylABC* operon as well as the *xylS* gene. Increased levels of xylS, with or without the inducer, directly promote transcription of the *xylDLEGFJKIH* operon.[38] In the presence of an aromatic acid such as benzoate, XylS activates transcription to high levels. The sigma factor ntrA has been implicated as participating in induction[39,40] and it has been suggested that host-encoded proteins may also play a role in induction.[41-43]

Molecular Analysis of the cis-*Toluene Dihydrodiol Pathway*

The *cis*-toluene dihydrodiol pathway of *P. putida* strain F1 has been extensively studied at the molecular level as well as at the biochemical level. The initial steps of the biochemical pathway for toluene degradation by *P. putida* F1 are shown in detail in Figure 4. The individual enzymes responsible for the first four catabolic reactions in the pathway (toluene to 2-hydroxypenta-2,4-dienoate and acetate) have been purified and characterized. The first step in the catabolic pathway, the conversion of toluene to *cis*-toluene dihydrodiol, is catalyzed by the multicomponent enzyme system designated toluene dioxygenase.[44] This enzyme system consists of three components: $reductase_{TOL}$,[45] $ferredoxin_{TOL}$,[46] and an iron sulfur protein (ISP_{TOL}).[47] Initially, electrons are accepted from NADH by the flavoprotein $reductase_{TOL}$. The electrons are then transferred to the 2Fe·2S ferredoxin which then transfers the electrons to ISP_{TOL}. The latter component then catalyzes the introduction of both atoms of molecular oxygen and the two electrons from NADH into the aromatic nucleus to form (+)-*cis*-1(S),2(R)-dihydroxy-3-methylcyclohexa-3,5-diene (*cis*-toluene dihydrodiol). A dehydrogenase[48] utilizes the latter compound as substrate to produce 3-methylcatechol and NADH. Thus, the NADH consumed in the first step in the catabolic pathway is regenerated by toluene *cis*-dihydrodiol dehydrogenase in the second step of the pathway. The 1,2-dihydroxylated compound formed, 3-methylcatechol, is the substrate for *meta*-ring-cleavage by 3-methylcatechol 2,3-dioxygenase to form 2-hydroxy-6-oxo-2,4-heptadienoate. The latter seven-carbon acid is subject to hydrolytic attack by 2-hydroxy-6-oxo-2,4-heptadienoate hydrolase to form acetate and 2-hydroxypenta-2,4-dienoate. Subsequent reactions in the pathway remain to be determined. The genes encoding for these proteins have been designated *tod* and the gene designations for each peptide are shown in Figure 4.

Mutants blocked in each of the initial steps of the *P. putida* F1 catabolic pathway have been isolated through ampicillin/D-cycloserine enrichment following exposure to the mutagen N-methyl-N′-nitro-N-nitrosoguanidine (NTG).[49,50] Mutant strains were readily identified through the inclusion of the redox dyes nitro blue tetrazolium and 2,3,5-triphenyl-2H-tetrazolium chloride in the plating medium used following NTG mutagenesis. This dye selection technique requires the inclusion of 1 m*M* succinate in the solid growth medium and exposure of the plates to toluene vapors. Rapidly growing colonies utilizing toluene as a carbon source will turn red due to reduction of the dye 2,3,5-triphenyl-2H-tetrazolium chloride, while slow growing colonies utilizing only succinate as carbon source will appear white. Inclusion of the second dye, nitro blue tetrazolium, allows characterization of mutants lacking components of the toluene dioxygenase enzyme system. This is due to differences in reduction of the dye by the redox

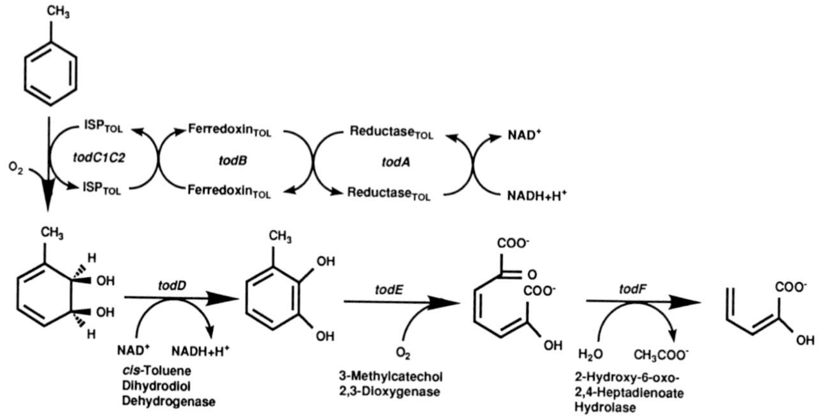

Figure 4. *P. putida* F1 *cis*-dihydrodiol catabolic pathway for toluene degradation. Gene designations for each enzyme are indicated.

components of toluene dioxygenase. Full characterization of the mutant strains included enzymatic assays and identification of accumulating intermediates. Mutants were also constructed using the suicide Tn*5* transposon donor pRKTV14. These mutants exhibited polar effects, suggesting that the genes for toluene degradation are operonic in *P. putida* F1. The polar effects of different transposon mutants allowed a preliminary gene order to be determined, although ambiguities were still present. The transposon mutants fell into four classes based on enzyme assays for the individual components of toluene dioxygenase, *cis*-toluene dihydrodiol dehydrogenase, 3-methylcatechol dioxygenase, and 2-hydroxy-6-oxo-2,4-heptadienoate hydrolase. Class I mutants were *todFC1C2BADE*, class II mutants were *todBADE*, class III mutants were *todDE*, and class IV mutants were *todE*. The relative gene order is thus *todFC1C2, todBA, todD*, and *todE*. Since the initial transposon mutagenesis data is not comprehensive, ambiguities exist in determining the order of the *todFC1C2* genes and the *todBA* genes. Many of the class I mutant strains were later shown to be insertions in a positive activator gene since they could not be complemented by the cloned structural genes.

The *tod* genes were cloned from *P. putida* strain F1 on two genomic *Eco*RI fragments using the broad host range mobilizable cosmid pLAFR1.[50] This clone was labeled pDTG301 and contains two *Eco*RI fragments (14 and 10.7 kb pairs) in addition to the cosmid vector. This plasmid could complement all of the NTG-generated mutations of PpF1, although the complementation results with PpF21 (*todF*) were ambiguous due to leakiness of the mutation. Subcloning experiments localized the position of each of the *todC1C2BADE* genes on the cloned fragment through complementation analysis.[50] The *todF* gene was located through Southern blotting experiments utilizing a synthetic degenerate oligonucleotide that was constructed based on the N-terminal amino acid sequence of the purified hydrolase.[51] The determined gene locations on the restriction map of the cloned *Eco*RI fragments is shown in Figure 5. This data confirms and extends the transposon

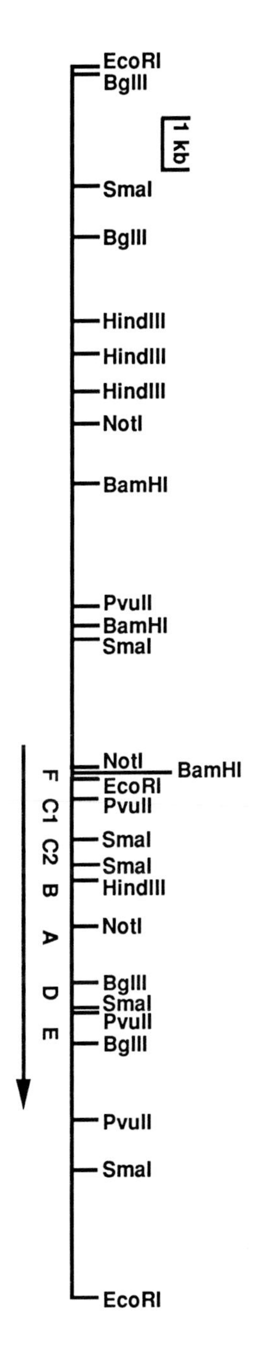

Figure 5. Restriction map of pDTG301 containing the *tod* genes of *P. putida* F1. The location of the *todFC1C2BADE* genes is indicated. The arrow indicates the direction of transcription as well as the extent of the sequenced region.

mutagenesis data in determining that the genes are transcribed in the order *todF*, *todC1*, *todC2*, *todB*, *todA*, *todD*, and *todE*. The entire 7.6 kb base pair region encoding for the *todFC1C2BADE* genes and flanking regions was sequenced.[51,52] The open reading frames were located and the genes identified through comparison to the N-terminal amino acid sequence derived from the purified protein or subunit. In addition, the peptide molecular weights determined by analysis of the purified proteins were compared to the molecular weights derived from the open reading frames to ensure the proper length of the open reading frame. This analysis was necessary in order to accurately locate the open reading frame responsible for each protein. This proved especially important since the start codon for *todD*, the gene coding for *cis*-toluene dihydrodiol dehydrogenase, is actually GTG rather than ATG. Additionally, this codon overlaps the stop codon TGA of *todA*, the previous gene. If the N-terminal amino acid sequence of the protein was not known, then the open reading frame would have been identified incorrectly (there is an in-frame ATG located 330 nucleotides downstream). The partially sequenced open reading frames at both ends of the sequenced region possess homology with genes of the TOL plasmid.[33] The individual *todC1C2BA* genes coding for toluene dioxygenase exhibit 57 to 68% homology at the nucleotide sequence level to the genes for biphenyl dioxygenase from *Pseudomonas* sp. LB400.[53] There is also significant homology (>55%) of the *todC1C2BADE* genes coding for toluene dioxygenase, *cis*-toluene dihydrodiol dehydrogenase, and 3-methylcatechol 2,3-dioxygenase to the genes coding for biphenyl dioxygenase, *cis*-biphenyl dihydrodiol dehydrogenase, and 2,3-dihydroxybiphenyl 1,2-dioxygenase from *P. pseudoalcaligenes* KF707.[54,55] (The cloned genes for biphenyl degradation are covered in more detail in Chapter 4.) One major difference between the toluene dioxygenase and biphenyl dioxygenase nucleotide sequences is the addition of a gene of unknown function between the genes analogous to *todC2* and *todB*.

Knowledge of the nucleotide sequence allowed the construction of several expression clones in *E. coli* that are designed to synthesize high levels of the enzymes. Clones were made that expressed each of the enzymes alone and in combination with others.[52,56] These clones were constructed using the expression vector pKK223-3 containing the hybrid *tac* promoter that is inducible by isopropyl-β-D-thiogalactopyranoside (IPTG). Clones were constructed that separately express each individual gene (either *todC1, todC2, todB, todA, todD,* or *todE*) and that express groups of genes (either *todC1C2, todC1C2BA, todC1C2BAD,* or *todC1C2BADE*). The latter set of clones was utilized in substrate range experiments to identify the intermediates formed during the degradation of a particular aromatic compound. For instance, it is possible to produce *cis*-toluene dihydrodiol from toluene using the *todC1C2BA* clone pDTG601, 3-methylcatechol from toluene using the *todC1C2BAD* clone pDTG602, and 2-hydroxy-6-oxo-2,4-heptadienoate from toluene using the *todC1C2BADE* clone pDTG603. Analogous intermediates can be produced using other aromatic hydrocarbons as substrates.[56-59] Compounds such as benzene, ethylbenzene, phenol, *o*-cresol, *m*-cresol, *p*-cresol, naphthalene, biphenyl, and certain chlorobiphenyls can be oxidized to the

Table 1
Aromatic Compounds Oxidized by Toluene Dioxygenase Cloned in *E. coli*

Aromatic hydrocarbons	Substituted aromatics	Heterocyclic and bicyclic compounds
Benzene	Chlorobenzene	Indan
Toluene	*o*-Dichlorobenzene	Indene
Ethylbenzene	*m*-Dichlorobenzene	Indole
Propylbenzene	*p*-Dichlorobenzene	Naphthalene
Isopropylbenzene	*o*-Chlorotoluene	Biphenyl
Butylbenzene	*m*-Chlorotoluene	2-Chlorobiphenyl
sec-Butylbenzene	*p*-Chlorotoluene	3-Chlorobiphenyl
tert-Butylbenzene	*p*-Flourotoluene	4-Chlorobiphenyl
o-Xylene	*p*-Iodotoluene	2,3-Dichlorobiphenyl
m-Xylene	*o*-Chlorophenol	2,4-Dichlorobiphenyl
p-Xylene	*m*-Chlorophenol	2,5-Dichlorobiphenyl
Phenol	*p*-Chlorophenol	3,4-Dichlorobiphenyl
o-Cresol	Nitrobenzene	
m-Cresol	*o*-Nitrotoluene	
p-Cresol	*m*-Nitrotoluene	
	p-Nitrotoluene	

corresponding ring-cleavage products. Compounds such as chlorobenzene, *o*-dichlorobenzene, *m*-dichlorobenzene, *p*-dichlorobenzene, *o*-chlorophenol, *m*-chlorophenol, and *p*-chlorophenol can be oxidized only to the corresponding chlorocatechols. This is due to the inability of the *todE*-encoded catechol 2,3-dioxygenase to cleave chlorocatechols. This wide substrate range is mainly due to the ability of toluene dioxygenase to oxidize the aromatic ring of a number of different aromatic compounds. A partial list of compounds known to be oxidized by toluene dioxygenase is given in Table 1.

A number of other microbial strains have been identified besides *P. putida* F1 that degrade monocyclic aromatic hydrocarbons by a *cis*-dihydrodiol type of catabolic pathway. *P. putida* strain RE204 degrades isopropylbenzene via a *cis*-dihydrodiol type of intermediate.[60] The isopropylbenzene degradation pathway is encoded by a large 105 kb plasmid contained by the strain. The gene positions were determined through analysis of Tn*5* insertion mutations and by cloning and restriction analysis. The benzene degradation pathway encoded by *P. putida* strain BE-81 also involves a dihydrodiol type of intermediate. The genes from this strain that encode the enzymes for the conversion of benzene to catechol have been cloned and sequenced.[61] The nucleotide sequence is practically identical to the same genes sequenced from *P. putida* F1. The genes for the conversion of toluene to 2-hydroxy-6-oxo-2,4-heptadienoate via the *cis*-dihydrodiol pathway were cloned from *P. putida* NCIB 11767.[62] The 20 kb partial *Sau*3A fragment may also contain regulatory genes since expression of the cloned genes in *E. coli* was enhanced by exposure to toluene. A benzene dioxygenase encoded by a 112 kb plasmid was cloned from *P. putida* ML2.[63] These genes showed homology in Southern blotting experiments to the *tod* genes discussed above. The genes for toluene degradation have recently been cloned on a 5.5 kb *Hin*dIII fragment from strain GJZ9, an uncharacterized isolate.[64] Initial nucleotide sequence analysis shows very little

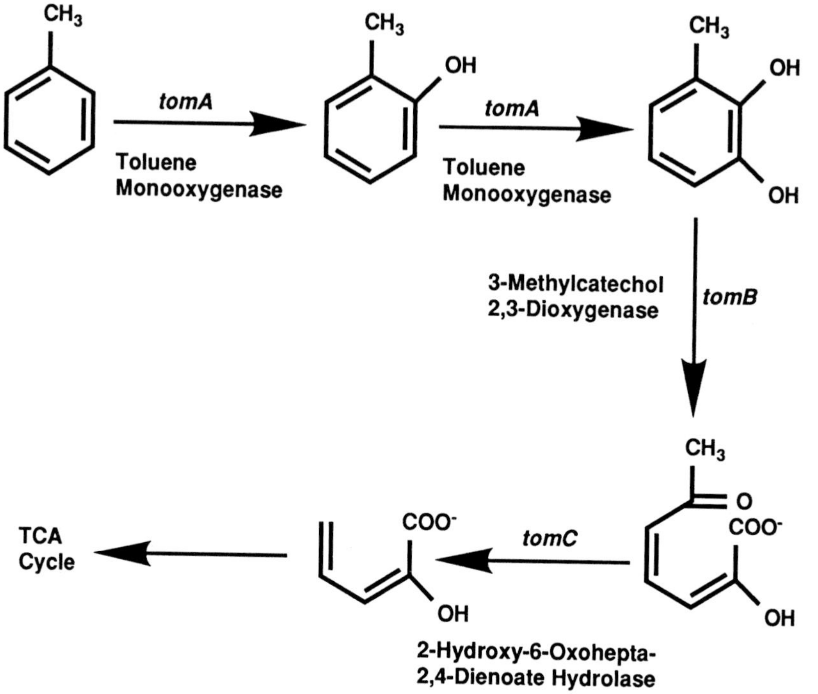

Figure 6. *P. cepacia* G4 *o*-cresol pathway for toluene degradation. Gene designations for each enzyme are indicated.

homology between the toluene dioxygenase encoded by this strain and that from *P. putida* F1.

Molecular Analysis of the o*-Cresol Pathway*

In contrast to the well-studied TOL plasmid and *cis*-dihydrodiol pathways for toluene degradation, not much is known about the molecular aspects of the cresol pathways for toluene degradation. This is due mainly to the fact that these metabolic pathways were only recently discovered. The *o*-cresol catabolic pathway for toluene degradation is encoded by *P. cepacia* strain G4.[9] The catabolic pathway for toluene in this strain (Figure 6) first involves the oxidation of toluene to *o*-cresol through the action of toluene 2-monooxygenase. A subsequent hydroxylation forms 3-methylcatechol which is the substrate for *meta*-cleavage of the aromatic ring. The product of this reaction, 2-hydroxy-6-oxo-2,4-heptadienoate, undergoes hydrolytic cleavage to form acetate and 2-hydroxypenta-2,4-dienoate. Further catabolic reactions lead to the formation of tricarboxylic acid cycle metabolites. The same catabolic pathway is utilized for the degradation of toluene, phenol, *o*-cresol, and *m*-cresol by this strain. The genes responsible for this pathway have been given the designation *tom*. Mutants in the catabolic pathway

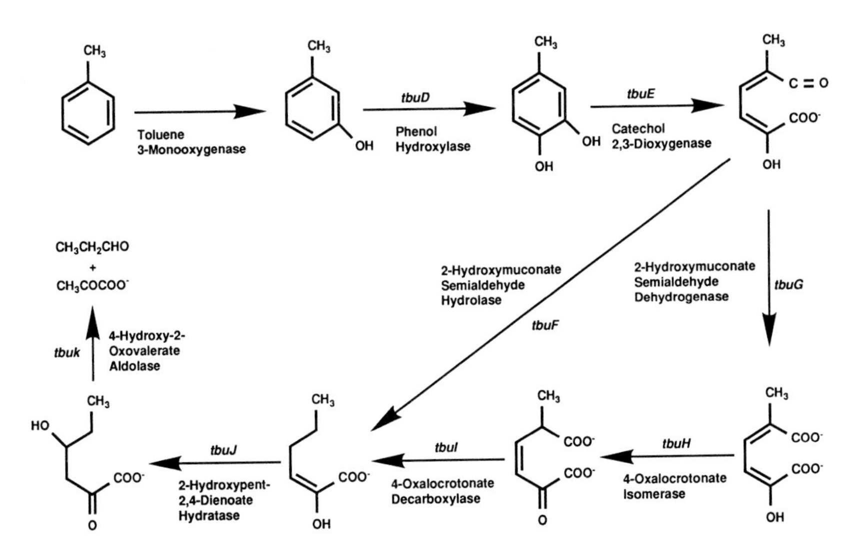

Figure 7. *P. pickettii* PKO1 *m*-cresol pathway for toluene degradation. Gene designations for each enzyme are indicated.

have been isolated[65] using traditional mutagenesis methods and also through the use of suicide substrates. Interestingly, mutants defective in the oxidation of toluene to *o*-cresol are also defective in the oxidation of *o*-cresol to 3-methylcatechol. Due to the large number of mutants examined (110 in this class), it is highly probable that a single enzyme is responsible for catalyzing both enzymatic steps. The genes responsible for this catabolic pathway have not yet been cloned from this strain.

Molecular Analysis of the m-*Cresol Pathway*

The *m*-cresol catabolic pathway for toluene degradation is encoded by *P. picketii* strain PKO1.[10] This strain is capable of growing on toluene, benzene, phenol, and *m*-cresol as the sole carbon source. The identical catabolic pathway is involved in the oxidation of each growth substrate (Figure 7). The catabolic pathway initially involves a toluene 3-monooxygenase that catalyzes the formation of *m*-cresol from toluene. A subsequent hydroxylation leads to the formation of 3-methylcatechol which is then the substrate for *meta*-cleavage of the aromatic ring to form 2-hydroxy-6-oxo-2,4-heptadienoate. Subsequent catabolic reactions are similar to those determined for the TOL pathway described above, with two metabolic routes leading to the production of the central metabolites pyruvate and propionaldehyde.

The genes responsible for the *m*-cresol catabolic pathway for toluene degradation of *P. pickettii* PKO1 were cloned on a 26.5 kb *Bam*HI restriction fragment.[10] This fragment of DNA was originally isolated through its ability to confer on *P. aeruginosa* PAO1c the ability to grow on toluene, benzene, phenol, and *m*-cresol

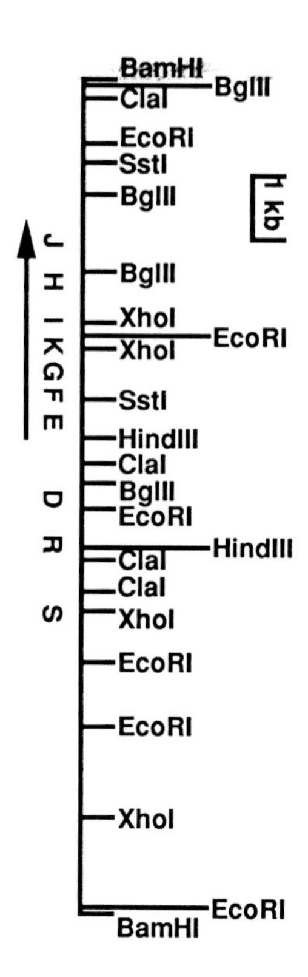

Figure 8. Restriction map of pRO1957 containing the *tbu* genes of *P. pickettii* PKO1. The locations of the *tbuEFGKIHJ, tbuD, tbuR, tbuS* genes are indicated. The arrow indicates the direction of transcription for the *tbuJHIKGFE* operon.

as carbon sources. The restriction map of this fragment of DNA is shown in Figure 8 along with the gene locations as determined through subcloning, expression, and enzyme assays. The location of the gene(s) involved in the initial hydroxylation of benzene and toluene to phenol and *m*-cresol, respectively, have not yet been determined. Two genes were shown to be required for the oxidation of phenol and *m*-cresol to catechol and 3-methylcatechol, respectively. These genes were originally named *phlA* and *phlR*,[10] but were subsequently renamed *tbuD* and *tbuR*, respectively.[66] The gene *tbuD* encodes for a phenol hydroxylase which is regulated in trans by the *tbuR* gene product. Since there is an absolute requirement for *tbuR* in *tbuD* induction, a positive regulatory system must be involved. Inducers of phenol hydroxylase activity include both phenol and *m*-cresol; however, the structurally similar compounds *o*-cresol and *p*-cresol were not inducers. In addition, the products of the phenol hydroxylase catabolic reaction, catechol and 3-methylcatechol, could not induce phenol hydroxylase activity. Even though the *ortho* and *para* isomers of *m*-cresol were not inducers of phenol hydroxylase, they were shown to be substrates of the enzyme, with the concomitant formation of 3-methylcatechol and 4-methylcatechol, respectively. In addition, both catechol and resorcinol could be hydroxylated to form pyrogallol, while 3-chlorophenol could be

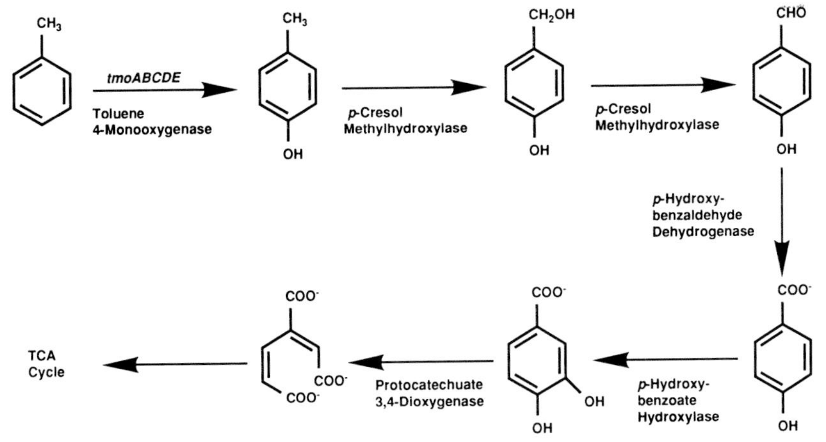

Figure 9. *P. mendocina* KR *p*-cresol pathway for toluene degradation. Gene designations for each enzyme are indicated.

hydroxylated to 4-chlorocatechol. The compounds 3,4-dimethylphenol and guaiacol act as nonsubstrate effectors since they stimulate NADPH oxidation by the phenol hydroxylase, with no formation of product. The nucleotide sequence of the *tbuD* gene has recently been determined.[67]

The genes for the further catabolism of catechol and 3-methylcatechol via a *meta*-cleavage pathway to central metabolites are also encoded by the 26.5 kb *Bam*HI fragment.[66] These genes have been given the designation *tbuEFGHIJK* (Figures 7 and 8). The genes are arranged in a single operon, with transcription proceeding in the order *tbuE, tbuF, tbuG, tbuK, tbuI, tbuH, tbuJ*. Although the enzymes encoded by these genes are similar in function to those of the TOL pathway, the gene order is not strictly conserved between the two operons. Another gene, designated *tbuS*, was shown to be required in trans for full induction of the *tbuEFGKIHJ* operon. The product of this gene acts both as a repressor and as an activator for gene transcription. Inducers for this *meta*-cleavage pathway were shown to be phenol and *m*-cresol; *o*-cresol, *p*-cresol, catechol, 3-methylcatechol, or 4-methylcatechol could not induce expression of this operon. Once induced, the *tbu* gene products were capable of metabolizing catechol, 3-methylcatechol, or 4-methylcatechol to the analogous central metabolites.

Molecular Analysis of the p-*Cresol Pathway*

The *p*-cresol catabolic pathway for toluene degradation[11,12] is encoded by *P. mendocina* strain KR (Figure 9). This strain has been the subject of extensive investigations into the biochemistry and genetics of the initial enzyme of the pathway, toluene 4-monooxygenase. Whited and Gibson demonstrated that three partially purified components were necessary for the enzymatic activity.[11] The reconstituted enzyme system involves a monooxygenase reaction in which one

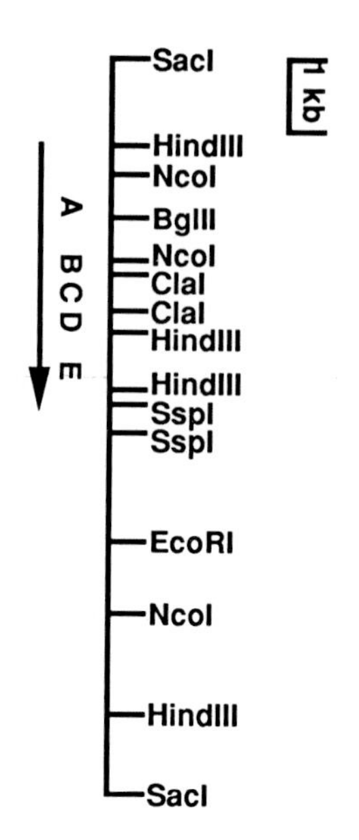

Figure 10. Restriction map of pKMY277 containing the *tmo* genes of *P. mendocina* KR. The locations of the *tmoABCDE* genes are indicated. The arrow indicates the direction of transcription.

atom of molecular oxygen is incorporated into the aromatic nucleus in an NAD(P)H-dependent reaction to form *p*-cresol. It is suggested that toluene-3,4-oxide is the first initial product of the enzymatic reaction since the reaction proceeds with a concomitant NIH shift. The three succeeding steps of the catabolic pathway involve the oxidation of the methyl group of *p*-cresol to a carboxylic acid, through the actions of *p*-cresol methylhydroxylase and *p*-hydroxybenzaldehyde dehydrogenase.[12] The *p*-hydroxybenzoate formed is then hydroxylated to protocatechuate through the action of *p*-hydroxybenzoate hydroxylase. Protocatechuate undergoes *ortho*-cleavage by protocatechuate 3,4-dioxygenase to form β-carboxy-*cis,cis*-muconate. Subsequent reactions proceed through the β-ketoadipate pathway to tricarboxylic acid cycle intermediates.[12]

The genes encoding for toluene 4-monooxygenase were originally cloned on two 10.2 kb *Sac*I fragments of DNA and further localized to a 3.6 kb *Hin*dIII-*Ssp*I fragment (Figure 10).[68] The nucleotide sequence of this region was subsequently determined.[68] Five open reading frames were detected that are required for toluene 4-monooxygenase activity. These genes were subsequently designated *tmoA*, *tmoB*, *tmoC*, *tmoD*, and *tmoE*. That each of these genes are required for toluene 4-monooxygenase activity was determined through site-directed mutagenesis of each of the genes, followed by *trans*-complementation of the mutant with the cloned wild-type gene. Two open reading frames were detected upstream from the *tmoABCDE* genes, although no function could be ascribed to them. Following *tmoE* is a region of dyad symmetry followed by a string of five thymidine residues, suggesting a factor-independent transcription terminator. The observed

G+C content of the DNA (48.9%) is much lower than the value reported for the *P. mendocina* genome (62.8 to 64.3%). This suggests that the genes may have originated from another species and were only recently transferred to *P. mendocina*. The deduced amino acid sequences of the *tmoA, tmoD*, and *tmoE* gene products possess some homology to the genes for the five-component phenol hydroxylase enzyme system.[69] The genes for the subsequent oxidation of the methyl substituent of *p*-cresol to *p*-hydroxybenzoic acid have recently been cloned on a separate DNA fragment by Wright and Olsen.[70]

Molecular Approaches to Polycyclic Aromatic Hydrocarbon Degradation

Polycyclic Aromatic Hydrocarbon Catabolic Pathways

Numerous microorganisms and consortia have been reported in the literature that have the ability to degrade polycyclic aromatic hydrocarbons. Pure cultures have been isolated that are capable of degrading naphthalene (reviewed by Yen and Serdar[71]), phenanthrene, anthracene, fluorene, fluoranthene, and pyrene. The catabolic pathways for the microbial degradation of the simplest polycyclic aromatic hydrocarbons (napthalene, anthracene, and phenanthrene) were elucidated over 25 years ago.[72] However, only in the last 5 years have the microbial degradation pathways for the more complex polycyclic compounds been determined. The microbial catabolic pathways involved almost invariably initiate attack on the aromatic nucleus to form a *cis*-dihydrodiol type of intermediate similar to that described above for the *P. putida* F1 toluene catabolic pathway. The two best characterized strains for the microbial degradation of naphthalene are *P. putida* G7 containing the NAH plasmid[73] and *P. putida* NCIB 9816-4 containing plasmid pDTG1.[74,75] Most of the genetics and regulation investigations have been performed on *P. putida* G7, while the most intense biochemical investigations have been performed on *P. putida* NCIB 9816-4. The initial enzyme in this pathway, naphthalene dioxygenase, has been purified and shown to consist of three components.[76] The components of this enzyme system function similarly to those described for toluene dioxygenase from *P. putida* F1. Initially, a reductase component,[77] designated $reductase_{NAP}$, accepts electrons from NAD(P)H. This flavoprotein differs from $reductase_{TOL}$ in that $reductase_{NAP}$ includes an iron sulfur cluster in addition to the flavin moiety. Electrons are subsequently transferred to a 2Fe·2S ferredoxin[78] and then to an iron sulfur protein, designated ISP_{NAP}.[79] The latter component of naphthalene dioxygenase catalyzes the addition of both atoms of molecular oxygen into the aromatic nucleus to form (+)-*cis*-(1R,2S)-dihydroxy-1,2-dihydronaphthalene (*cis*-naphthalene dihydrodiol).[80] The *cis*-dihydrodiol is dehydrogenated to form 1,2-dihydroxynaphthalene by a dehydrogenase with NAD as the electron acceptor, thus replenishing the NADH utilized in the first enzymatic step of the catabolic pathway. The dihydroxylated intermediate is the substrate for *meta*-cleavage of the aromatic ring. Subsequent reactions result in the formation of salicylate (Figure 11). Salicylate has been shown to be further metabolized by three different pathways, depending on the microbial strain. The most common pathway is the metabolism of

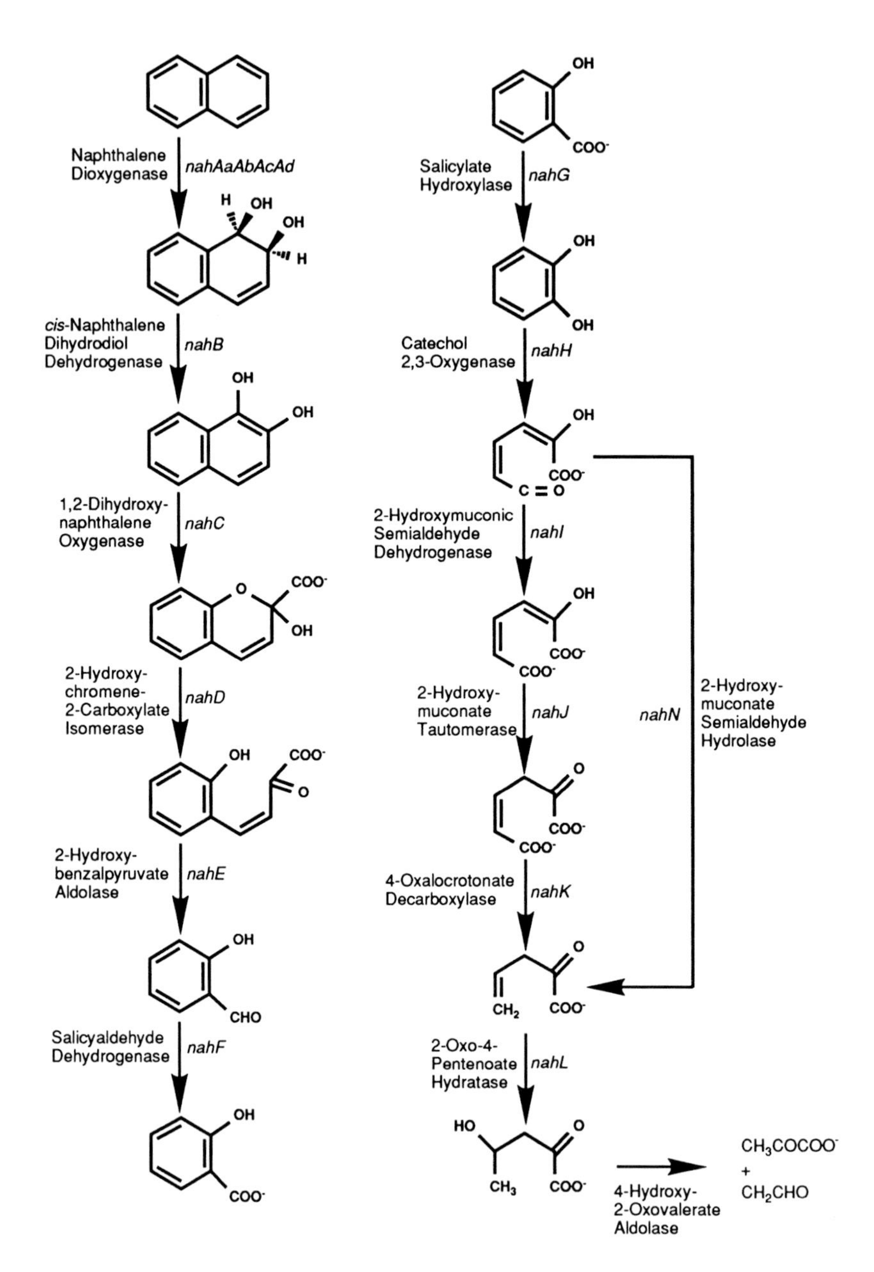

Figure 11. *P. putida* G7 pathway for naphthalene degradation. Gene designations for each enzyme are indicated.

salicylate to catechol by salicylate hydroxylase. Catechol is then subject either to *meta* ring cleavage and the eventual formation of pyruvate and acetaldehyde or to *ortho* ring cleavage and the eventual formation of succinate and acetate. The complete catabolic pathway encoded by the NAH7 plasmid for salicylate degradation via *meta* cleavage is shown in Figure 11. Certain microorganisms metabolize salicylate via gentisate and the eventual formation of pyruvate and fumarate.[81] Degradation of the three-ring polycyclic aromatic hydrocarbons anthracene and phenanthrene proceeds in a similar fashion.[82]

Recently, several microorganisms have been isolated that have the ability to degrade the four-ring polycyclic aromatic hydrocarbon pyrene.[83,84] The proposed catabolic pathway for this compound[85] is both similar to and different from the naphthalene catabolic pathway described above. Pyrene degradation by a *Mycobacterium* species proceeds initially through the formation of pyrene *cis*-4,5-dihydrodiol. The *cis*-pyrene dihydrodiol is dehydrated to a 1,2-dihydroxylated intermediate which is the substrate for *ortho* ring cleavage. Decarboxylation leads to the formation of 4-phenanthroic acid. A series of subsequent metabolic reactions lead to the single aromatic ring intermediates phthalic acid and cinnamic acid. Further metabolism leads to the formation of tricarboxylic acid cycle intermediates. The major difference between this pathway and that for naphthalene is the *ortho* cleavage of 4,5-dihydroxypyrene and the subsequent decarboxylation to form 4-phenanthroic acid.

Molecular Analysis of Naphthalene Degradation

A number of self-transmissible, large catabolic plasmids have been isolated that encode for the ability to degrade naphthalene (reviewed by Yen and Serdar[71]). These plasmids range in size from approximately 80 kb pairs to 175 kb pairs, of which only about 30 kb pairs are responsible for the degradation of naphthalene. The best studied of these plasmids is the NAH7 plasmid of *P. putida* G7.[73] The genes for the catabolic pathway are arranged in two operons, similar in function to the two operons of the TOL plasmid. The "upper" operon encodes for the conversion of naphthalene to salicylate, while the "lower" operon encodes for the conversion of salicylate through the *meta*-cleavage pathway to pyruvate and acetaldehyde. Thus, both the TOL and NAH plasmid "upper" pathway operons convert aromatic substrates to aromatic acids, while the "lower" pathway operons convert the aromatic acids to central metabolites. The *nah* region of plasmid NAH7 was shown to be a defective transposon, having the ability to transpose in the presence of the transposase encoded by the Tn*1721* subgroup of class II transposons.[86]

The location of each gene in the pathway on the NAH7 plasmid was determined through transposon mutagenesis and mapping of the subsequent insertion.[87] In addition, the determined polarity effects of the transposon insertions indicated the orientation of transcription of each operon.[87] The *nah* genes have been cloned from NAH7.[88] A restriction map of the region is shown in Figure 12 with the gene locations. Transcription of the genes in the "upper" operon proceeds from *nahA* to *nahF*, while transcription in the "lower" operon proceeds from *nahG* to *nahK*.

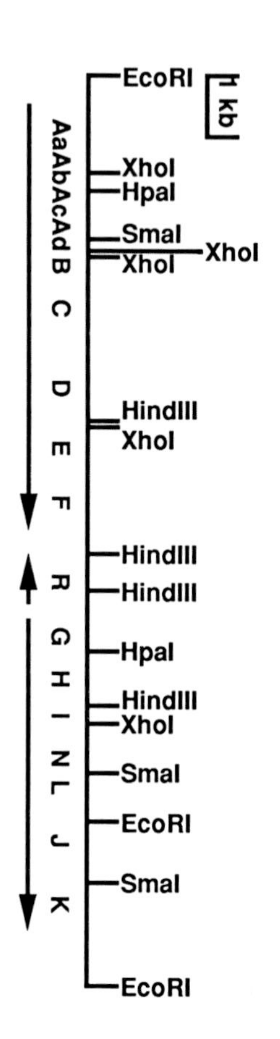

Figure 12. Restriction map of the segment of the NAH7 plasmid containing the *nah* genes. The locations of the *nah* genes in each of the three operons is indicated. Arrows indicate the direction of transcription.

Transcription of the regulatory gene *nahR* proceeds in the opposite direction from that of the lower operon. The *nah* genes were also cloned from *P. putida* NCIB 9816-4[74,75] and were shown to have a similar, although not identical, restriction map to those cloned from the NAH7 plasmid.

Regulation of the naphthalene catabolic system encoded by the NAH7 plasmid is mediated by the product of the *nahR* gene.[89] This single protein mediates an increase in transcription of both the upper and lower pathway operons in the presence of salicylate. Thus, low levels of the enzymes encoded by the upper pathway must be present in order to convert naphthalene to salicylate to effect full induction of both operons. The promoters of both the lower and upper operons have been sequenced and a detailed investigation of the ability of NahR to promote transcription has been performed.[90] A transcription activation site at –70 has been identified, along with an RNA polymerase binding site at the –35 region. The corresponding region of the upper pathway operon has been sequenced from *P.*

putida NCIB 9816-4.[91] This putative promoter region shows 67% homology to that of the NAH7 plasmid, with the most homology around the transcription activation site, RNA polymerase binding site, and transcription starting point. Upstream of this region, the homology between the two sequences diverges rapidly.

Although the genes for naphthalene degradation have been cloned and their locations determined, only a few of the genes have been sequenced. The two catechol 2,3-dioxygenase genes of NAH7 have been sequenced: the *nahC* gene encoding 1,2-dihydroxynaphthalene dioxygenase[92] and the *nahH* gene encoding catechol 2,3-dioxygenase.[93,94] The deduced amino acid sequences for the enzymes encoded by the two genes are structurally quite different from each other. The 1,2-dihydroxynaphthalene dioxygenase is similar to 2,3-dihydroxybiphenyl dioxygenases from biphenyl degrading pathways, while the catechol 2,3-dioxygenase gene sequence is almost identical to that described for catechol 2,3-dioxygenase encoded by the *xylE* gene of the TOL plasmid. The gene for salicylate hydroxylase, *nahG*, from the NAH plasmid has also been sequenced.[95]

The genes responsible for the individual components of naphthalene dioxygenase have been sequenced from both *P. putida* strains G7 and NCIB 9816-4.[91,96] Four genes are responsible for encoding the three components of this enzyme. The *nahAa* gene encodes for reductase$_{NAP}$, the *nahAb* gene encodes for ferredoxin$_{NAP}$, and the *nahAc* and *nahAd* genes encode for the large and small subunits of the iron sulfur protein (ISP$_{NAP}$). The nucleotide sequences of the *nahAaAbAcAd* genes from the two sources show 93% homology at the nucleotide sequence level and 94% homology at the deduced amino acid sequence level. The most divergence is seen upstream of the structural gene sequences. There is 77% homology between the two nucleotide sequences in the 100 base pair region immediately upstream from *nahAa*. However, past this region, the nucleotide sequences diverge rapidly. Comparisons between the deduced amino acid sequences determined for the three components of naphthalene dioxygenase from *P. putida* strains G7 and NCIB 9816-4 and the three components of toluene dioxygenase from *P. putida* F1 show very few similarities. This is even though both enzymes catalyze the formation of *cis*-dihydrodiol intermediates from their respective substrates. The most homology can be seen between the two small 2Fe·2S electron transfer components, ferredoxin$_{TOL}$ and ferredoxin$_{NAP}$. However, there are similarities in motifs between the two sequences, around the flavin binding site and iron sulfur binding site regions. Despite these differences in enzyme structure, it has been shown that toluene dioxygenase cloned in *E. coli* can oxidize naphthalene to *cis*-naphthalene dihydrodiol.[56] The rate of oxidation of naphthalene by toluene dioxygenase is almost the same as the rate of oxidation of toluene. The *nahAaAbAcAd* genes have been used as molecular probes by various researchers to detect similarities between polycyclic aromatic hydrocarbon degrading microorganisms. Many naphthalene degrading strains isolated from terrestrial environments show homology to the *nahA* genes; however, many estuarine and marine isolates do not show homology to the *nahA* genes.[97] In addition, many microbial isolates capable of degrading polycyclic aromatic hydrocarbons of more than three rings do not show cross-hybridization to the *nahA* genes.[98]

BIODEGRADATIVE APPLICATIONS OF THE CLONED GENES FOR AROMATIC HYDROCARBON DEGRADATION

Molecular Tools for Biodegradation Analysis

Knowledge of the genetic structure of the catabolic pathways for aromatic hydrocarbon degradation supplies needed information for applications in the environmental field. Besides the obvious construction of genetically engineered strains that may possibly have improved performance over indigenous microorganisms, molecular techniques can benefit biodegradation monitoring and optimization.[99] The cloned structural genes can be used as gene probes for specific microorganisms and catabolic pathways. The cloned regulatory genes are especially useful since they allow probing for specific regulatory features and thus immediately demonstrate the type of inducers that will be needed to activate expression of the desired degradative ability. Nucleotide sequences determined for genes encoding key enzymes or regulatory proteins allow the design of specific PCR primers for rapid detection and quantification of the types of microorganisms present at a particular location. Substrate range information gained from experiments using cloned genes expressed in heterologous microorganisms can be utilized to predict the rate of disappearance of complex mixtures of xenobiotics by a single catabolic pathway as well as the appearance of any dead end products resulting from incomplete degradation occurring through cometabolic transformations.

One of the most powerful tools of molecular biology is the use of gene probes for monitoring populations. The use of gene probes for monitoring microbial populations in the environment has been reviewed recently.[100-102] It has been shown that in colony blotting procedures, one could easily detect *P. putida* colonies containing the TOL plasmid even when the nonhybridizing background population was high (10^6 bacteria).[103] Genes for naphthalene degradation have been utilized to monitor the number of naphthalene-degrading bacteria in activated sludge.[104] This study showed that an increase in gene frequency correlated with an increase in naphthalene-degrading bacteria. The gene probe technique was almost two orders of magnitude more sensitive than a standard plate count for naphthalene-degrading bacteria. Gene probes have been utilized to detect gene expression in the environment by direct extraction of mRNA from soils followed by specific detection of *nah* gene transcription by hybridization.[105] The gene probe technique does have limitations in monitoring certain types of catabolic populations. For instance, in a mixed population of toluene degraders one could expect to detect the genes for any or all of the five different catabolic pathways for toluene degradation. The gene probe(s) to be used in monitoring must be chosen carefully, with knowledge of the specific microbial population to be monitored. It has been shown, for instance, that only 10% of naphthalene degraders isolated from an estuarine environment hybridize to a *nahAB* gene probe.[97]

One of the more important factors to consider in monitoring microbial populations is an on-line measurement of biodegradative activity or even of population density. One of the simplest and most sensitive measurements is that of light.

Recently, the ability of certain enzymes to produce light has been combined with biodegradative operons to produce bioluminescent reporters. The *lux* genes from *Vibrio fischeri*, encoding for the light-producing luciferase enzyme system, have been inserted into the *nahG* gene encoding for salicylate hydroxylase of a naphthalene catabolic plasmid in *P. fluorescens*.[106] The new plasmid formed expresses luciferase, and therefore produces light, in response to exposure to naphthalene. The strain has been utilized in soil slurries to detect naphthalene biodegradation *in situ* with instantaneous measurements.[106] In addition, the luciferase reporter system can be utilized as a biosensor to quantitate the amount of naphthalene that is bioavailable in contaminated soils.[107] Measurement of naphthalene concentration by this system is rapid and quantitative down to at least 45 ppb.

Degradation of Trichloroethylene

A number of microorganisms have been shown to possess the ability to completely degrade trichloroethylene under aerobic conditions.[108] This degradative ability is dependent on cometabolic processes. The enzymes ammonia monooxygenase,[109] methane monooxygenase,[110] propane monooxygenase,[111] isoprene monooxygenase,[112] toluene dioxygenase,[113] toluene 2-monooxygenase,[114] toluene 3-monooxygenase,[108] and toluene 4-monooxygenase[115] have all been shown to degrade trichloroethylene. Each of these enzyme systems must be induced using appropriate substrates in order to obtain full activity for trichloroethylene degradation. This poses a problem in that the inducing substrate must often be removed from the culture medium before trichloroethylene can be rapidly degraded. This is due to competition between the inducing substrate and trichloroethylene for the active site of the enzyme. One method of avoiding this problem is to utilize a gratuitous inducer, if one is available, that is not a substrate for the enzyme and thus will not competitively inhibit trichloroethylene degradation. Another method would be to construct constitutive mutants that continually express the desired enzyme system. However, perhaps the best way to obtain high levels of enzyme activity for the degradation of trichloroethylene is to express the cloned genes to high levels using foreign promoters.

The genes for two enzyme systems capable of degrading trichloroethylene have been cloned and expressed in *E. coli*. These are the *todC1C2BA* genes for toluene dioxygenase from *P. putida* F1[50,52] and the *tmoABCDE* genes for toluene 4-monooxygenase from *P. mendocina* KR.[115] Expression of these genes in appropriate *E. coli* hosts under the control of promoters such as *tac* or λP_L results in the high-level expression of toluene dioxygenase or toluene 4-monooxygenase activity, respectively. Each *E. coli* recombinant strain is then capable of degrading trichloroethylene at high rates and to low levels.[115,116] In the case of toluene dioxygenase cloned in *E. coli*, the recombinant strain shows a lower initial rate of trichloroethylene degradation than the parent strain *P. putida* F1 induced with toluene.[116] However, the rate of degradation is sustained over long periods of time (>6 hr) for the recombinant *E. coli*, while with *P. putida* F1 the degradation rate rapidly declines. Wackett has shown that the decline in degradative activity of toluene dioxygenase toward trichloroethylene is due to inactivation of the enzyme by the highly reactive degradation products formed from trichloroethylene.[117] The initial products of toluene dioxygenase-mediated degradation of trichloroethylene are formate and glyoxylate, and a mechanism for their formation has been proposed.[117] Experiments with *P. mendocina* KR have shown that toluene-induced toluene 4-monooxygenase activity against trichloroethylene rapidly declines when toluene is removed from the culture medium. In the presence of toluene, sustained degradation of trichloroethylene is observed, although degradative ability against trichloroethylene tapers off due to competition between toluene and trichloroethylene. However, using the recombinant *E. coli* strain expressing toluene 4-monooxygenase, trichloroethylene can be degraded to below detectable limits at sustained rates comparable to that obtained for toluene-induced *P. mendocina* KR. Products detected from toluene 4-monooxygenase-mediated trichloroethylene degradation also include formate and glyoxylate.[115]

Construction of New Biodegradative Pathways

Although nature is infinitely variable and has shown versatility in the variety of ways by which aromatic compounds can be degraded by microorganisms, many aromatic compounds are recalcitrant or only slowly degraded. The possibility exists that one could construct by recombinant means or by novel breeding techniques new microorganisms that have enhanced capabilities to degrade aromatic compounds. This enhancement could include the construction of new catabolic pathways for the degradation of recalcitrant compounds, the removal of metabolic blocks for compounds that are only partially degraded, or the expansion of the range of substrates that can be degraded by a single microorganism. Methods of obtaining this end include long-term chemostat enrichment culture; *in vivo* genetic engineering where natural processes such as transduction, transformation, or conjugation are utilized in a defined way to obtain a desired result; and *in vitro* genetic engineering, selectively introducing cloned genes or constructed operons into a host bacterium. The latter method often produces the quickest results and a totally defined genetic system is produced.

One of the earliest examples of this microbial "improvement" was the construction of halobenzoate degrading strains using elements of the TOL plasmid and *Pseudomonas* sp. B13.[118] The latter strain has the capability of utilizing 3-chlorobenzoate as carbon source by a modified *ortho*-cleavage pathway. The purpose of these experiments was to utilize the broad substrate specificity of toluate 1,2-dioxygenase (*xylXYZ*; see Figure 2) and dihydroxycyclohexadiene carboxylate dehydrogenase (*xylL*) from the TOL plasmid to catalyze the catabolic transformation of 4-chlorobenzoate to 4-chlorocatechol. The latter substrate could then be utilized as a carbon source by *Pseudomonas* sp. B13. Initially, this was accomplished by natural genetic means simply by mating the TOL plasmid into *Pseudomonas* sp. B13. At first, such strains were not capable of utilizing 4-chlorobenzoic acid as a carbon source. However, spontaneous mutants were obtained that did have this ability. It was subsequently determined that such mutations inactivated the TOL plasmid catechol 2,3-dioxygenase gene *(xylE)*. This *meta*-cleavage catechol dioxygenase had been acting to funnel 4-chlorocatechol into the nonproductive *meta* pathway. Inactivation of this enzyme by spontaneous mutation allowed funneling of 4-chlorocatechol through the productive *ortho*-cleavage catabolic route encoded by *Pseudomonas* sp. B13. This experiment was repeated years later through the use of genetic engineering techniques.[119] The structural genes *xylXYZL* were cloned from the TOL plasmid along with the regulatory gene *xylS* into a broad host range cloning vector. This recombinant plasmid allowed growth of *Pseudomonas* sp. B13 on 4-chlorobenzoate in a regulated fashion.

The example given in the previous paragraph is of expansion of the substrate range of a catabolic pathway in a so-called "horizontal" fashion. This means that the catabolic pathway is expanded in such a fashion that more structural analogs or isomers of a single class of compounds can be degraded. In the previous example, this was accomplished through the use of an isofunctional enzyme introduced from a second microorganism by *in vivo* or *in vitro* genetic techniques. This could also be accomplished in a more random fashion through mutation and selection for the modification of preexisting enzymes so that they have an expanded substrate range. Another manner by which the substrate range for growth of a microorganism may be altered is by the so-called "vertical" fashion. This approach involves the addition of new enzymatic steps to a preexisting catabolic pathway in such a manner that the new enzymes act to feed new substrates down to the existing biochemical pathway. An example of this technique is the addition of the ability to grow on chlorosalicylates to *Pseudomonas* sp. B13.[118] This was accomplished through the cloning of the structural gene for salicylate hydroxylase (*nahG*) and the regulatory gene (*nahR*) responsible for its induction from the NAH plasmid (see Figure 11) into *Pseudomonas* sp. B13. The artificially constructed microbial strain was then able to utilize 3-, 4-, and 5-chlorosalicylates as carbon sources.

One problem that is always inherent in the construction of new strains by genetic techniques is that of regulation of gene expression. An enzyme is often available that can catalyze a given transformation, but the regulatory system that controls its expression does not recognize the substrate or its catabolic products as inducers. In such cases, an enzyme is often utilized by expressing the genes constitutively in a host organism or by placing the gene(s) under artificial control (as seen for trichloroethylene degradation, described above). Both of these methods, valuable in laboratory studies or in controlled bioreactors, have distinct disadvantages under real-world environmental situations. A solution is to place the genes under the control of promoters which are activated in response to the desired substrate to be degraded. The easiest way to obtain this result is to alter the specificity of an existing activator through mutation of the coding gene. This is the approach taken by Ramos and co-workers in altering the specificity of the *xylS* gene product to recognize 4-ethylbenzoate.[120] The alteration involved random mutagenesis, followed by selection using a two-plasmid construct that would express tetracycline resistance in response to induction mediated by *xylS* in response to 4-ethylbenzoate. The mutant *xylS*-encoded activator thus produced not only still recognized the normal inducers 2-, 3-, and 4-methylbenzoate, but now additionally recognized 4-ethylbenzoate as an inducing compound. In such a manner, mutants of *xylS* were also produced that recognized a number of other aromatic compounds.[121]

ACKNOWLEDGMENTS

This work was performed under the sponsorship of the U.S. Department of Energy, Environmental Restoration and Waste Management Young Faculty Award Program administered by Oak Ridge Associated Universities. The assistance of Mel Scala in preparing the manuscript is gratefully acknowledged.

REFERENCES

1. **Blumer, M.,** Polycyclic aromatic compounds in nature, *Sci. Am.,* 234, 34, 1976.
2. **Bazylinski, D. A., Wirsen, C. O., and Jannasch, H. W.,** Microbial utilization of naturally occuring hydrocarbons at the Guaymas Basin hydrothermal vent site, *Appl. Environ. Microbiol.,* 55, 2832, 1989.
3. **Blumer, M., and Youngblood, W. W.,** Polycyclic aromatic hydrocarbons in soils and recent sediments, *Science,* 188, 53, 1975.
4. **Hites, R. A., LaFlamme, R. E., and Farrington, J. W.,** Sedimentary polycyclic aromatic hydrocarbons: the historical record, *Science,* 198, 829, 1977.
5. **Altenschmidt, U., and Fuchs, G.,** Anaerobic toluene oxidation to benzyl alcohol and benzaldehyde in a denitrifying *Pseudomonas* strain, *J. Bacteriol.,* 174, 4860, 1992.
6. **Evans, P. J., Ling, W., Goldschmidt, B., Ritter, E. R., and Young, L. Y.,** Metabolites formed during anaerobic transformation of toluene and *o*-xylene and their proposed relationship to the initial steps of toluene mineralization, *Appl. Environ. Microbiol.,* 58, 496, 1992.
7. **Williams, P. A., and Murray, K.,** Metabolism of benzoate and the methylbenzoates by *Pseudomonas putida (arvilla)* mt-2: evidence for the existence of a TOL plasmid, *J. Bacteriol.,* 120, 416, 1974.
8. **Worsey, M. J., and Williams, P. A.,** Metabolism of toluene and xylenes by *Pseudomonas putida (arvilla)* mt-2: evidence for a new function of the TOL plasmid, *J. Bacteriol.,* 124, 7, 1975.
9. **Shields, M. S., Montgomery, S. O., Chapman, P. J., Cuskey, S. M., and Pritchard, P. H.,** Novel pathway of toluene catabolism in the trichloroethylene-degrading bacterium G4, *Appl. Environ. Microbiol.,* 55, 1624, 1989.
10. **Kukor, J. J., and Olsen, R. H.,** Molecular cloning, characterization, and regulation of a *Pseudomonas pickettii* PKO1 gene encoding phenol hydroxylase and expression of the gene in *Pseudomonas aeruginosa* PAO1c, *J. Bacteriol.,* 172, 4624, 1990.
11. **Whited, G. M., and Gibson, D. T.,** Separation and partial characterization of the enzymes of the toluene-4-monooxygenase catabolic pathway in *Pseudomonas mendocina* KR1, *J. Bacteriol.,* 173, 3017, 1991.

12. **Whited, G. M., and Gibson, D. T.,** Toluene-4-monooxygenase, a three-component enzyme system that catalyzes the oxidation of toluene to *p*-cresol in *Pseudomonas mendocina* KR1, *J. Bacteriol.,* 173, 3010, 1991.
13. **Gibson, D. T., Koch, J. R., and Kallio, R. E.,** Oxidative degradation of aromatic hydrocarbons by microorganisms. I. Enzymatic formation of catechol from benzene, *Biochemistry,* 7, 2653, 1968.
14. **Gibson, D. T., Hensley, M., Yoshioka, H., and Mabry, T. J.,** Formation of (+)-*cis*-2,3-dihydroxy-1-methylcyclohexa-4,6-diene from toluene by *Pseudomonas putida, Biochemistry,* 9, 1626, 1970.
15. **Kobal, V. M., Gibson, D. T., Davis, R. E., and Garza, A.,** X-ray determination of the absolute stereochemistry of the initial oxidation product formed from toluene by *Pseudomonas putida* 39/D, *J. Am. Chem. Soc.,* 95, 4420, 1973.
16. **Ziffer, H., Jerina, D. M., Gibson, D. T., and Kobal, V. M.,** Absolute stereochemistry of the dextro *cis*-1,2-dihydroxy-3-methylcyclohexa-3,5-diene produced from toluene by *Pseudomonas putida, J. Am. Chem. Soc.,* 95, 4048, 1973.
17. **Gibson, D. T., Yeh, W.-K., Liu, T.-N., and Subramanian, V.,** Toluene dioxygenase: a multicomponent enzyme system from *Pseudomonas putida,* in *Oxygenases and Oxygenase Metabolism,* Nozaki, M., Yamamoto, S., Ishimura, Y., Coon, M. J., Ernster, L., and Estabrook, R. W., Eds., Academic Press, New York, 1982, 51.
18. **Cruden, D. L., Wolfram, J. H., Rogers, R. D., and Gibson, D. T.,** Physiological properties of a *Pseudomonas* strain which grows with *p*-xylene in a two-phase (organic-aqueous) medium, *Appl. Environ. Microbiol.,* 58, 2723, 1992.
19. **Kunz, D. A., and Chapman, P. J.,** Catabolism of pseudocumene and 3-ethyltoluene by *Pseudomonas putida (arvilla)* mt-2: evidence for new functions of the TOL (pWW0) plasmid, *J. Bacteriol.,* 146, 179, 1981.
20. **Harayama, S., Mermod, N., Rekik, M., Lehrbach, P. R., and Timmis, K. N.,** Roles of the divergent branches of the *meta*-cleavage pathway in the degradation of benzoate and substituted benzoates, *J. Bacteriol.,* 169, 558, 1987.
21. **Jacoby, G. A., Rogers, J. E., Jacob, A. E., and Hedges, R. W.,** Transposition of *Pseudomonas* toluene degrading genes and expression in *Escherichia coli, Nature,* 234, 220, 1979.
22. **Tsuda, M., and Iino, T.,** Genetic analysis of transposon carrying toluene degrading genes on a TOL plasmid pWWO, *Mol. Gen. Genet.,* 210, 270, 1987.
23. **Harayama, S., Leppik, R. A., Rekik, M., Mermod, N., Lehrbach, P. R., Reineke, W., and Timmis, K. N.,** Gene order of the TOL catabolic plasmid upper pathway operon and oxidation of both toluene and benzyl alcohol by the *xylA* product, *J. Bacteriol.,* 167, 455, 1986.
24. **Harayama, S., Lehrbach, P. R., and Timmis, K. N.,** Transposon mutagenesis of *meta*-cleavage pathway operon genes of the TOL plasmid of *Pseudomonas putida* mt-2, *J. Bacteriol.,* 160, 251, 1984.
25. **Franklin, F. C. H., Bagdasarian, M., Bagdasarian, M., and Timmis, K. N.,** Molecular and functional analysis of the TOL plasmid pWWO from *Pseudomonas putida* and cloning of genes for the entire regulated aromatic ring *meta* cleavage pathway, *Proc. Natl. Acad. Sci. U.S.A.,* 78, 7458, 1981.
26. **Harayama, S., Rekik, M., and Timmis, K. N.,** Genetic analysis of a relaxed substrate specificity aromatic ring dioxygenase, toluate 1,2-dioxygenase, encoded by TOL plasmid pWW0 of *Pseudomonas putida, Mol. Gen. Genet.,* 202, 226, 1986.

27. **Harayama, S., Rekik, M., Wubbolts, M., Rose, K., Leppik, R. A., and Timmis, K. N.,** Characterization of five genes in the upper-pathway operon of TOL plasmid pWW0 from *Pseudomonas putida* and identification of the gene products, *J. Bacteriol.,* 171, 5048, 1989.
28. **Harayama, S., and Rekik, M.,** The *meta* cleavage operon of TOL degradative plasmid pWW0 comprises 13 genes, *Mol. Gen. Genet.,* 221, 113, 1990.
29. **Benjamin, R. C., Voss, J. A., and Kunz, D. A.,** Nucleotide sequence of *xylE* from the TOL pDK1 plasmid and structural comparison with isofunctional catechol-2,3-dioxygenase genes from TOL pWWO and NAH7, *J. Bacteriol.,* 173, 2724, 1991.
30. **Nakai, C., Kagamiyama, H., Nozaki, M., Nakazawa, T., Inouye, S., Ebina, Y., and Nakazawa, A.,** Complete nucleotide sequence of the metapyrocatechase gene on the TOL plasmid of *Pseudomonas putida* mt-2, *J. Biol. Chem.,* 258, 2923, 1983.
31. **Harayama, S., Rekik, M., Bairoch, A., Neidle, E. L., and Ornston, L. N.,** Potential DNA slippage structures acquired during evolutionary divergence of *Acinetobacter calcoaceticus* chromosomal *benABC* and *Pseudomonas putida* TOL pWWO plasmid *xylXYZ*, genes encoding benzoate dioxygenases, *J. Bacteriol.,* 173, 7540, 1991.
32. **Neidle, E., Hartnett, C., Ornston, L. N., Bairoch, A., Rekik, M., and Harayama, S.,** *cis*-Diol dehydrogenases encoded by the TOL pWW0 plasmid *xylL* gene and the *Acinetobacter calcoaceticus* chromosomal *benD* gene are members of the short-chain alcohol dehydrogenase superfamily, *Eur. J. Biochem.,* 204, 113, 1992.
33. **Horn, J. M., Harayama, S., and Timmis, K. N.,** DNA sequence determination of the TOL plasmid pWW0 *xylGFJ* genes of *Pseudomonas putida*: implications for the evolution of aromatic catabolism, *Mol. Microbiol.,* 5, 2459, 1991.
34. **Suzuki, M., Hayakawa, T., Shaw, J. P., Rekik, M., and Harayama, S.,** Primary structure of xylene monooxygenase: similarities to and differences from the alkane hydroxylation system, *J. Bacteriol.,* 173, 1690, 1991.
35. **Inouye, S., Nakazawa, A., and Nakazawa, T.,** Determination of the transcription initiation site and identification of the protein product of the regulatory gene *xylR* for *xyl* operons on the TOL plasmid, *J. Bacteriol.,* 163, 863, 1985.
36. **Inouye, S., Nakazawa, A., and Nakazawa, T.,** Nucleotide sequence of the regulatory gene *xylS* on the *Pseudomonas putida* TOL plasmid and identification of the protein product, *Gene,* 44, 235, 1986.
37. **Worsey, M. J., Franklin, F. C. H., and Williams, P. A.,** Regulation of the degradative pathway enzymes coded for by the TOL plasmid (pWW0) from *Pseudomonas putida* mt-2, *J. Bacteriol.,* 134, 757, 1978.
38. **Mermod, N., Ramos, J. L., Bairoch, A., and Timmis, K. N.,** The *xylS* gene positive regulator of TOL plasmid pWW0: identification, sequence analysis, and overproduction leading to constitutive expression of *meta* cleavage operon, *Mol. Gen. Genet.,* 207, 349, 1987.
39. **Dixon, R.,** The *xylABC* promoter from the *Pseudomonas putida* TOL plasmid is activated by nitrogen regulatory genes in *Escherichia coli*, *Mol. Gen. Genet.,* 203, 129, 1986.
40. **Inouye, S., Gomada, M., Sangodkar, U. M. X., Nakazawa, A., and Nakazawa, T.,** Upstream regulatory sequence for transcriptional activator Xy1R in the first operon of xylene metabolism on the TOL plasmid, *J. Mol. Biol.,* 216, 251, 1990.

41. **DeLorenzo, V., Herrero, M., Metzke, M., and Timmis, K. N.,** An upstream *xylR*-induced and IHF-induced nucleoprotein complex regulates the sigma-54-dependent PU promoter of TOL plasmid, *EMBO J.*, 10, 1159, 1991.
42. **Holtel, A., Timmis, K. N., and Ramos, J. L.,** Upstream binding sequences of the *xylR* activator protein and integration host factor in the *xylS* gene promoter region of the *Pseudomonas* TOL plasmid, *Nucleic Acids Res.*, 20, 1755, 1992.
43. **Cuskey, S. M., and Sprenkle, A. M.,** Benzoate-dependent induction from the OP2 operator-promoter region of the TOL plasmid pWW0 in the absence of known plasmid regulatory genes, *J. Bacteriol.*, 170, 3742, 1988.
44. **Yeh, W.-K., Gibson, D. T., and Liu, T.-N.,** Toluene dioxygenase: a multicomponent enzyme system, *Biochem. Biophys. Res. Commun.*, 78, 401, 1977.
45. **Subramanian, V., Liu, T.-N., Yeh, W.-K., Narro, M., and Gibson, D. T.,** Purification and properties of NADH-ferredoxin$_{TOL}$ reductase, *J. Biol. Chem.*, 256, 2723, 1981.
46. **Subramanian, V., Liu, T.-N., Yeh, W.-K., Serdar, C. M., Wackett, L. P., and Gibson, D. T.,** Purification and properties of ferredoxin$_{TOL}$: a component of toluene dioxygenase from *Pseudomonas putida* F1, *J. Biol. Chem.*, 260, 2355, 1985.
47. **Subramanian, V., Liu, T.-N., Yeh, W.-K., and Gibson, D. T.,** Toluene dioxygenase: purification of an iron sulfur protein by affinity chromatography, *Biochem. Biophys. Res. Commun.*, 91, 1131, 1979.
48. **Rogers, J. E., and Gibson, D. T.,** Purification and properties of *cis*-toluene dihydrodiol dehydrogenase from *Pseudomonas putida*, *J. Bacteriol.*, 130, 1117, 1977.
49. **Finette, B. A., Subramanian, V., and Gibson, D. T.,** Isolation and characterization of *Pseudomonas putida* PpF1 mutants defective in the toluene dioxygenase enzyme system, *J. Bacteriol.*, 160, 1003, 1984.
50. **Zylstra, G. J., McCombie, W. R., Gibson, D. T., and Finette, B. A.,** Toluene degradation by *Pseudomonas putida* F1: genetic organization of the *tod* operon, *Appl. Environ. Microbiol.*, 54, 1498, 1988.
51. **Menn, F.-M., Zylstra, G. J., and Gibson, D. T.,** Location and sequence of the *todF* gene encoding 2-hydroxy-6-oxohepta-2,4-dienoate hydrolase in *Pseudomonas putida* F1, *Gene*, 104, 91, 1991.
52. **Zylstra, G. J., and Gibson, D. T.,** Toluene degradation by *Pseudomonas putida* F1, nucleotide sequence of the *todC1C2BADE* genes and their expression in *Escherichia coli*, *J. Biol. Chem.*, 264, 14940, 1989.
53. **Erickson, B. D., and Mondello, F. J.,** Nucleotide sequencing and transcriptional mapping of the genes encoding biphenyl dioxygenase, a multicomponent polychlorinated-biphenyl-degrading enzyme in *Pseudomonas* strain LB400, *J. Bacteriol.*, 174, 2903, 1992.
54. **Taira, K., Hirose, J., Hayashida, S., and Furukawa, K.,** Analysis of *bph* operon from the polychlorinated biphenyl-degrading strain of *Pseudomonas pseudoalcaligenes* KF707, *J. Biol. Chem.*, 267, 4844, 1992.
55. **Furukawa, K., Arimura, N., and Miyazaki, T.,** Nucleotide sequence of the 2,3-dihydroxybiphenyl dioxygenase gene of *Pseudomonas pseudoalcaligenes*, *J. Bacteriol.*, 169, 427, 1987.
56. **Zylstra, G. J., and Gibson, D. T.,** Aromatic hydrocarbon degradation: a molecular approach, in *Genetic Engineering: Principles and Methods*, Vol. 13, Setlow, J. K., Ed., Plenum Press, New York, 1991.
57. **Robertson, J. B., Spain, J. C., Haddock, J. D., and Gibson, D. T.,** Oxidation of nitrotoluenes by toluene dioxygenase: evidence for a monooxygenase reaction, *Appl. Environ. Microbiol.*, 58, 2643, 1992.

58. **Zylstra, G. J., Chauhan, S., and Gibson, D. T.,** Degradation of chlorinated biphenyls by *Escherichia coli* containing cloned genes of the *Pseudomonas putida* F1 toluene catabolic pathway, in *Proc. Sixteenth Annu. Hazardous Waste Res. Symp.: Remedial Action, Treatment, and Disposal of Hazardous Waste,* USEPA, 1990, 290.
59. **Brand, J. M., Cruden, D. L., Zylstra, G. J., and Gibson, D. T.,** Stereospecific hydroxylation of indan and indene by *Escherichia coli* containing the cloned toluene dioxygenase genes from *Pseudomonas putida* F1, *Appl. Environ. Microbiol.,* 58, 3407, 1992.
60. **Eaton, R. W., and Timmis, K. N.,** Characterization of a plasmid-specified pathway for catabolism of isopropylbenzene in *Pseudomonas putida* RE204, *J. Bacteriol.,* 168, 123, 1986.
61. **Irie, S., Doi, S., Yorifuji, T., Takagi, M., and Yano, K.,** Nucleotide sequencing and characterization of the genes encoding benzene oxidation enzymes of *Pseudomonas putida, J. Bacteriol.,* 169, 5174, 1987.
62. **Stephens, G. M., Sidebotham, J. M., Mann, N. H., and Dalton, H.,** Cloning and expression in *Escherichia coli* of the toluene dioxygenase gene from *Pseudomonas putida* NCIB11767, *FEMS Microbiol. Lett.,* 57, 295, 1989.
63. **Tan, H. M., and Mason, J. R.,** Cloning and expression of the plasmid-encoded benzene dioxygenase genes from *Pseudomonas putida* ML2, *FEMS Microbiol. Lett.,* 72, 259, 1990.
64. **Zylstra, G. J., Sandoli, R. L., Biegel, S. D., and Didolkar, V. A.,** Isolation and characterization of microbial strains with aromatic dioxygenase pathways, *J. Cellular Biochem.,* 17C, 194, 1993.
65. **Shields, M. S., Montgomery, S. O., Cuskey, S. M., Chapman, P. J., and Pritchard, P. H.,** Mutants of *Pseudomonas cepacia* G4 defective in catabolism of aromatic compounds and trichloroethylene, *Appl. Environ. Microbiol.,* 57, 1935, 1991.
66. **Kukor, J. J., and Olsen, R. H.,** Genetic organization and regulation of a *meta* cleavage pathway for catechols produced from catabolism of toluene, benzene, phenol, and cresols by *Pseudomonas pickettii* PKO1, *J. Bacteriol.,* 173, 4587, 1991.
67. **Kukor, J. J., and Olsen, R. H.,** Complete nucleotide sequence of *tbuD*, the gene encoding phenol/cresol, hydroxylase from *Pseudomonas pickettii* PKO1, and functional analysis of the encoded enzyme, *J. Bacteriol,* 174, 6518, 1992.
68. **Yen, K. M., Karl, M. R., Blatt, L. M., Simon, M. J., Winter, R. B., Fausset, P. R., Lu, H. S., Harcourt, A. A., and Chen, K. K.,** Cloning and characterization of a *Pseudomonas mendocina* KR1 gene cluster encoding toluene-4-monooxygenase, *J. Bacteriol.,* 173, 5315, 1991.
69. **Nordlund, I., Powlowski, J., and Shingler, V.,** Complete nucleotide sequence and polypeptide analysis of multicomponent phenol hydroxylase from *Pseudomonas* sp. strain CF600, *J. Bacteriol.,* 172, 6826, 1990.
70. **Wright, A., and Olsen, R. H.,** Genetic organization and regulation of the *p*-cresol regulon of *Pseudomonas mendocina* KR1, Paper K33, *Abstr. Annu. Meet. Am. Soc. Microbiol.,* New Orleans, 1992.
71. **Yen, K.-M., and Serdar, C. M.,** Genetics of naphthalene catabolism in pseudomonads, *Crit. Rev. Microbiol.,* 15, 247, 1988.
72. **Davies, J. I., and Evans, W. C.,** Oxidative metabolism of naphthalene by soil *Pseudomonads, Biochem. J.,* 91, 251, 1964.
73. **Dunn, N., and Gunsalus, I. C.,** Transmissable plasmids coding early enzymes of naphthalene oxidation in *Pseudomonas putida, J. Bacteriol.,* 114, 974, 1973.

74. **Serdar, C. M., and Gibson, D. T.,** Isolation and characterization of altered plasmids in mutant strains of *Pseudomonas putida* NCIB 9816, *Biochem. Biophys. Res. Commun.,* 164, 764, 1989.
75. **Serdar, C. M., and Gibson, D. T.,** Studies of nucleotide sequence homology between naphthalene-utilizing strains of bacteria, *Biochem. Biophys. Res. Commun.,* 164, 772, 1989.
76. **Ensley, B. D., Gibson, D. T., and Laborde, A. L.,** Oxidation of naphthalene by a multicomponent enzyme system from *Pseudomonas* sp. strain NCIB 9816, *J. Bacteriol.,* 149, 948, 1982.
77. **Haigler, B. E., and Gibson, D. T.,** Purification and properties of NADH-ferredoxin$_{NAP}$ reductase, a component of naphthalene dioxygenase from *Pseudomonas* sp. strain NCIB 9816. *J. Bacteriol.,* 172, 457, 1990.
78. **Haigler, B. E., and Gibson, D. T.,** Purification and properties of ferredoxin$_{NAP}$, a component of naphthalene dioxygenase from *Pseudomonas* sp. strain NCIB 9816, *J. Bacteriol.,* 172, 465, 1990.
79. **Ensley, B. D., and Gibson, D. T.,** Naphthalene dioxygenase: purification and properties of a terminal oxygenase component. *J. Bacteriol.,* 155, 505, 1983.
80. **Jeffrey, A. M., Yeh, H. J. C., Jerina, D. M., Patel, T. R., Davey, J. F., and Gibson, D. T.,** Initial reactions in the oxidation of naphthalene by *Pseudomonas putida, Biochemistry,* 14, 575, 1975.
81. **Grund, E., Denecke, B., and Eichenlaub, R.,** Naphthalene degradation via salicylate and gentisate by *Rhodococcus* sp. strain B4, *Appl. Environ. Microbiol.,* 58, 1874, 1992.
82. **Evans, W. C., Fernley, H. N., and Griffiths, E.,** Oxidative metabolism of phenanthrene and anthracene by soil pseudomonads, *Biochem. J.,* 95, 819, 1965.
83. **Heitkamp, M. A., and Cerniglia, C. E.,** Mineralization of polycyclic aromatic hydrocarbons by a bacterium isolated from sediment below an oil field, *Appl. Environ. Microbiol.,* 54, 1612, 1988.
84. **Heitkamp, M. A., Franklin, W., and Cerniglia, C. E.,** Microbial metabolism of polycyclic aromatic hydrocarbons: isolation and characterization of a pyrene-degrading bacterium, *Appl. Environ. Microbiol.,* 54, 2549, 1988.
85. **Heitkamp, M. A., Freeman, J. P., Miller, D. W., and Cerniglia, C. E.,** Pyrene degradation by a *Mycobacterium* sp.: identification of ring oxidation and ring fission products, *Appl. Environ. Microbiol.,* 54, 2556, 1988.
86. **Tsuda, M., and Iino, T.,** Naphthalene degrading genes on plasmid NAH7 are on a defective transposon, *Mol. Gen. Genet.,* 223, 33, 1990.
87. **Yen, K.-M., and Gunsalus, I. C.,** Plasmid gene organization: naphthalene/salicylate oxidation, *Proc. Natl. Acad. Sci. U.S.A.,* 79, 874, 1982.
88. **Schell, M. A.,** Cloning and expression in *Esherichia coli* of the naphthalene degradation genes from plasmid NAH7, *J. Bacteriol.,* 153, 822, 1983.
89. **Schell, M. A.,** Transcriptional control of the *nah* and *sal* hydrocarbon-degradation operons by the *nahR* gene product, *Gene,* 36, 301, 1985.
90. **Huang, J., and Schell, M. A.,** *In vivo* interactions of the nahR transcriptional activator with its target sequences, *J. Biol. Chem.,* 266, 10830, 1991.
91. **Simon, M. J., Osslund, T. D., Saunders, R., Ensley, B. D., Suggs, S., Harcourt, A., Suen, W.-c., Cruden, D. L., Gibson, D. T., and Zylstra, G. J.,** Sequences of genes encoding naphthalene dioxygenase in *Pseudomonas putida* strains G7 and NCIB 9816-4, *Gene,* 127, 31, 1993.
92. **Harayama, S., and Rekik, M.,** Bacterial aromatic ring-cleavage enzymes are classified into two different gene families, *J. Biol. Chem.,* 264, 15328, 1989.

93. **Ghosal, D., You, I.-S., and Gunsalus, I. C.,** Nucleotide sequence and expression of gene *nahH* of plasmid NAH7 and homology with gene *xylE* of TOL pWWO, *Gene,* 55, 19, 1987.
94. **Harayama, S., Rekik, M., Wasserfallen, A., and Bairoch, A.,** Evolutionary relationships between catabolic pathways for aromatics: conservation of gene order and nucleotide sequences of catechol oxidation genes of pWWO and NAH7 plasmids, *Mol. Gen. Genet.,* 210, 241, 1987.
95. **You, I.-S., Ghosal, D., and Gunsalus, I. C.,** Nucleotide sequence analysis of the *Pseudomonas putida* PpG7 salicylate hydroxylase gene *(nahG)* and its 3′-flanking region, *Biochemistry,* 30, 1635, 1991.
96. **Kurkela, S., Lehvaslaiho, H., Palva, E. T., and Teeri, T. H.,** Cloning, nucleotide sequence and characterization of genes encoding naphthalene dioxygenase of *Pseudomonas putida* strain NCIB9816, *Gene,* 73, 355, 1988.
97. **Smith, C., and Shiaris, M. P.,** Distribution of *nahAB* genes among naphthalene-transforming bacteria isolated from soil, freshwater sediments, and marine sediments, Paper Q70, *Abstr. Annu. Meet. Am. Soc. Microbiol.,* New Orleans, 1989.
98. **Wang, X. P., and Zylstra, G. J.,** Isolation and characterization of polycyclic aromatic hydrocarbon degrading bacteria from a coal tar contaminated soil, Paper K50, *Abstr. Annu. Meet. Am. Soc. Microbiol.,* New Orleans, 1992.
99. **Sayler, G. S.,** Contribution of molecular biology to bioremediation, *J. Hazard. Mater.,* 28, 13, 1991.
100. **Olson, B. H.,** Tracking and using genes in the environment, *Environ. Sci. Technol.,* 25, 604, 1991.
101. **Sayler, G. S., and Layton, A. C.,** Environmental applications of nucleic acid hybridization, *Annu. Rev. Microbiol.,* 44, 625, 1990.
102. **Atlas, R. M., Sayler, G., Burlage, R. S., and Bej, A. K.,** Molecular approaches for environmental monitoring of microorganisms, *Biotechniques,* 12, 706, 1992.
103. **Sayler, G. S., Shields, M. S., Tedford, E. T., Breen, A., Hooper, S. W., Sirotkin, K. M., and Davis, J. W.,** Application of DNA-DNA colony hybridization to the detection of catabolic genotypes in environmental samples, *Appl. Environ. Microbiol.,* 49, 1295, 1985.
104. **Blackburn, J. W., Jain, R. L., and Sayler, G. S.,** Molecular microbial ecology of a naphthalene-degrading genotype in activated sludge, *Environ. Sci. Technol.,* 21, 884, 1987.
105. **Tsai, Y. L., Park, M. J., and Olson, B. H.,** Rapid method for direct extraction of messenger RNA from seeded soils, *Appl. Environ. Microbiol.,* 57, 765, 1991.
106. **King, J. M. H., DiGrazia, P. M., Applegate, B., Burlage, R., Sanseverino, J., Dunbar, P., Larimer, F., and Sayler, G. S.,** Rapid, sensitive bioluminescent reporter technology for naphthalene exposure and biodegradation, *Science,* 249, 778, 1990.
107. **Heitzer, A., Webb, O. F., Thonnard, J. E., and Sayler, G. S.,** Specific and quantitative assessment of naphthalene and salicylate bioavailability by using a bioluminescent catabolic reporter bacterium, *Appl. Environ. Microbiol.,* 58, 1839, 1992.
108. **Ensley, B. D.,** Biochemical diversity of trichloroethylene metabolism, *Annu. Rev. Microbiol.,* 45, 283, 1991.
109. **Arciero, D., Vannelli, T., Logan, N., and Hooper, A. B.,** Degradation of trichloroethylene by the ammonia oxidizing bacterium *Nitrosomonas europaea, Biochem. Biophys. Res. Commun.,* 159, 640, 1989.

110. **Little, C. D., Palumbo, A. V., Herbes, S. E., Lidstrom, M. V., Tyndall, R. L., and Gilmen, P. J.,** Trichloroethylene biodegradation by a methane-oxidizing bacterium, *Appl. Environ. Microbiol.,* 54, 951, 1988.
111. **Wackett, L. P., Brusseau, G. A., Householder, S. R., and Hanson, R. S.,** Survey of microbial oxygenases: trichloroethylene degradation by propane oxidizing bacteria, *Appl. Environ. Microbiol.,* 55, 2960, 1988.
112. **Ewers, J., Freirer, S. D., and Knackmuss, H. J.,** Selection of trichloroethene TCE degrading bacteria that resist inactivation by TCE, *Arch. Microbiol.,* 154, 410, 1990.
113. **Wackett, L. P., and Gibson, D. T.,** Degradation of trichoroethylene by toluene dioxygenase in whole cell studies with *Pseudomonas putida* F1, *Appl. Environ. Microbiol.,* 54, 1703, 1988.
114. **Nelson, M. J. K., Montgomery, S. O., O'Neill, E. J., and Pritchard, P. H.,** Aerobic metabolism of trichloroethylene by a bacterial isolate, *Appl. Environ. Microbiol.,* 52, 383, 1986.
115. **Winter, R. B., Yen, K. M., and Ensley, B. D.,** Efficient degradation of trichloroethylene by a recombinant *Escherichia coli, Bio/Technology,* 7, 282, 1989.
116. **Zylstra, G. J., Wackett, L. P., and Gibson, D. T.,** Trichloroethylene degradation by *Escherichia coli* containing the cloned *Pseudomonas putida* F1 toluene dioxygenase genes, *Appl. Environ. Microbiol.,* 55, 3162, 1989.
117. **Li, S., and Wackett, L. P.,** Trichloroethylene oxidation by toluene dioxygenase, *Biochem. Biophys. Res. Commun.,* 185, 443, 1992.
118. **Reineke, W., and Knackmuss, H.-J.,** Construction of haloaromatics utilizing bacteria, *Nature,* 277, 385, 1979.
119. **Lehrbach, P. R., Zeyer, J., Reineke, W., Knackmuss, W., and Timmis, K. N.,** Enzyme recruitment *in vitro*: use of cloned genes to extend the range of haloaromatics degraded by *Pseudomonas* sp. B13, *J. Bacteriol.,* 158, 1025, 1984.
120. **Ramos, J. L., Wasserfallen, A., Rose, K., and Timmis, K. N.,** Redesigning metabolic routes: manipulation of TOL plasmid pathway for catabolism of alkylbenzoates, *Science,* 235, 593, 1987.
121. **Ramos, J. L., Stolz, A., Reineke, W., and Timmis, K. N.,** Altered effector specificities in regulators of gene expression: TOL plasmid *xylS* mutants and their use to engineer expansion of the range of aromatics degraded by bacteria, *Proc. Natl. Acad. Sci. U.S.A.,* 83, 8467, 1986.

6

Assessment of Environmental Degradation by Molecular Analysis of a Sentinel Species: Atlantic Tomcod

Isaac I. Wirgin and Seymour J. Garte

Nelson Institute of Environmental Medicine, New York University Medical Center, New York, NY

NEOPLASIA IN FERAL FISH POPULATIONS

In recent years, reports of epizootics of neoplasia in North American feral fish populations have increased dramatically. These episodes have been observed in marine, estuarine, and freshwater species. In all cases, these outbreaks have afflicted populations in waterways impacted from the anthropogenic discharges from highly urban or industrialized areas. Some of the more publicized and studied cases have involved winter flounder from Boston Harbor,[1] English sole from Puget Sound,[2] brown bullhead from the Great Lakes,[3] and Atlantic tomcod from the Hudson River.[4,5] Hepatocellular carcinomas have been the predominant lesion and, given the liver's proclivity to accumulate and process xenobiotics, this tissue specificity is suggestive of an environmental etiology for the disease. Follow-up studies have correlated the incidence of neoplasia in these fish with elevated sediment, tissue, or metabolite concentrations of organic pollutants, most often polycyclic aromatic hydrocarbons (PAHs).[6] Attempts to correlate neoplasia

0-87371-631-0/94/$0.00+$.50

with tissue concentrations of PAHs have, however, been thwarted by the extremely rapid metabolism of this class of xenobiotics by fishes.[7] The apparent sensitivity of selected finfish species to environmental insult and increased concern over environmental degradation have led to the suggestion that sentinel aquatic species may serve as surrogates to monitor the health of our nation's waterways.

POLLUTION AND HUDSON RIVER FISHERIES

The Hudson River serves as a domestic water supply, supports commercial fisheries, provides recreational opportunities, and is in close proximity to a vast urban population. The Hudson has been subject to environmental perturbation from a myriad of xenobiotic sources for decades. These have included organic pollutants such as dibenzofurans and dioxins,[8] PAHs,[9] polychlorobiphenyls (PCBs),[10] and heavy metals.[11] Illustrative of the problem is the designation of two Superfund sites within the confines of the Hudson River estuary. These include the Hudson River PCB Superfund Site which extends from Hudson Falls, NY to the Battery in New York City, a distance of over 200 river miles, and the Foundry Cove Superfund site in Cold Spring, NY (river mile 55) because of remarkably high levels of sediment-borne cadmium and nickel.

Historically, the Hudson River has supported lucrative commercial fisheries for three species of anadromous fishes: striped bass, American shad, and Atlantic sturgeon. Currently, commercial fishermen ply their trade from off upper Manhattan Island to at least Catskill, NY (river mile 125). At the present time, American shad and striped bass populations in the Hudson are at or near record levels of abundance, while sturgeon abundance is probably in moderate decline. Yet, despite the availability and potential lucrative commercial values of these fisheries, commercial fishing is prohibited for resident Hudson River species and striped bass due to their elevated tissue levels of contaminants, primarily PCBs and heavy metals. Recently, PCB concentrations have declined in some Hudson River finfish species; however, overall levels within populations have not reached acceptable standards for human consumption.

USE OF BIOMARKERS IN AQUATIC SYSTEMS

The development and validation of biomarkers in sentinel species has become popular in recent years due to their potential ability to quantify actual exposure history, provide a temporal framework for this exposure, and determine a threshold biological response and possibly detrimental biological consequences. The biomarker approach provides several advantages over the measurement of environmental or tissue concentrations of contaminants. In aquatic systems, these might include integration of bioavailable sediment-borne contaminant levels within a river or estuarine system, possible hypersensitivity of the bioresponse, reduced cost, and most importantly evaluation of an endpoint of a quantifiable biological effect.

Biomarkers which have been evaluated in fish span the range of biological organization from the molecular to the community level. Biomarkers on the molecular level offer the advantage of rapid and often sensitive responses to xenbiotics and a degree of specificity in the response. Molecular biomarkers quantify the level of expression or structural alterations in environmentally responsive genes at the level of their messenger RNA (mRNA). The applicability of this approach to aquatic organisms is currently limited by the number of gene probes available; however, this battery is increasing rapidly. Our studies have focused on an evaluation of molecular biomarkers in Atlantic tomcod as indicators of exposure to Hudson River-borne contaminants and point sources of pollution in other northwest Atlantic estuarine systems.

CHOICE OF SENTINEL SPECIES: THE ATLANTIC TOMCOD

The Atlantic tomcod is a very common bottom-dwelling fish species in the Hudson River whose overall distribution extends from Labrador to New Jersey.[12] The Hudson River supports their southernmost spawning population and, as a result, at times they may be thermally stressed. Tomcod are not a commercial species; however, recreational fishermen frequently target them within the Hudson River. Atlantic tomcod are anadromous and spend their entire life cycles within the confines of their natal estuaries.[13,14] Within these estuaries, adult tomcod undergo winter spawning migrations to the upper reaches of the estuary, and juveniles and surviving adults then drop back to more saline reaches of the rivers for summer feeding. Tomcod are opportunistic feeders, their diet consisting of many available bottom-dwelling food species. Atlantic tomcod also have unusually high liver lipid concentrations which serve to bioaccumulate many xenobiotics of concern due to their lipophilic nature. Additionally, tomcod collected from the Hudson River have increased hepatic liver lipid levels, perhaps indicative of elevated detoxification of organic xenobiotics.[15] As a result of these factors, their exposure history and physiology potentially integrates the contaminant levels throughout the estuary and its food chain.

The Atlantic tomcod population in the Hudson River exhibits a truncated age class structure in comparison with tomcod from other rivers. More than 90% of the population is comprised of fish up to 1 year of age;[5] 3-year-old tomcod are exceedingly rare in the Hudson (0.1% of the population), although tomcod up to 7 years of age are reported from more northern populations.[16]

Atlantic tomcod collected from the Hudson River have one of the highest incidences of neoplasia in any feral population. For example, >50% of 1-year-old Hudson River collected tomcod exhibit hepatocellular carcinomas and >90% of 2-year-olds display this disease.[5] In contrast, <5% of tomcod from the Pawcatuck River, Rhode Island, or the Saco River, Maine, exhibited liver lesions.[17] To date, liver lesions have not been observed in tomcod from more pristine rivers in the Canadian Maritime Provinces, although other diseases have been observed.[18] The frequency and severity of liver tumors in Hudson River tomcod increases directly with the age and length of the fish such that within an age class, the frequency of tumors is significantly higher in larger fish.[5] All of these factors suggest an environmental etiology to the elevated level of neoplasia in these fish. Thus, it has been suggested that Atlantic tomcod may serve as an excellent sentinel species of tidal river systems in the northeastern U.S. and in the Canadian Maritime Provinces.[17]

GENETIC ALTERATIONS IN HUDSON RIVER TOMCOD

Initially, we sought to determine if Hudson River tomcod constitute a separate population, perhaps genetically predisposed to neoplasia; or was the population subjected to insult from environmentally borne carcinogenic agents? The NIH3T3 transfection assay was used to determine that tomcod liver tumor DNA could transform these mouse fibroblasts *in vitro*. Soft agar assays confirmed the anchorage independence of these transformed cells and their tumorigenicity was demonstrated in nude mouse tumor assays. The injection of primary transfectant cells from the NIH3T3 assay into athymic mice resulted in the rapid generation of very large tumors from all cell lines tested. Furthermore, in Southern blot analysis, DNA isolated from these nude mouse tumor DNAs exhibited an exogenous tomcod K-*ras* oncogene.[19] This was the first demonstration of an activated oncogene in any feral population of fish and was followed by reports of activated *ras* oncogenes in winter flounder from a highly PAH polluted site in Boston Harbor[20] and in aflatoxin-treated rainbow trout.[21] Polymerase chain reaction (PCR) analysis of panels of tumor DNAs from both the winter flounder and rainbow trout revealed mutations at the 12th codon of the *ras* gene, a sequence consistently mutated in tumor DNA from animal models treated with chemical carcinogens and in human populations. Surprisingly, the same rodent-derived PCR primers used successfully on flounder and trout DNA did not amplify *ras* in tomcod. As a result, we have sequenced the tomcod *ras* gene from a cDNA library.

In addition to this somatic mutation in the *ras* oncogene in tomcod tumor DNA, normal tomcod liver DNA revealed a restriction fragment length polymorphism (RFLP) in the c-*abl* oncogene.[22] Allelic variation was observed at two of the three c-*abl* domains scored. Hardy-Weinberg analysis confirmed the Mendelian inheritance of these c-*abl* hybridizable genotypes. Surprisingly, the frequencies of c-*abl* genotypes differed significantly between the cancer-prone Hudson River and cancer-free Maine tomcod populations. While the relationship between this RFLP in the *abl* oncogene and tumorigenesis in tomcod has yet to be investigated, this was among the first demonstrations of a difference between feral populations in the frequency of variant oncogene genotypes. This difference was even more surprising in view of the results of an RFLP analysis of mitochondrial DNA (mtDNA) between the same tomcod populations.[22] Mitochondrial DNA evolves approximately an order of magnitude more rapidly than single copy nuclear DNA genes[23] (see Chapter 2), and thus it would be expected that mtDNA polymorphisms would be numerous between the Hudson and Maine tomcod populations given the difference in oncogene genotypes. To the contrary, we found no variation in the mitochondrial genome within or between these two tomcod populations, a finding consistent with the short evolutionary time elapsed since their divergence. Considered against this background of mtDNA monomorphism, the divergence in c-*abl* oncogene genotypes between the two populations probably was a very recent event and highlights the hypermutability of the tomcod nuclear genome or a barrage of environmental insult within northeast estuarine systems.

USE OF CYP1A INDUCTION IN ENVIRONMENTAL MONITORING

For the past 15 years, studies conducted on levels of cytochrome P450 gene products in fish from contaminated environments have generally shown a positive correlation between overall levels of contamination and expression of the gene.[24,25] Thus, it has been proposed that induction of the cytochrome P450 system could serve as an effective biomarker of exposure and effect to selected organic pollutants. The cytochrome P450 system encodes for protein products responsible for the metabolism of both endogenous and exogenous substrates. Phase I enzymes encoded for by this system oxidize xenobiotic agents to an inactive form in which they may be excreted from the body by the action of Phase II conjugative enzymes. Alternatively, classes of environmental procarcinogens are activated to their penulimate carcinogenic state, diolepoxides, by CYP1A enzymes.[7] Activated carcinogens then form DNA adducts at critical genetic loci, protooncogenes or tumor suppressor genes, and initiate the neoplastic process. Thus, it may be argued that induction of CYP1A gene expression is not only indicative of xenobiotic exposure, but also detrimental biological effect.

Levels of CYP1A transcription in fishes may be induced by several classes of organic pollutants. These include polycyclic aromatic hydrocarbons, coplanar PCB congeners, and dibenzofurans and dioxins.[26] Therefore, it has been suggested that levels of CYP1A gene products can serve as effective biological markers of exposure to these environmental toxicants.[27] Levels of CYP1A induction may be quantified by measurement of its enzyme activities (EROD or AHH), quantification of CYP1A protein concentration by western blotting, or analysis of CYP1A mRNA by northern or slot blot analyses. Given the incidence of liver cancer in Hudson River tomcod, the presumed role of CYP1A enzymes in the neoplastic process, and their responsiveness to organic pollutants of concern, we proceeded to evaluate CYP1A gene expression as a potentially sensitive marker of exposure and early biological effect.

To date, most studies using CYP1A induction to monitor environmental levels of these xenobiotics have focused on the protein level. Potential use of CYP1A mRNA levels as a biomarker of exposure requires an examination of several aspects of the response. These include quantification of the sensitivity of the CYP1A mRNA response in comparison to levels of induction detected of the CYP1A proteins. What are the kinetics of induction and clearance of CYP1A mRNA in fish treated with known inducers? Is induction of CYP1A mRNA sufficiently persistent to serve as an effective biomarker? Can differences among inducers in temporal aspects of induction and clearance of CYP1A mRNA be used to identify environmental inducing agents? How much genetic variability is there in the inducibility of CYP1A mRNA? This can be measured between species, populations, and among individuals within a species. Background levels of intraspecific variability in CYP1A mRNA inducibility in feral populations need to be quantified.

Levels of CYP1A mRNA in Tomcod Collected from Natural Systems

Initially, levels of hepatic CYP1A mRNA were quantified in tomcod collected from the Hudson River and other more pristine estuarine systems to determine if levels of gene expression were sufficiently sensitive, but also followed gradients in known levels of contamination among systems and therefore validate its use as a suitable biomarker of exposure. Detectable levels of CYP1A mRNA were observed in almost all individuals, even those collected from pristine rivers. Not unexpectedly, levels of hepatic CYP1A mRNA were significantly higher in tomcod collected from the Hudson River than seen in fish from all other rivers. The extent of variability in the response among Hudson River fish was substantial, whereby some individuals exhibited a highly induced phenotype and other fish showed very low levels of CYP1A mRNA.[28] This may have reflected genetic variability among individuals in terms of inducibility of the CYP1A gene or difference among individuals in their recent exposure history. The Margaree River drainage, Nova Scotia, has no urbanization or industrialization, and Margaree-collected tomcod showed extremely low levels of CYP1A mRNA. No variability was seen in the response among the Margaree-collected tomcod; all fish showed low levels of gene expression. Tomcod collected from the St. Lawrence River, Quebec, the Saco River, Maine, and the Miramichi River, New Brunswick, showed intermediary levels of CYP1A mRNA consistent with their moderate levels of anthropogenic influences.[29] These levels were significantly higher than levels of gene expression detected in Margaree tomcod, yet lower than levels detected in Hudson River-collected fish. Once again, interindividual variability in the response was considerable.

Additionally, levels of CYP1A mRNA expression were found to be very low in two other species of Hudson River-collected fish, hogchokers and striped bass.[30] Hogchokers are a bottom-dwelling flatfish and share a similar ecological niche and exposures as tomcod and would therefore be expected to exhibit comparable levels of CYP1A expression. Striped bass are a mid-water feeding species and therefore their low levels of CYP1A mRNA expression were not unexpected.

To validate the responsiveness of enhanced CYP1A mRNA expression to levels of environmental contaminants, two other biological markers were measured in a subset of these samples, including levels of fluorescent aromatic compounds in their bile and levels of hepatic DNA adducts. Levels of fluorescent aromatic compounds (FAC) determined by HPLC with fluorometric detection at a wavelength pair appropriate for higher molecular weight PAC such as benzo[a]pyrene and normalized for bile protein concentrations were significantly higher in tomcod from the Hudson River than in St. Lawrence or Miramichi River fish. For example, levels of FACs were approximately eightfold higher in the Hudson River than in Miramichi River tomcod. Levels of hepatic DNA adducts as detected by [^{32}P] postlabeling analysis followed the same general pattern. For example, levels of DNA adducts (nmol adducts/mol DNA) were between 10- and 40-fold higher in Hudson River tomcod than seen in Margaree or Miramichi River fish. Interestingly, levels of hepatic DNA adducts did not decrease significantly in Hudson River-collected tomcod that were depurated for more than 20 days in clean water. Additionally, hogchokers collected from the Hudson River, which did not exhibit induced CYP1A mRNA expression or liver cancer, had levels of hepatic DNA adducts that were comparable to those in Hudson River tomcod.[29] Not only did these results with alternative biomarkers confirm the exposure history of these fish, they also helped in identifying the CYP1A mRNA inducing agents in these systems. Both the FAC and DNA adduct analyses are only sensitive to PAH exposures and thus suggested that PAH exposure contributed to the elevated levels of CYP1A mRNA in Hudson River tomcod.

We also examined the rate of clearance of induced CYP1A mRNA in tomcod collected from a Hudson River site and transferred to clean laboratory water. Levels of expression in fish sacrificed immediately after collection were high. Surprisingly, levels of CYP1A mRNA decreased very rapidly upon transfer: by 4 hr a slight decline in expression (15%) was observed, and by 8 hr a 75% reduction; by 1 to 5 days, levels of gene expression approached basal levels for all individuals.[28] This rapid rate of CYP1A clearance in Hudson River-exposed fish indicates their likely exposure to rapidly metabolized compounds, such as PAHs. Levels of CYP1A mRNA remained at uniformly low basal levels for tomcod maintained in laboratory water for up to 120 days after collection. All depurated fish showed low levels of CYP1A mRNA expression.

Genetic Polymorphism in the CYP1A Gene in Atlantic Tomcod

During the course of quantifying levels of Hudson River tomcod CYP1A mRNA in northern blot analysis, it became evident that there was variability in the number of CYP1A hybridizable mRNA bands among individual fish.[31] All tomcod displayed a 3.0 kb CYP1A mRNA band, while variant individuals exhibited an additional 2.2 kb CYP1A mRNA band. Southern blot analysis was conducted on genomic DNA from these same individuals using a battery of six different restriction enzymes to determine if tomcod with the variant CYP1A mRNA also displayed a polymorphism in the structure of the CYP1A gene. DNA from all tomcod with the normal single 3.0 kb CYP1A mRNA band exhibited a single CYP1A hybridizable DNA band with all six restriction enzymes. All tomcod with the second variant mRNA band exhibited an extra DNA fragment in relation to tomcod with the common CYP1A genotype with five of the six restriction enzymes tested. The molecular size of the second variant DNA fragment was approximately 500 to 600 bp smaller than the normal DNA fragment. Thus, the size difference among DNA fragments was consistent with that observed among mRNA bands. The fact that five of six restriction enzymes tested revealed this smaller DNA fragment in variant individuals suggests that this polymorphism did not represent a single base substitution. These results suggested that this polymorphic DNA fragment represented a second CYP1A allele that contains a 500 bp deletion in comparison to the common CYP1A allele.

Densitometry revealed approximately equal concentrations of both CYP1A hybridizable mRNA bands in polymorphic individuals. Additionally, the optical densities of both CYP1A hybridizable DNA fragments in variant individuals were equivalent following high stringency wash conditions, suggesting equal homology between these two DNA sequences and the 3-MC induced rainbow trout CYP1A cDNA probe.[31] Thus, it is likely that both mRNA transcripts are equally expressed and are both products of the CYP1A gene.

Restriction fragment length polymorphisms in the structure of the CYP1A gene have also been detected in humans, and the frequency of variant genotypes differs among racial groups.[32] It has been hypothesized that the presence of the variant CYP1A genotypes may affect the susceptibility of individuals to cancer. Studies to date have provided mixed results suggesting that these CYP1A polymorphisms may impact significantly on genetic susceptibility to some cancers and not others.

We sought to determine if the CYP1A polymorphism in tomcod was ubiquitous throughout the species' distribution or restricted to the cancer-prone Hudson River population. Southern blot analyses of tomcod genomic DNA demonstrated that more than 10% of over 200 Hudson River tomcod analyzed displayed the CYP1A polymorphism. In all cases, these individuals also displayed the variant CYP1A mRNA. Further analysis determined that this polymorphism was absent in greater than 40 tomcod, each from four other populations: the St. Lawrence River, Quebec; Saco River, Maine; Margaree River, NS; and Miramichi River, NB.[33] Thus, this CYP1A polymorphism was restricted to the cancer-prone Hudson River population. Although it is tempting to speculate that this polymorphism plays a role in the neoplastic process, the functional significance of this variation has yet to be demonstrated. The fact that the variant allele is transcribed and harbors a large deletion suggests a possible selective disadvantage for this genotype. Furthermore, preliminary western blot studies have indicated that the variant CYP1A mRNA is translated to a second CYP1A protein recognizable by a CYP1A monoclonal antibody. Additional studies underway to assess the significance of this polymorphism include a comparison of EROD activity and CYP1A mRNA inducibility in beta-naphthoflavone (β-NF) treated fish and a comparison of the DNA sequence of the variant and normal tomcod CYP1A alleles.

The Effects of Prior Exposure or Genetic Variability in CYP1A mRNA Inducibility

To gain information concerning the identity of environmental inducing agents, Hudson River-collected tomcod were depurated for 20 to 30 days and subsequently treated with pure chemicals to determine which were able to induce CYP1A gene expression. After this period of depuration and prior to treatment, a subset of these fish were sacrificed and expression of their CYP1A mRNA was determined to be at basal levels. A single i.p. injection of β-NF induced a rapid and strong response; levels of gene expression approximated that seen in other species of fish treated with the same compound.[34,35] Unexpectedly, similar treatment of Hudson River-collected tomcod with two halogenated aromatic hydrocarbons, 2,3,7,8-TCDD (dioxin) and 3,3′,4,4′-TCB (coplanar PCB congener 77) did not result in induction of CYP1A mRNA.[36] In contrast, studies have demonstrated the inducibility of the CYP1A gene in other species of fish by i.p. injection of equivalent doses of these two compounds.[37,38] To determine if more extensive depuration, and thus additional opportunity for metabolism of resident hepatic xenobiotics, would permit induction of the CYP1A gene, a second set of Hudson River-collected tomcod was depurated for 120 days and again treated with 3,3′,4,4′-TCB. Despite extensive depuration, these fish did not exhibit induced CYP1A mRNA.

To determine if the noninducibility of CYP1A mRNA in Hudson River tomcod is a function of prior exposure history or a genetic difference between tomcod and other fish species, CYP1A gene inducibility was evaluated in tomcod from a second population. In this case, tomcod were collected from the Miramichi River, New Brunswick, depurated for 20 to 30 days, and treated with doses of 2,3,7,8-TCDD and 3,3′,4,4′-TCB equivalent to those used on Hudson River fish. Levels of CYP1A mRNA were highly induced in these Miramichi River tomcod collected subsequent to treatment with these two halogenated aromatic hydrocarbons, demonstrating that the CYP1A gene could be induced in tomcod by these compounds. The question still remains as to whether the failure of the Hudson River tomcod to respond to these treatments resulted from genetic differences among tomcod populations or their prior exposure history. Monosson and Stegeman[39] observed a similar phenomenon in winter flounder collected from two sites, pristine offshore Georges Bank and a moderately polluted river in Narragansett Bay. Flounder collected from Georges Bank had very low P4501A protein levels, but were inducible by subsequent i.p. injection of 3,3′,4,4′-TCB. In contrast, flounder collected from Narragansett Bay had 80-fold higher levels of P4501A protein, but TCB treatment resulted in no significant change in their P4501A protein content. They concluded that this lack of response was due to the maximal or near-maximal induction in these Narragansett Bay fish as a result of environmental exposure to inducers.

In winter flounder, prior exposure to some environmentally borne compounds resulting in strong CYP1A protein induction may have acted to inhibit further induction of gene expression. However, in tomcod, levels of CYP1A mRNA were not induced, and in fact were at very low levels prior to TCB treatment. This result at the transcriptional level would argue against prior CYP1A mRNA induction serving to inhibit additional increases in gene expression. Most intriguing is the observation that CYP1A mRNA could be induced in depurated Hudson River tomcod by treatment with β-NF, but not with the two halogenated aromatic hydrocarbons. CYP1A mRNA was inducible in Hudson River tomcod by β-NF despite the fact that hepatic levels of total PCBs and coplanar PCB congeners were very high, at levels comparable to those observed in Narragansett Bay flounder. For example, hepatic levels of TCB in Hudson River-collected tomcod ranged from 11 to 16 ng/g wet weight,[36,40] whereas the level of TCB in Narragansett Bay flounder livers was 8.0 ng/g wet weight.[39] Total coplanar PCB concentrations in Hudson River-collected tomcod ranged from 1.4 to 1.9 μg/g wet weight. These observations lead to the suggestion that perhaps separate molecular pathways lead to CYP1A mRNA induction in halogenated vs non-halogenated hydrocarbon treated tomcod.

Caging of Tomcod in Natural Environments

We tested the efficacy of CYP1A mRNA expression as a monitor of the effects of point sources of pollution or the general levels of contaminants in estuarine systems by reintroducing depurated caged tomcod into natural systems. For example, levels of CYP1A mRNA were measured in tomcod caged on two separate occasions in bleached mill kraft effluents on the Miramichi River, New Brunswick and compared to levels in fish caged at two sites downstream. Control tomcod were caged at a pristine upriver site not impacted by the mill's effluents. A second group of control fish consisted of unexposed tomcod maintained in the laboratory. Levels of CYP1A mRNA were low and did not differ between the laboratory maintained tomcod and those caged at the upriver site. During a winter exposure, with low flow levels in the river, tomcod caged in the mill's effluents exhibited a significant fourfold induction in levels of CYP1A mRNA over controls. A gradient in levels of induction was observed at the two downstream sites, three- and twofold induction, respectively, at stations 5 and 10 km downriver from the mill site. The observed gradient in levels of CYP1A mRNA at two downstream sites is consistent with a single point source of inducers in the system and subsequent dilution of the effluents with ambient river water. Significant 11-fold induction of CYP1A mRNA was also detected in tomcod caged at the mill site in the spring months; levels of gene expression were higher than those observed during the earlier winter exposure.[41] This seasonal difference in levels of CYP1A gene expression in fish caged at the same site could have resulted from several factors, both biotic and abiotic. Several studies have demonstrated that sex and maturational levels significantly impact on inducibility of CYP1A proteins in fishes[42] and, considering that tomcod spawn in midwinter, it is possible that seasonal differences in caging time may have impacted on levels of CYP1A mRNA. It is likely that spawning would have reduced levels of CYP1A mRNA in females during the winter exposure, thus lowering mean levels of gene expression in that group. Experiments are currently underway to quantify the effects of sex, maturation level, and season on levels of CYP1A mRNA inducibility in chemically treated tomcod. Additionally, the exposure and bioavailability of inducers in the effluents may vary seasonally due to differences in river flow which impacts on dilution factors and resuspension of sediment-borne xenobiotics. These results illustrate that seasonal variability must be considered when interpreting the results of caging experiments of feral fishes in natural environments.

We proposed that the rate of clearance of induced CYP1A mRNA in tomcod exposed at the mill site and subsequently depurated in clean water might permit identification of the inducing agents in the mill's effluents. To test this hypothesis, tomcod caged at the mill site for 14 days were transferred to clean water and subsets of these fish were sacrificed 1, 3, and 10 days later. CYP1A mRNA expression in these exposed and depurated tomcod was compared to levels in tomcod caged at the mill and upriver sites and sacrificed immediately after exposure. Levels of CYP1A mRNA in tomcod transferred from the mill site to clean water remained significantly induced for at least 3 days post transfer and even increased over time despite their depuration in clean water. Levels of CYP1A mRNA remained above control values (fourfold) for the entire 10-day duration of the experiment. How does this rate of clearance of CYP1A mRNA in these environmentally exposed tomcod compare to that observed in fish treated with pure chemicals? The rates of induction and clearance of CYP1A mRNA in killifish[34] and rainbow trout[35] treated with single i.p injections of b-NF were very rapid. Maximum gene induction was reached within 40 and 18 hr, respectively, with no secondary peak in gene expression evident. Basal levels of CYP1A mRNA expression in trout were reached by 48 hr post treatment. Clearly, the kinetics of clearance of CYP1A mRNA in tomcod caged at the mill site were very different from that seen in fish treated with a single dose of a rapidly metabolized PAH. In contrast, single treatments of rodents with the halogenated hydrocarbons 2,3,7,8-TCDD and the PCB Aroclor 1254, revealed persistently elevated levels for up to 30 weeks and secondary peaks in hepatic and lung CYP1A gene expression.[42,43] In combination, these results suggest that the temporal characterization of the kinetics of induction and clearance of CYP1A gene expression in exposed tomcod should provide insight into the identity of environmental inducers.

Characterization of Rates of Induction and Clearance of CYP1A mRNA in Tomcod Treated with Three Model Chemicals

Tomcod were collected from natural populations, depurated for 20 to 60 days in the laboratory, and i.p. injected with single doses of three model chemicals — 2,3,7,8,-TCDD (dioxin), 3,3′,4,4′-TCB (PCB congener 77), and β-NF — and then allowed to depurate in clean laboratory water.[44] Doses of each chemical were selected based on their ability to significantly induce CYP1A mRNA in earlier dose-response studies. Subsets of these chemically treated tomcod were sacrificed at various times following treatment and levels of their CYP1A mRNA were determined. Not surprisingly, single treatment with the two halogenated hydrocarbons yielded a very different profile of CYP1A mRNA induction and clearance than seen in tomcod treated with β-NF. For example, initial significant induction (4- to 19-fold) of CYP1A gene expression in 2,3,7,8-TCDD-treated fish was not observed until 5 days post treatment. Induction levels remained constant from day 5 through day 14, then declined at day 18, followed by a secondary increase to maximum 7.4-fold induction at day 25. In contrast, in β-NF-treated fish, twofold induction of gene expression was detected by 8 hr, maximum induction was observed at 72 hr and expression returned to near basal levels by day 5. In several respects, treatment with the PCB congener 3,3′,4,4′-TCB resulted in a response similar to that observed with dioxin. For example, induction was persistent for the duration of the experiment with both halogenated hydrocarbons, with no decrease in levels of gene expression observed for the durations of the experiments (10 and 25 days). Additionally, treatments with both halogenated hydrocarbons resulted in an initial induction in gene expression, then decline followed by a second and more pronounced peak. Thus, the response with both halogenated hydrocarbons was characterized by persistent induction and a secondary increase in gene expression. We believe that these very different kinetics of mRNA induction and clearance in halogenated vs nonhalogenated hydrocarbon exposures provide signature profiles that could be used to identify environmental inducing agents.

Dose Response of CYP1A mRNA to Three Model Chemicals

In order to calibrate the response of CYP1A mRNA in environmentally exposed tomcod, fish were treated in the laboratory with environmentally relevant concentrations of three model chemicals; β-NF; 3,3′,4,4′-teterachlorobiphenyl (TCB, a coplaner PCB congener); and 2,3,7,8-TCDD (dioxin). In all cases, fish were i.p. injected with these agents and sacrificed at times determined to provide at least 50% of maximum induction. Maximum induction ranged from approximately 50-fold in the dioxin-treated fish to 160-fold in the PCB-injected tomcod. Maximum induction in the β-NF-injected tomcod was 100-fold. In both β-NF- and dioxin-injected tomcod, saturation in levels of CYP1A mRNA was observed at concentrations of 50 ppm and 5 ppb, respectively. Levels of CYP1A was still increasing in TCB-injected tomcod at a concentration of 10 ppm. Significant induction with these compounds was initially observed at the following concentrations: β-NF, 1 ppm; TCB, 100 ppb; and TCDD, 100 ppt. In all three cases, no threshold concentration of inducing agent was required before CYP1A mRNA induction was observed. These studies illustrated that significant induction of CYP1A mRNA was possible in tomcod exposed to environmentally relevant concentrations of these xenobiotic agents. Additionally, it was demonstrated that via this route of exposure and for all three model chemicals, no threshold concentration of inducer was required to elicit an induced response at the transcriptional level.

CONCLUSIONS AND RECOMMENDATIONS

Clearly, CYP1A mRNA induction in Atlantic tomcod is a sensitive marker of exposure and early biological effect. Monitoring can be conducted via three alternative strategies: surveying levels of CYP1A mRNA in fish collected from feral populations, *in situ* caging exposures at suspected contaminated sites, or laboratory exposures to field collected sediments. Field studies showed that levels of CYP1A mRNA expression correlated well with environmental levels of aromatic hydrocarbon contamination, and concordance was seen between relative levels of CYP1A gene expression and a second biological marker of exposure (bile metabolites detected by FACs) and a biological marker of advanced effect (levels of hepatic DNA adducts). Furthermore, elevated levels of DNA adducts, oncogene activation, and genomic DNA mutability in tomcod highlight the probable detrimental consequences of this xenobiotic exposure and processing.

Laboratory studies demonstrated that expression of CYP1A mRNA is sensitive to extremely low levels of exposure to xenobiotics and a dose response was observed between extremely low and saturating levels of inducers. Temporal characteristics of induction and clearance of CYP1A mRNA varied widely among classes of inducers and suggested that this relationship could be used to identify environmental inducers. In most cases, the persistence of the mRNA response was sufficient to validate its use as a marker of exposure in field studies.

However, our results highlight several precautions which should be considered when interpreting results from environmental monitoring programs. First, genetic differences in the inducibility and activity of CYP1A mRNAs should be quantified and evaluated before instituting environmental monitoring programs. This variability exists at varying levels of taxonomic comparisons, including among individuals within a population, among populations, and among species. We observed significant variation among individuals from a single genetically defined population and between species in fish exposed under laboratory controlled conditions to pure model inducers. Furthermore, levels of gene expression varied widely among induced individuals from feral populations exposed to environmentally borne inducing agents. Because of the large interspecific variability in CYP1A inducibility, selection of the appropriate sentinel species is critical. Additionally, variation in the structure of the CYP1A mRNA and protein suggests that the biological consequences of induction may vary widely among individuals. However, a definitive answer to this question awaits characterization of the variant proteins, quantification of their enzymatic activities, and evaluation of more advanced biological effects such as levels of DNA adducts or oncogene activation.

Second, prior exposure history may be a significant factor impacting on levels of CYP1A gene expression in exposed populations of feral fishes. In two fish species, prior exposure to environmental inducers apparently inhibited subsequent increases in levels of CYP1A gene expression. In the case of Hudson River tomcod, this inhibition of further gene induction at the mRNA level was observed despite low levels of CYP1A mRNA. In the case of winter flounder, inhibition of additional CYP1A gene expression and even reduction of CYP1A enzymatic activity was detected in fish which already exhibited induced levels of CYP1A proteins. Thus, it is possible that different mechanisms of inhibition of CYP1A induction are operative in different species. Alternatively, these interspecific differences in levels of gene expression in fish from contaminated sites prior to subsequent treatment with inducing agents may have resulted from the gene products examined. *De novo* CYP1A mRNA and protein synthesis may have shut down, and induced protein levels may have resulted from their decreased lability. In either case, it is clear that further characterization and understanding of this inhibitory effect is needed to interpret the results obtained from environmental monitoring programs.

REFERENCES

1. **Murchelano, R. A., and Wolke, R. E.,** Epizootic carcinoma in the winter flounder, *Science*, 228, 587, 1985.
2. **McCain, B. B., Pierce, K. V., Wellings, S. R., and Miller, B. S.,** Hepatomas in marine fish from an urban estuary, *Bull. Environ. Contamin. Toxicol.*, 18, 1, 1977.
3. **Baumann, P. C., Smith, W. D., and Parland, W. K.,** Tumor frequencies and contaminant concentrations in brown bullheads from an industrialized river and a recreational lake, *Trans. Am. Fish. Soc.*, 116, 79, 1987.
4. **Smith, C. E., Peck, T. H., Klauda, R. J., and McLaren, J. B.,** Hepatomas in Atlantic tomcod *Microgadus tomcod* (Walbaum) collected from the Hudson River estuary in New York, *J. Fish Dis.*, 2, 313, 1979.
5. **Dey, W., Peck, T., Smith, C., Cormier, S., and Kreamer, G.-L.,** A study of the occurrence of liver cancer in Atlantic tomcod *(Microgadus tomcod)*, Final Report to the Hudson River Foundation, New York, 1986.
6. **Malins, D. C., McCain, B. B., Landahl, J. T., Myers, M. S., Krahn, M. M., Brown, D. W., Chan, S.-L., and Roubal, W. T.,** Neoplastic and other diseases in fish in relation to toxic chemicals: an overview, *Aquat. Toxicol.*, 11, 43, 1988.
7. **Varanasi, U., Nishimoto, M., Reichert, W. L., and Eberhart, B.-T. L.,** Comparative metabolism of benzo[a] pyrene and covalent binding to hepatic DNA in English sole, starry flounder, and rat, *Cancer Res.*, 46, 3817, 1986.
8. **Pruell, R. J., Rubinstein, N. I., Taplin, B. K., LiVolsi, J. A., and Norwood, C. B.,** 2,3,7,8-TCDD, 2,3,7,8-TCDF and PCBs in marine sediments and biota: laboratory and field studies, Final Report to U.S. Army Corps of Engineers, March 12, 1990.
9. National Status and Trends Program for Marine Environmental Quality, A summary of selected data on chemical contaminants in sediments collected during 1984, 1985, 1986, and 1987, NOAA Technical Memorandum NOS OMA 44, Rockville, MD, 1988.
10. **Sloan, R. J., and Armstrong, R. W.,** PCB patterns in Hudson River fish: II. Migrant and marine species, in *Fisheries Research in the Hudson River*, Smith, C. L., Ed., State University of New York Press, 1988, chap. 13.

11. **Koepp, S. J., Santoro, E. D., and DiNardo, G.,** Heavy metals in finfish and selected macroinvertebrates of the lower Hudson River estuary, in *Fisheries Research in the Hudson River,* Smith, C. L., Ed., State University of New York Press, 1988, chap. 10.
12. **Bigelow, H., and Schroeder, W.,** Fishes of the Gulf of Maine, *U.S. Fish Wildl. Serv. Fish. Bull.,* 74, 1, 1953.
13. **McLaren, J. B., Peck, T. H., Dey, W. P., and Gardiner, M.,** Biology of the Atlantic tomcod in the Hudson River estuary, in *Science, Law, and Hudson River Power Plants,* Barnthouse, L. W., Klauda, R. J., Vaughan, D. S., and Kendall, R. L., Eds., Vol. 4, American Fisheries Society Monograph, 1988, 102–112.
14. **Klauda, R. J., Peck, T. H., and Rice, G. K.,** Accumulation of polychlorinated biphenyls in Atlantic tomcod *(Microgadus tomcod)* collected from the Hudson River estuary, New York., *Bull. Environ. Contamin. Toxicol.,* 27, 829, 1981.
15. **Cormier, S. M, Racine, R. N., Smith, C. E., Dey, W. P., and Peck, T. H.,** Hepatocellular carcinoma and fatty infiltration in the Atlantic tomcod, *Microgadus tomcod* (Walbaum), *J. Fish Dis.,* 12, 105, 1989.
16. **Cormier, S. M.,** Fine structure of hepatocytes and hepatocellular carcinomas of the Atlantic tomcod, *Microgadus tomcod* (Walbaum), *J. Fish Dis.,* 9, 179, 1990.
17. **Cormier, S. M., and Racine, R. N.,** Histopathology of Atlantic tomcod: a possible monitor of xenobiotics in northeast tidal rivers and estuaries, in *Biological Markers of Environmental Contamination,* McCarthy, J. F., and Shugart, L. R., Eds., Lewis Publishers, Boca Raton, FL, 1990.
18. **S. Courtenay,** personal communication.
19. **Wirgin, I., Currie, D., and Garte, S. J.,** Activation of the K-*ras* oncogene in liver tumors of Hudson River tomcod, *Carcinogenesis,* 10, 2311, 1989.
20. **McMahon, G., Huber, L. J., Moore, M., Stegeman, J. J., and Wogan, G. N.,** Mutations in c-Ki-*ras* oncogenes in diseased livers of winter flounder from Boston Harbor, *Proc. Natl. Acad. Sci. U.S.A.,* 87, 841, 1990.
21. **Chang, Y.-J., Mathew, C., Mangold, K., Marien, K., Hendricks, J., and Bailey, G.,** Analysis of *ras* gene mutations in rainbow trout liver tumors initiated by aflatoxin B1, *Mol. Carcin.,* 4, 112, 1991.
22. **Wirgin, I. I., D'Amore, M., Grunwald, C., Goldman, A., and Garte, S. J.,** Genetic diversity at an oncogene locus and in mitochondrial DNA between populations of cancer-prone Atlantic tomcod, *Biochem. Genet.,* 28, 459, 1990.
23. **Brown, W. M., George, M., Jr., and Wilson, A. C.,** Rapid evolution of animal mitochondrial DNA, *Proc. Natl. Acad. Sci. U.S.A.,* 18, 1967, 1979.
24. **Payne, J. F., Fancey, L. L., Rahimtula, A. D., and Porter, E. L.,** Review and perspective on the use of mixed-function oxygenase enzymes in biological monitoring, *Comp. Biochem. Physiol.,* 86C, 233, 1987.
25. **Stegeman, J. J., Brouwer, M., Di Giulio, R. T., Förlin, L., Fowler, B. A., Sanders, B. M., and Van Weld, P. A.,** Enzyme and protein synthesis as indicators of contaminant exposure and effect, in *Biomarkers: Biochemical, Physiological, and Histological Markers of Anthropogenic Stress,* Huggett, R. J., Kimerle, R. A., Mehrle, P. M., Jr., and Bergman, H. L., Eds., Lewis Publishers, Boca Raton, FL, 1992.
26. **Stegeman, J. J.,** Cytochrome P450 forms in fish: catalytic, immunological and sequence similarities, *Xenobiotica,* 19, 1989.
27. **Goksøyr, A., and Förlin, L.,** The cytochrome P-450 system in fish, aquatic toxicology and environmental monitoring, *Aquat. Toxicol.,* 22, 287, 1992.

28. **Kreamer, G.-L., Squibb, K., Gioeli, D., Garte, S. J., and Wirgin, I.,** Cytochrome P450IA mRNA expression in feral Hudson River tomcod, *Environ. Res.*, 55, 64, 1991.
29. **Wirgin, I. I., Grunwald, C., Courtenay, S., Reichert, W. L., and Stein, J.,** Assessment of chemical contaminant exposure in Atlantic tomcod from five rivers using a suite of biomarkers, in *Pollution Response in Marine Organisms,* Gotëberg, Sweden, 1993.
30. **Wirgin, I. I.,** unpublished data.
31. **Wirgin, I., Kreamer, G.-L., and Garte, S. J.,** Genetic polymorphisms of cytochrome P-450IA in cancer-prone Hudson River tomcod, *Aquat. Toxicol.*, 19, 205, 1991.
32. **Cosma, G., Crofts, F., Currie, D., Wirgin, I., Toniolo, P., and Garte, S. J.,** Racial differences in restriction fragment length polymorphisms and messenger RNA inducibility of the human CYP1A1 gene, *Cancer Epidemol. Biomark. Prev.,* 2, 53, 1992.
33. **Wirgin, I. I., and Van Cleef, K.,** unpublished data.
34. **Kloepper-Sams, P. J., and Stegeman, J. J.,** The temporal relationships between P450 protein content, catalytic activity, and mRNA levels in the teleost *Fundulus heteroclitus* following treatment with b-napthoflavone, *Arch. Biochem. Biophys.*, 268, 525, 1989.
35. **Haasch, M. L., Wejksnora, P. J., Stegeman, J. J., and Lech, J. J.,** Cloned rainbow trout liver P1450 complementary DNA as a potential environmental monitor, *Toxicol. Appl. Pharmacol.,* 98, 422, 1989.
36. **Wirgin, I. I., Kreamer, G.-L., Grunwald, C., Squibb, K., Garte, S. J., and Courtenay, S.,** Effects of prior exposure history on cytochrome P450IA mRNA induction by PCB congener 77 in Atlantic tomcod, *Marine Environ. Res.*, 34, 1, 1992.
37. **Gooch, J., Elskus, A. A., Kloepper-Sams, P. J., Hahn, M. E., and Stegeman, J. J.,** Effects of ortho- and non-ortho substituted polychlorinated biphenyl congeners on the hepatic monooxygenase systems in scup *(Stenotomus chrysops), Toxicol. Appl. Pharmacol.,* 908, 422, 1989.
38. **Haasch, M. L., Quardokus, E. M., Sutherland, L. A., Goodrich, M. S., Prince, R., Cooper, K. R., and Lech, J. J.,** CYP1A1 protein and mRNA in teleosts as an environmental bioindicator: laboratory and environmental studies, *Marine Environ. Res.,* 34, 139, 1992.
39. **Monosson, E., and Stegeman J. J.,** Cytochrome P450E (P450IA) induction and inhibition in winter flounder by 3,3′,4,4′-tetrachlorobiphenyl: comparison of response in fish from Georges Bank and Narragansett Bay, *Environ. Toxicol. Chem.*, 10, 765, 1991.
40. **Wirgin, I. I.,** unpublished data.
41. **Courtenay, S., Grunwald, C., Alexander, R., Kreamer, G.-L., and Wirgin, I. I.,** Induction of CYP1A mRNA in Atlantic tomcod caged in a bleached kraft mill effluent in the Miramichi River, *Aquat. Toxicol,* in press.
42. **Beebe, L. E., Fox, S. D., Issaq, H. J., and Anderson, L. M.,** Biological and biochemical effects of retained halogenated hydrocarbons, *Environ. Toxicol. Chem.*, 10, 757, 1991.
43. **Beebe, L., Fox, S. D., Riggs, C. W., Park, S. S., Gelboin, H. V., Issaq, H. J., and Anderson, L. M.,** Persistent effects of a single dose of Aroclor 1254 on cytochrome P450IA1 and IIB1 in mouse lung, *Toxicol. Appl. Pharamacol.*, 11, 16, 1992.
44. **Grunwald, C., et al.,** unpublished data.

7

Molecular Epidemiology of Common Diseases

Kari Hemminki, M.D.

Center for Nutrition and Toxicology, Karolinska Institute, Huddinge, Sweden

INTRODUCTION

Etiology-oriented epidemiologic research has been successful in many areas; for example, many risk factors of cardiovascular diseases and carcinogenic agents have been established, which has helped to design preventive measures.[1,2] However, even modern epidemiological methods have limitations when multiple etiologic factors and small risk ratios are involved. Furthermore, epidemiologic findings rarely establish the mechanism of the disease. The lack of mechanistic handles has been called "the black box" (Figure 1). It has been suggested that the opening of the black box by the exploitation of molecular biology will be a key to future success in etiology-oriented epidemiological research.[3,4]

Over the years, epidemiology and biochemical research have conceptually interacted with each other. Epidemiologic research, the identification of causes and risk factors, has provided clues for biochemists to study, and vice versa. Examples of studies where biochemical parameters have been extensively used to refine epidemiological studies include those where blood lipids, vitamins, and microelements have been correlated to risks of cardiovascular disease and cancer.

0-87371-631-0/94/$0.00+$.50

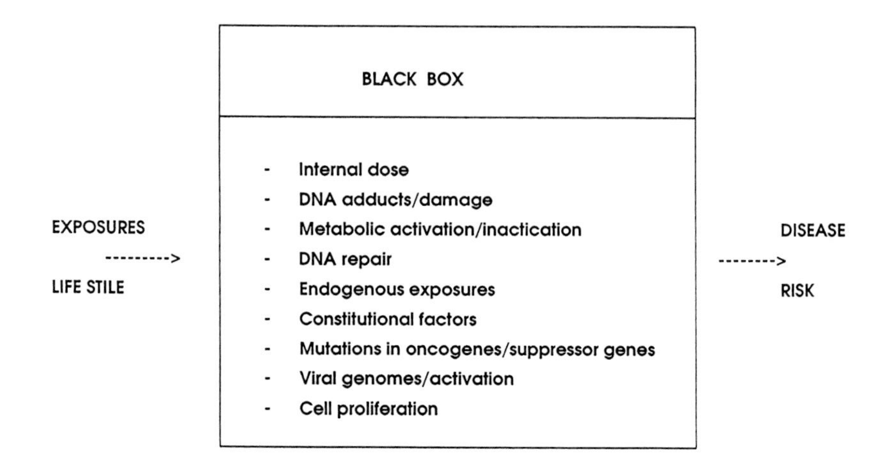

Figure 1. Traditional epidemiology examines disease frequency in relation to exposure/lifestyle variable. Molecular epidemiology involves mechanistic aspects described in the "black box", specifically in cancer studies.

More recently, epidemiologic study designs have been exploited in molecular biological studies. Such studies have a unique possibility to unravel disease mechanisms at the gene level.[5,6] Optimally, the combination of molecular biological and epidemiologic research (molecular epidemiology) can establish biological disease indicators, sometimes called "markers", which could substitute for actual epidemiological endpoints.[7,8]

Molecular biology can contribute to epidemiologic research by providing measures of environmental exposure, effect, and susceptibility to disease. Reciprocally, epidemiology can contribute to molecular biology by contributing guidance in study design and analysis. In this chapter, possibilities for interaction between molecular biologists and epidemiologists are discussed in the context of common human diseases.

HERITABLE, GENETIC, AND ENVIRONMENTAL DISEASES

Humans inherit one set of chromosomes (alleles) from the maternal and the other from the paternal side.[9,10] Genes in chromosomes code for proteins that are enzymes and structural components. Enzymes metabolize nucleic acids, lipids, carbohydrates, and other proteins. The whole system is regulated by a complex network of nutrients, hormones, and growth factors through feedback mechanisms.

Even in the modern medical literature, diverse and confusing opinions are presented on the role of genetic and environmental factors in disease etiology. These are partially terminological differences: distinction should be made between inheritance through germ cells (= hereditary) and genetic constitution (combination of hereditary and somatic events). It is now well understood that the inherited DNA is under continuous change through recombinations and mutations, which may be endogenous or environmentally caused. Some of the confusion in clinical genetics and other fields of medicine is caused by the idea of "heritability" being defined as "the proportion of the total variation of a character attributable to genetic as opposed to environmental factors".[9] This definition considers environmental and genetic events to be distinct, and fails to take into account the influence of environmental factors on genes. Furthermore, molecular biology has revealed considerable heterogeneity in the structure and function of the human genome relating to polymorphisms, the interplay of alleles, and the types and expression of mutations, illustrating the difference between phenotype and genotype.

In medical practice, hereditary diseases are important for genetic counseling, whereby disease risk may be lowered in the next generation. Also, many hereditary disorders are associated with special sensitivities to environmental agents, and these diseases may be alleviated or prevented by intervention.[11,12] When a disease is caused by environmental factors, primary prevention is possible. For this reason, much of analytic epidemiology has been directed toward detection of these factors.

MEDICAL GENETICS

The area of medicine where biochemistry and epidemiology have had a traditional, fruitful interaction is medical genetics. Hundreds of genetic and metabolic diseases have been characterized and essential concepts have been established. It would thus be instructive to analyze why the marriage of epidemiology and biochemistry has been so successful in medical genetics.[9,10]

One undoubted reason for this success has been the thorough foundation of genetics established since the days of Mendel. Another reason is that for some genetic diseases, it has been easy to establish family pedigrees. In many such diseases, there has been found to be a clear deficiency of a functional enzyme in affected individuals. The symptoms of disease have often guided scientists to search for certain types of enzymes. Finally, many genetic diseases have manifested with severe chromosomal changes, detectable even with early karyotyping methods, which has subsequently led to the mapping of the affected gene.

The success of clinical genetics in unraveling the causes of many heritable diseases and inborn errors of metabolism has provided a model for molecular epidemiology (Table 1). After identification of the disease or syndrome, a family pedigree is constructed for the inspection of familiarity and mode of inheritance. In inborn errors of metabolism, the clinical manifestation prompts a search for the affected metabolic pathway and enzyme. More recently, karyotype analysis of the affected family members has been used to find the chromosomal location of the candidate locus, which is then probed with known adjacent

Table 1
Search for Pathogenesis in Genetic Diseases

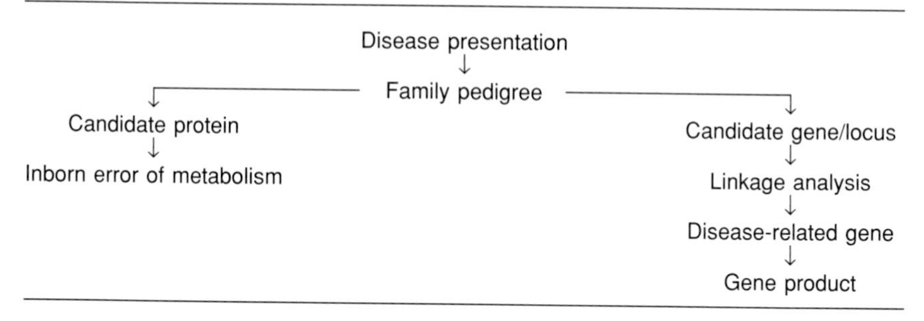

chromosomal sequences. Linkage analysis with DNA from several affected individuals helps to identify the disease-related gene, followed by characterization of the encoding protein.

In an encyclopedia of gene loci in man, about 4600 genes are represented out of the estimated total of 50,000 to 100,000,[10] and about 1000 of them have been cloned and sequenced. The ultimate sequencing of the total human genome (HUGO) has been launched and, once completed, the genes of all the proteins and regulatory sequences will be known. This will provide an immense base for hypothesis testing in disease etiology.

CANCER

Cancer is thought to be a multistage process in which cells lose their normal growth control through a number of successive stages.[13-15] The idea of a multistage process arose mainly from epidemiological analyses of age-dependence and latency periods in the onset of cancer.[16-19] The early observations of malignant growth in *in vitro* systems led Boveri[20] to propose that permanent changes underlie malignant growth. It was noted later in human populations that many cancers presented a hereditary component.[21,22] Knudson and Strong[23] analyzed heredity of retinoblastoma, Wilm's tumor, neuroblastoma, and pheochromocytoma, postulating mutational etiology. In the case of retinoblastoma, it was noted that the bilateral and unilateral forms of the disease differed in heritability, age of onset, and number of tumor clones. It was supposed that the bilateral form required one mutation, whereas the unilateral form required two mutations in the same locus. It was reasoned that in the bilateral case, a mutated allele in the retinoblastoma locus was inherited and thus only one somatic mutation was required.[24] Such epidemiological analyses have recently been confirmed by molecular genetics, leading to a marked advancement in the understanding of the genetic bases of certain cancers and refinement of the critical genes, oncogenes, and antioncogenes interplaying in malignant transformation.[13,25]

Parallel to the studies on the genetic basis of cancer, other lines of cancer research were also proliferative and pointing to the same direction as the genetic studies. Extensive work with chemically induced cancers led to an overall understanding that a large proportion of chemical carcinogens are electrophiles, able to react covalently with DNA.[26] These chemicals were often mutagenic;[27] when the genotoxicity of well-documented human and animal carcinogens have been summarized, an overwhelming proportion (80 to 90%) have been found to be positive in various genotoxicity assays.[28] Taken together, these data indicated that DNA is the target for many environmental carcinogens, and their mechanism of action is to cause somatic mutations in some of the critical genes outlined above.

A third line of cancer research used epidemiological data on the incidence of cancer in various countries and in immigrant populations to conclude that the causes of cancer are mainly environmental.[29,30] This excludes dominant inheritance as being the cause of the most common cancers, but does not exclude the role of somatic mutations.

The multistage process, where constitutional factors and environmental exposures interact, will be discussed below.

Constitutional Factors

A number of constitutional factors underlie susceptibility to cancer. These include, among others, levels of enzymes that activate and inactivate chemical carcinogens and of DNA repair enzymes.[31,32] The activities of several of these enzymes have been studied in human tissue explants *in vitro* and using model substrates *in vivo;* marked interindividual differences, up to several hundred-fold, have been noted. To what extent these polymorphisms in human populations actually predispose individuals to cancer has been a field of active research and some suggested relationships are presented in Table 2. It is not established that the carcinogens indicated in the table are the sole factors. Furthermore, many controversial data exist. In a recent study, extensive metabolizers of debrisoquine were found to be at a fourfold higher risk of lung cancer, when the data were adjusted for age, sex, and smoking. As a comparison, in the same study, occupational exposure of men to asbestos and PAH carried a 2.9- and 2.4-fold risk, respectively.[33] There have been extensive methodological advances in the area of drug-metabolizing enzymes recently, as genotyping instead of phenotyping has become available (see Chapter 11). Genotyping allows mapping of the types of mutations or polymorphisms involved.

A number of genetic diseases are known to carry an increased risk of cancer.[11,12] Among these, xeroderma pigmentosum has defects in DNA repair; and in ataxia telangiectasia and Fanconi's anemia, somatic chromosomal aberrations are frequently found in the patients. Down's syndrome and polyposis coli are associated with an increased risk of leukemia and colon carcinoma, respectively, but the molecular mechanisms have not been unraveled.

Measurement of Exposure

Direct measurement of a chemical or its metabolite in ambient air or in biological fluids is the most common means of assessing exposure in epidemiologic

Table 2
Suggested Relationships Between Xenobiotic Metabolism and Risk of Cancer

Metabolism	Xenobiotics suspected	Relationship
Acetylation	Aromatic amines	Bladder cancer risk in slow acetylators
Debrisoquine	PAH, nitrosamines	Lung cancer risk in extensive metabolizers
Aryl hydrocarbon hydroxylase	PAH	Lung cancer risk in highly induced induction
Glutathione transferase	Many carcinogens	Lung cancer risk in low activity

Note: Modified from Reference 32.

studies. However, usually concurrent representative individual measurements are lacking and exposure categories are constructed from sparse data combined with other hygienic knowledge.

Recently, methods have been developed to measure the covalent binding products of environmental carcinogens with DNA and protein. The rationale for these methods is that binding of a chemical to DNA is thought to be the initiating event triggering the cascade of biochemical changes eventually leading to malignancy.[14] In target tissues of experimental animals, the levels of adducts correlate with tumor responses.[34,35] This has not been directly studied in humans, but inferences can be made from animal studies and from the indirect data, discussed later, on groups at high risk of cancer. Because target tissues are usually not available in humans and the DNA binding studies resort to surrogate tissue, the above inference is less compelling. Generally, binding to protein, such as hemoglobin and albumin, correlates with binding to DNA, and protein adducts can be used as a surrogate measure of DNA binding.[36,37]

The development of sensitive methods based on ^{32}P-postlabeling[38-40] or immunoassay[41] have laid the groundwork for the determination of DNA adducts in humans (see Chapter 8). The technology is new and ever improving and expanding.[42] Due caution must be exercised in the interpretation of the data obtained as to accuracy and identity of the measurements. Recently, interlaboratory comparisons have been carried out for both postlabeling technique and immunoassay methods, and reasonable correlations have been found.[43] Still unknown variables exist, and it is not yet clear that in human DNA the two assays measure identical adducts.

Postlabeling and immunoassays have been applied in occupational and environmental studies.[44,45] One population has been foundry workers that are known to be at a risk of lung cancer.[46] In a series of studies, it has been shown that the levels of aromatic adducts in white blood cell DNA of the workers correlate with the estimated exposure to PAHs.[43,47,48]

Measurement of Effect

Cytogenetic endpoints, chromosomal aberrations, sister chromatid exchanges, and micronuclei have been used extensively in occupational monitoring.[49] In addition to some 10 to 20 occupational exposures, several anticancer agents have been shown to increase one or several of the cytogenetic endpoints in humans.[28] Many of these exposures also increase the risk of cancer. At a group level, an increase in chromosomal aberrations is usually associated with a risk of cancer.[50] This is additionally supported by the increased incidence of cancer syndromes which involve chromosome instability, such as ataxia telangiectasia and Fanconi's anemia. A collaborative follow-up study has recently been initiated in the Nordic countries in which cytogenetic data on the frequency of chromosomal aberrations, sister chromatid exchanges, and micronuclei are correlated with the risk of all cancers.[51]

Karyotyping of cancer cells has revealed many consistent changes. By now, more than 10,000 abnormal karyotypes have been cataloged,[52] and many are related to a risk of cancer. The first of these changes was the translocation between chromosomes 22 and 8 (the Philadelphia chromosome) present in some 98% of the patients with chronic myeloid leukemia.[52] This translocation is sometimes observed before the clinical disease, which appears, however, with about 95% probability. It has been shown by molecular analyses that the translocation involves the transfer of *abl* oncogene to chromosome 22.[53]

Point mutations can be scored by polymerase chain reaction (PCR), where the sequence of interest is amplified thousands of times.[54] PCR can be run after gradient denaturing gel electrophoresis if rare mutants are detected among normal cells. This has been applied to human hypoxanthine guanine phosphoribosyltransferase (HPRT) genes.[55] Ionizing radiation[56] and anticancer agents[57] have been shown to cause mutations in the HPRT gene in humans.

carcinogen-related mutational hot spots have been noted.[59] With the new powerful techniques, extensive advances have been made in the mapping of mutations in a number of growth-controlling genes, such as the p53 tumor suppressor gene, the most common target of genetic change found in human cancers. The p53 protein has been found to be a DNA-binding protein recognizing a 33-base pair region in DNA. The common mutant proteins found in human tumors were unable to bind to this sequence.[60] The mutational spectrum differs among the tissues where cancers are found; for example, G to T transitions predominate in colon, brain, and lymphoid malignancies, while transversions are commonly observed in lung and liver cancer.[61] Mutations at the A-T base pair are common in esophageal carcinomas. A mutational hot spot in colorectal carcinomas, brain tumors, leukemias, and lymphomas is a CpG sequence. In liver cancers from areas where aflatoxin B_1 and hepatitis B virus are risk factors, most mutations are at codon 249, a frequent target of aflatoxin B_1-induced genetic change.[61]

Genes involved in familial adematous polyposis, an autosomal dominant disease posing a high risk of colorectal cancer, have been cloned[62] and germ line point mutations in one of them have been described.[63]

p21 POSITIVE SAMPLES AND CANCER DIAGNOSES (↓)

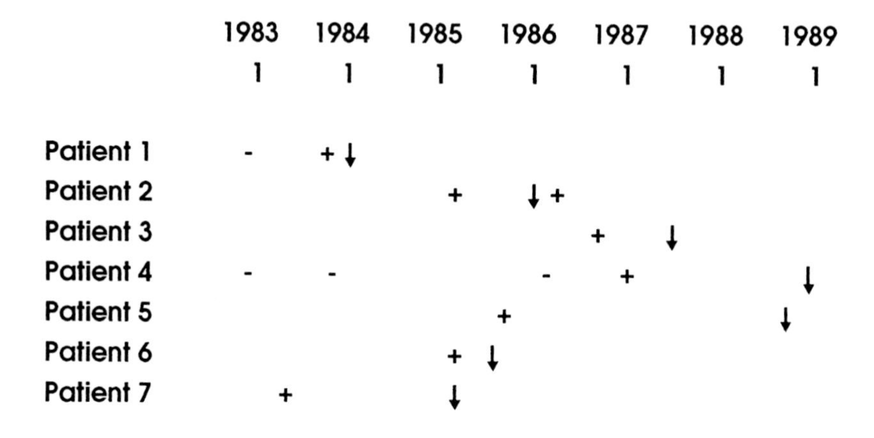

Mean : p21 + 16 months prior diagnosis

Figure 2. Pneumoconiosis (asbestosis and silicosis) patients sampled at indicated times for oncoprotein analysis from plasma: – = no p21 protein present; + = p21 protein detectable; ↓ = diagnosis of cancer.

Two new techniques have been used to assay for proteins associated with cancer risk. In the glycophorin A assay, antibodies specific to the M and N alleles are used with flow cytometry and cell sorting. The variants lacking one of the alleles are isolated and the presumed mutants are scored. The frequency of variants is about 1×10^5 in normal blood donors and increases in individuals exposed to anticancer agents and radiation.[64] The other protein assay measures the amount and size of oncoproteins by immunoblotting. *Ras* and *fes* oncoproteins are detected in plasma in some forms of cancer,[65] and even some occupational groups exposed to polychlorinated biphenyls[66] and PAHs.[67]

In a retrospective follow-up study of pneumoconiosis patients (asbestosis and silicosis patients), plasma samples were analyzed for a panel of oncoproteins. Interestingly, many of the pneumoconiosis patients were positive for PDGE (platelet-derived growth factor), which was correlated with the degree of fibrosis.[68] Moreover, significantly more cancer cases were positive in the *ras*-oncogene coding p21 protein as compared to noncancer controls. Among seven patients, the sequence of appearance of p21 in plasma could be inspected in relation to the diagnosis of cancer (Figure 2). The appearance of detectable levels of p21 in plasma preceded the diagnosis of cancer by 16 months.[68] Other oncogenes, such as c-*myc*, have been used as tools for prognosis in cancer cases. It will be of interest whether the p21 protein or other tumor-related proteins can be used in the early diagnosis and prevention of cancer.

CARDIOVASCULAR DISEASES

Some of the largest and most sophisticated studies of analytical epidemiology have concerned environmental and hereditary risk factors of cardiovascular diseases.[2] Many of them involved several biological parameters and the most concrete outcome has been the demonstration of blood cholesterol — particularly, low-density, lipoprotein-bound cholesterol — as a risk factor of atherosclerosis.

Molecular biologic approaches to atherosclerosis have resembled those used in cancer studies in the sense that the affected individuals have been compared to healthy controls. Interest has focused on the protein components and the related genes of blood lipid carrier proteins, lipid metabolizing enzymes, and cellular receptor proteins.[69-71]

Hypercholesterolemias

In familial hypercholesterolemia, blood total and low-density, lipoprotein-bound cholesterol levels are elevated two to three times, and some five times in very rare homozygotes. Typically, atherosclerosis appears in mid-age in the affected individuals, unless cholesterol-lowering drugs are used.

Familial hypercholesterolemia is a dominant monogenic disease, where a low-density lipoprotein (LDL) receptor is lacking or deficient.[70] The gene has been cloned and the deficient protein is either due to a point mutation or deletion.[72,73] So far, over 40 different mutations have been described.

Many other genes related to lipid metabolism are under intensive study, including the one coding for apolipoproteins A, B, C, and E and lipases.[74-79] Some of the detected gene defects appear to be risk factors for atherosclerotic disease, although the risk is usually lower than in familial hypercholesterolemia. The diseases of lipid metabolism, where considerable understanding at the molecular level has been gained, in addition to familial hypercholesterolemia, include hypobetalipoproteinemia with mutations causing truncated proteins or interfering with LDL receptor function[80,81] and deficiencies or lack of apolipoproteins A,[82] C, and E.[83] In these diseases, the hereditary component is large, but dietary and other factors contributed to the risk of cardiovascular disease. Overall, the familial forms of the disease account for a minor proportion of cardiovascular diseases.

Atherogenesis

Hypercholesterolemia is an important cause of coronary heart disease, and clinical intervention studies have demonstrated the therapeutic value of correcting hypercholesterolemia.[84,85] However, no matter how successfully hypercholesterolemias

Table 3
Steps Involved in Atherogenesis

Low-density lipoprotein (LDL)
↓ Active oxygen species
Modified LDL
↓
Internalization of modified LDL by macrophages in vascular endothelium
↓
Formation of fatty streaks
↓
Endothelial injury
↓
Accumulation of platelets
↓
Secretion of PDGF
↓
Cell proliferation, thickening of atheroma, plaque formation
↓
Rupture of plaque, infarction

are dealt with, coronary heart disease may appear because a high cholesterol level is not the only causative factor. At any given level of hypercholesterolemia, there is considerable variation in the clinical expression of the disease. Siblings with familial hypercholesterolemia and very closely matched cholesterol levels can have clinical coronary heart disease at very different ages. One basis for such variation lies in the biologic responses of cells in the artery wall to the presence of a given level of plasma cholesterol. Advances in the understanding of the metabolism of lipoproteins by the artery wall have yielded new insights into the factors that may be involved in the arterial response. Certain postsecretory modifications in the structure of lipoproteins appear to affect their atherogenic potential. There is a clear-cut mechanism proposed by Steinberg et al.[86] for atherogenesis. LDL-containing apolipoprotein B transports two thirds of plasma cholesterol in humans, and individuals with high plasma levels of this lipoprotein are at a risk of coronary heart disease.[81] LDL is normally taken up by liver and peripheral tissues by the LDL receptor. LDL modified by lipid peroxidation is a poor ligand to the receptor, but is a good ligand to a "scavenger receptor" present (e.g., in macrophages and endothelial cells). The first vascular lesion, the fatty streak, is an accumulation of macrophages loaded with cholesterol, beneath the vascular endothelium (Table 3). The lesion attracts platelets which secrete PDGF, leading to the growth of endothelial cells and thickening of atheroma to an atherosclerotic plaque. Modified LDL is also immunogenic, which may be an additional mechanism of accumulation of macrophages in the arteries. Triggered by physiological stress or other mechanisms, the plaque may rupture, causing infarction.

Growth factors are thought to contribute to the formation of atheromas in the vascular epithelium, as discussed above. There is evidence that in a cardiac arrhythmia (the long QT syndrome), which is not linked to abnormalities of lipid metabolism, H-*ras* oncogenes would be involved. The rationale for this finding is that *ras* proteins residing on the inner surface of cell membranes may take part in the regulation of potassium channels in the heart.[87]

There are several points where environmental factors may influence the process. Plasma LDL levels are partially controlled by dietary intake. Oxidative modification of LDL is an interplay of a number of endogenous and exogenous factors. Furthermore, the plaque formation and triggering of its rupture are multicausal and may depend on exogenous factors. Thus, there are many opportunities for input from epidemiology to unravel the environmental and constitutional contributing factors in atherogenesis.

MUSCULOSKELETAL DISEASES

Rheumatoid Arthritis

Rheumatoid arthritis is an autoimmune disease characterized by long-term inflammation of multiple joints. Mononuclear cell infiltration of the synovial membrane can eventually lead to the destruction of articular cartilage and surrounding structures. Because of its high frequency and potentially severe nature, this disease is a major cause of long-term disability in adults. Although the pathogenesis of rheumatoid arthritis and other similar autoimmune disease remains unknown, genetic and environmental factors have been implicated. A high incidence of chronic arthritis has been reported in patients with human T cell leukemia virus type I (HTLV-I) associated myelopathy, suggesting a possible involvement of HTLV-I in the disease.[88]

Several lines of evidence suggest that T cells specific for self-antigens may play a critical role in the initiation of these diseases. In the case of rheumatoid arthritis, the linkage of the disease to the DR4 and DR1 alleles of the class II genes of the major histocompatibility complex and the finding of sometimes oligoclonal, activated $CD4^+$ T cells in synovial fluid and tissue of affected joints suggest the involvement of $CD4^+$, ab T cell receptor-bearing, class II-restricted T cells in the disease. This view is supported by the finding that partial elimination or inhibition of T cells by a variety of techniques can lead to an amelioration of disease in certain patients.[89]

Diseases of the Connective Tissue

Connective tissue is distributed in most parts of the human body and hence collagen diseases may affect many organ systems. The collagen gene family is complex, constituting at least 13 different types of collagens composed of 25 distinct polypeptide chains. For example, the most common type I collagen is present in bone, skin, tendons, teeth, and blood vessels. All the collagens are composed of three polypeptides (type I collagen is composed of two α chains and one β-chain) wound around each other in a highly structured, triple helical manner, which explains some of the unique features of the genetic diseases of collagen.[90] The structure is highly sensitive to any amino acid perturbation; i.e., mutations anywhere along the polypeptides, particularly in glycine amino acids of the glycine-X-Y repeats, may render the molecule nonfunctional (Figure 3[91,92]). Moreover, there is an unusual amount of posttranslational modification in the collagen triple helix, making it particularly susceptible to mutations.

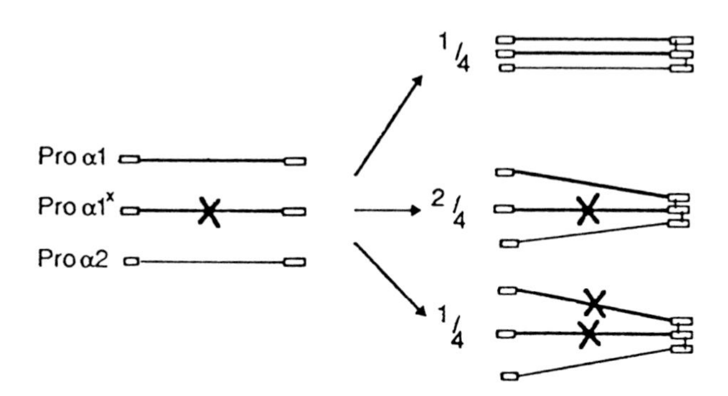

Figure 3. The effect of mutation (X) in one procollagen chain on the assembly of collagen fibrils, composed of three procollagen chains. A mutation in the gamma chain results in only $^1/_4$ of Type I collagen fibrils being normal, while $^1/_2$ have one mutated alpha chain. (Modified from Reference 91.)

Mutations in other collagen types have been described in a number of inherited collagen diseases. Mutations in type II collagen have been found in chondrodysplasias and some families with osteoarthritis. Type III collagen is a main component of arteries. Mutations have been noted in families with aortic aneurysms.[92]

Based on the results from heritable collagen diseases, it has been postulated that mild aberrations of collagen structure may contribute to many common diseases of the musculoskeletal and cardiovascular system such as osteoarthritis and aneurysms. Exposure to mutagenic agents may contribute to the accumulation of defunct collagen types. In the investigation of such multietiological common diseases, interactions with epidemiologic research are critically important.

ALLERGY AND ASTHMA

Atopic persons have an inherited oversensitivity to allergic reactions such as rhinitis and asthma.[93] Atopic reactions involve an antigen-induced sensitization and formation of IgE-type antibodies.[94] The affected individuals may have highly elevated serum IgE levels or raised IgE types specific toward the sensitizing antigens. The sensitive individuals may respond to tiny amounts of the antigen (μg/year). The mechanisms of allergic reactions are not uniformly understood. The theories invoke mast cells and certain classes of leucocytes that associate with mucous membranes and can bind IgE-type antibodies. When antigens bind to the sensitized cells, chemical mediators such as histamine, prostaglandins, and leucotrienes are released.[95] These influence the nearby muscle cells to contract and contribute to increased mucous secretion and vascular permeability. The reactions may be immediate and chronic, and they may gradually lead to damage in the epithelial cells.

Preliminary attempts have been made to search for candidate genes involved. Family studies have led to a localization of the candidate gene to chromosome 11.[96]

The families were defined by skin tests or serum total or specific IgE levels; 85% of the individuals with the candidate gene had symptoms, 60% rhinitis, and 20% asthma. Understanding the pathogenic mechanisms leading to allergy requires information on the structures of allergenic proteins. Cloned allergens have been used to characterize important allergens such as birch, grass and weed pollen, white-faced hornet venom, and house dust mite allergens.[97] A birch pollen allergen, profilin, is recognized by IgE antibodies from individuals allergic to pollen. Interestingly, humans have a profilin-like protein that also reacts with IgE antibodies and causes a histamine release from blood basophils. Thus, a self-antigen is suggested as a mechanism to maintain high IgE levels in individuals allergic to profilins.[97]

Even though atopy seems to have an inherited disposition, it is obvious that environmentally modifying factors exist, particularly affecting the immune system early in life.[98] The prevalence of allergy has been assumed to be increasing and a number of contributing factors have been proposed, ranging from breast feeding to food additives and environmental contaminants. It appears particularly appropriate in the case of allergic diseases that sound epidemiology, clinical investigations, and molecular studies are mounted to explore the field.

DIABETES

Insulin-dependent (type I) diabetes mellitus usually affects young individuals, perhaps with an increasing frequency in many European countries.[99] Twin studies show a concordance between 30 and 50% indicating hereditary, polygenic disposition to the disease. Some 50% of the heredity cosegregates with the major histocompatibility complex (HLA-D locus), suggesting autoimmune etiology.[100,101]

The autoimmune process associated with the disease includes massive lymphocytic infiltration of the pancreatic islets and circulating autoantibodies to B-cells. A 64,000 molecular weight (64 K) B-cell autoantigen is a target of autoantibodies in this disease. The 64-K autoantibodies are present in up to 80% of newly diagnosed type 1 diabetes patients and have been detected up to several years before clinical onset of the disease concomitant with a gradual loss of B-cells. The 64-K antigen was found to be B-cell-specific in an analysis of several tissues which did not include the brain.

Recent work has identified the 64-K autoantigen as glutamic acid carboxylase, the enzyme synthesizing the inhibitory neurotransmitter, γ-aminobutyric acid.[102] Pancreatic B-cells and a subpopulation of central nervous system neurons express high levels of this enzyme. Autoantibodies against glutamic acid decarboxylase with a higher titer are also found in stiff-man syndrome, a rare neurological disorder characterized by a high coincidence with type 1 diabetes.[102]

The effect of environmental factors in type 1 diabetes is suggested by a number of epidemiologic findings; the incidence varies extensively between populations and sex.[99] Other factors suggested to be involved are the duration of breast feeding and intake of high-protein diet or nitrosamines.[103] An early intake of foreign proteins has been speculated to trigger the autoimmune process leading to the

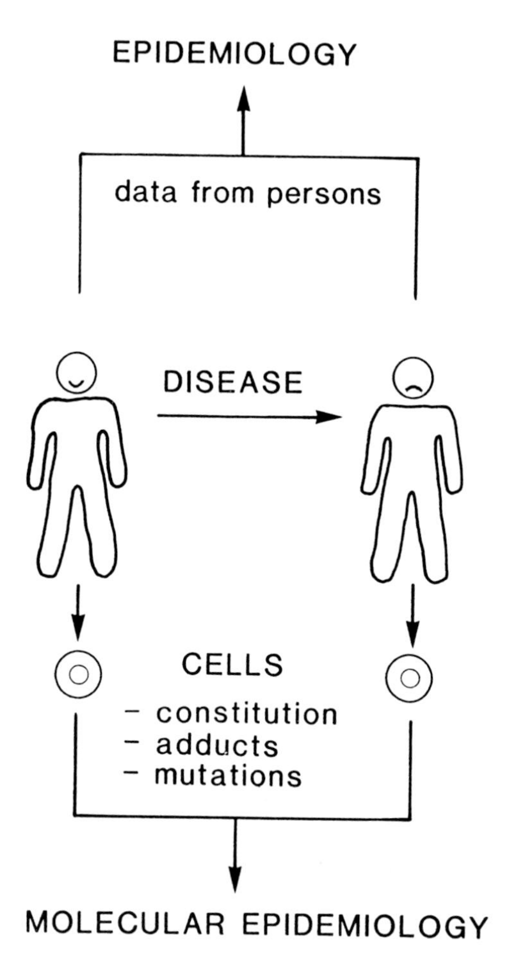

Figure 4. Differences in epidemiology and molecular epidemiology.

destruction of B-cells. Perinatal factors that can affect the immune system are of interest in disease etiology and should be tackled by a combined effort of epidemiologists, clinicians, and molecular biologists.

CONCLUSIONS

Biochemical and epidemiological studies have benefited each other conceptually in forming research hypotheses (Figure 4). The design of a typical biochemical study has been to assay samples from diseased and healthy individuals. This can lead to conclusive results when the factors under study produce large risks. At smaller risks, a formal case-control approach could be necessary.

Molecular epidemiology is of greatest use when applied to a multifactorial chronic disease. Biochemical analysis can reveal constitutional exposure or effect indicators that predict the disease with a high probability. Many such indicators or markers have been listed in this chapter for a number of diseases. The predictivity of the risk factor can then be tested epidemiologically. If high predictivity markers can be validated, they will serve the purpose of disease prevention or delay of disease appearance.

REFERENCES

1. **Miettinen, O. S.,** *Theoretical Epidemiology*, John Wiley & Sons, New York, 1985.
2. **Rothman, K. J.,** *Modern Epidemiology*, Little Brown & Co., Boston, 1986.
3. **Vandenbroucke, J. P.,** Is 'the causes of cancer' a miasma theory for the end of the twentieth century?, *Int. J. Epidemiol.,* 17, 708, 1988.
4. **Loomis, D., and Wing, S.,** Is molecular epidemiology a germ theory for the end of the twentieth century?, *Int. J. Epidemiol.,* 19, 1, 1990.
5. **Schulte, P. A.,** Methodologic issues in the use of biologic markers in epidemiologic research, *Am. J. Epidemiol.,* 126, 1006, 1987.
6. **Muir, C. S.,** Epidemiology, basic science, and the prevention of cancer: implications for the future, *Cancer Res.,* 50, 6441, 1990.
7. **Perera, F. P., and Weinstein, I. B.,** Molecular epidemiology and carcinogen-DNA adduct detection: new approaches to studies of human cancer causation, *J. Chronic Dis.,* 35, 581, 1982.
8. **Perera, F. P.,** Molecular cancer epidemiology: a new tool in cancer prevention, *J. Natl. Cancer Inst.,* 78, 887, 1982.
9. **Emery, A. E. H.,** *Elements of Medical Genetics,* Churchill Livingstone, London, 1979.
10. **McKusick, V. A.,** *Mendelian inheritance in Man*, 8th ed., Johns Hopkins University Press, Baltimore, 1988.
11. **Calabrese, E. J.,** *Ecogenetics: Genetic Variation in Susceptibility to Environmental Agents*, Wiley Interscience, New York, 1984.
12. **Calabrese, E. J.,** Ecogenetics: historical foundation and current status, *J. Occup. Med.,* 10, 1096, 1986.
13. **Peto, J.,** Genetic predisposition to cancer. Cancer incidence in defined populations, Cold Spring Harbor Laboratory, Cold Spring Harbor, NY, 1980 (Banbury Report 4).

13a. **Hansen, M. F., and Cavenee, W. K.,** Genetics of cancer predisposition, *Cancer Res.,* 47, 5518, 1987.

14. **Weinstein, I. B.,** The origins of human cancer: molecular mechanisms of carcinogenesis and their implications for cancer prevention and treatment, *Cancer Res.,* 48, 4135, 1988.
15. **Weinberg, R. A.,** Oncogenes, antioncogenes and the molecular bases of multistep carcinogenesis, *Cancer Res.,* 49, 3713, 1989.
16. **Stocks, P.,** Cancer of the stomach in the large towns of England and Wales 1921-39, *Br. J. Cancer,* 4, 147, 1950.
17. **Stocks, P.,** Studies of cancer death rates at different ages in England and Wales in 1921 to 1950. Uterus, breast and lung, *Br. J. Cancer,* 7, 283, 1953.
18. **Armitage, R., and Doll, R.,** The age distribution of cancer and a multistage theory of cancer, *Br. J. Cancer,* 8, 1, 1954.
19. **Armitage, R., and Doll, R.,** A two-stage theory of carcinogenesis in relation to the age distribution of human cancer, *Br. J. Cancer,* 11, 161, 1957.

20. **Boveri, T.,** *The Origin of Malignant Tumors,* Williams & Wilkins, Baltimore, 1929.
21. **Lynch, H. T., Krush, A. J., Mulcahy, G. M., and Reed, W. B.,** Familial occurrences of a variety of premalignant disease and uncommon malignant neoplasms, *Cancer (Phila.),* 38, 1474, 1979.
22. **Li, F. P., and Fraumeni, J. F.,** Soft-tissue sarcomas, breast cancers and other neoplasms. A familial syndrome, *Ann. Intern. Med.,* 71, 747, 1969.
23. **Knudson, A. G., and Strong, L. C.,** Hereditary and cancer in man, *Prog. Med. Genet.,* 9, 113, 1973.
24. **Knudson, A. G.,** Mutation and cancer: statistical study of retinoblastoma, *Proc. Natl. Acad. Sci. U.S.A.,* 68, 820, 1971.
25. **Knudson, A. G.,** Hereditary cancer, oncogenes, and antioncogenes, *Cancer Res.,* 45, 1437, 1985.
26. **Miller, J. A., and Miller, E. C.,** Ultimate chemical carcinogens as reactive mutagenic electrophiles, in *Origins of Human Cancer*, Hiatt, H. H., Watson, J. D., and Winsten, J. A., Eds., Cold Spring Harbor Laboratory, Cold Spring Harbor, NY, 1977, 605-627.
27. **Ames, B. N., Durston, W. E., Yamasaki, E., and Lee, F. D.,** Carcinogens are mutagens: a simple test system combining liver homogenates for activation and bacteria for detection, *Proc. Natl. Acad. Sci. U.S.A.,* 70, 2281, 1973.
28. **International Agency for Research on Cancer,** *Genetic and related effects: an updating of selected IARC monographs from volumes 1-42,* International Agency for Research on Cancer, Lyon, 1987 (IARC Monographs on the evaluation of carcinogenic risks to humans; Suppl. 6).
29. **Higginson, J.,** Present trends in cancer epidemiology, *Proc. Can. Cancer Congr.,* 8, 40, 1969.
30. **Doll, R., and Peto, R.,** The causes of cancer — quantitative estimates of avoidable risks of cancer in the United States today, *J. Natl. Cancer Inst.,* 66, 1191, 1981.
31. **Vahakangas, K., and Pelkonen, O.,** Host variations in carcinogen metabolism and DNA repair, in *Genetic-Epidemiology of Cancer,* Lynch, H. T., and Hirayama, R., Eds., CRC Press, Boca Raton, FL, 1989, 6-40.
32. **Law, M. R.,** Genetic predisposition to lung cancer, *Br. J. Cancer,* 61, 195, 1990.
33. **Caporaso, N., Hayes, R. B., Dosemeci, M., Hoover, R., Ayesh, R., Hetzel, M., and Idle, J.,** Lung cancer risk, occupational exposure, and the debrisoquine metabolic phenotype, *Cancer Res.,* 49, 3675, 1989.
34. **Beland, F. A., Fullerton, N. F., Kinouchi, T., and Poirier, M. C.,** DNA adducts formation during continuous feeding of 2-acetylaminogluorene at multiple concentrations, in *Methods for Detecting DNA Damaging Agents in Humans: Applications in Cancer Epidemiology and Prevention,* Bartsch, H., Hemminki, K., and O'Neill, I. K., Eds., International Agency for Research on Cancer, Lyon, 1988, 175-180.
35. **Wogan, G. N.,** Detection of DNA damage in studies on cancer etiology and prevention, in *Methods for Detecting DNA Damaging Agents in Humans: Applications in Cancer Epidemiology and Prevention,* Bartsch, H., Hemminki, K., and O'Neill, I. K., Eds., International Agency for Research on Cancer, Lyon, 1988, 32-51.
36. **Ehrenberg, L., Moustacchi, E., and Osterman-Golkar, S.,** Dosimetry of genotoxic agents and dose-response relationships in their effects, *Mutation Res.,* 123, 121, 1983.
37. **Neumann, H.-G.,** Analysis of hemoglobin as a dose monitor for alkylating and arylating agents, *Arch. Toxicol.,* 56, 1, 1984.
38. **Randerath, K., Reddy, M. V., and Gupta, R. C.,** ^{32}P-postlabeling test for DNA damage, *Proc. Natl. Acad. Sci. U.S.A.,* 78, 6126, 1981.

39. **Gupta, R. C., Reddy, M. V., and Randerath, K.,** ^{32}P-postlabeling analysis of non-radioactive aromatic carcinogen-DNA adducts, *Carcinogenesis,* 3, 1081, 1982.
40. **Reddy, M. V., and Randerath, K.,** Nuclease P_1-mediated enhancement of sensitivity of ^{32}P-postlabeling test for structurally diverse DNA adducts, *Carcinogenesis,* 7, 1543, 1986.
41. **Poirier, M. C.,** Antibodies to carcinogens — DNA adducts. *J. Natl. Cancer Inst.,* 67, 515, 1981.
42. **Bartsch, H., Hemminki, K., and O'Neill, I. K., Eds.,** *Methods for Detecting DNA Damaging Agents in Humans: Applications in Cancer Epidemiology and Prevention,* International Agency for Research on Cancer, Lyon, 1988.
43. **Hemminki, K., Perera, F. P., Phillips, D. H., Randerath, K., Reddy, M. V., and Santella, R. M.,** Aromatic DNA adducts in white blood cells of foundry workers, in *Methods for Detecting DNA Damaging Agents in Humans: Applications in Cancer Epidemiology and Prevention,* Bartsch, H., Hemminki, K., and O'Neill, I. K., Eds., International Agency for Research on Cancer, Lyon, 1988, 190-195.
44. **Hemminki, K., Grzybowska, E., Chorazy, M., Twardowska-Sancha, K., Sroczynski, J. W., Putman, K. L., Randerath, K., Phillips, D. H., Hewer, A., Santella, R. M., Young, T. L., and Perera, F. P.,** DNA adducts in humans environmentally exposed to aromatic compounds in an industrialized area of Poland, *Carcinogenesis,* 11, 1229, 1990b.
45. **Hemminki, K., Randerath, K., Reddy, M. V., Putman, K. L., Santella, R. M., Perera, F. P., Young, T.-L., Phillips, D. H., Hewer, A., and Savela, K.,** Postlabeling and immunoassay analysis of polycyclic aromatic hydrocarbons — adducts of deoxyribonucleic acid in white blood cells of foundry workers, *Scand. J. Work Environ. Health,* 16, 158, 1990a.
46. **International Agency for Research on Cancer,** Polynuclear aromatic compounds, Part 3: Industrial exposure in aluminum production, coal gasification, coke production and iron and steel foundring, International Agency for Research on Cancer, Lyon, 1984. (IARC monographs on the evaluation of the carcinogenic risk of chemicals to humans.)
47. **Phillips, D. H., Hemminki, K., Alhonen, A., Hewer, A., and Grover, P. L.,** Monitoring occupational exposure to carcinogens: detection by ^{32}P-postlabeling of aromatic DNA adducts in white blood cells from iron foundry workers, *Mutation Res.,* 204, 531, 1988.
48. **Perera, F. P., Hemminki, K., Young, T. L., Brenner, D., Kelly, G., and Santella, R. M.,** Detection of polycyclic aromatic hydrocarbon-DNA adducts in white blood cells of foundry workers, *Cancer Res.,* 48, 2288, 1988.
49. **Sorsa, M.,** Monitoring of sister chromatid exchanges and micronuclei as biological endpoints, in *Monitoring Human Exposure to Carcinogenic and Mutagenic Agents,* Berlin, A., Draper, M., Hemminki, K., and Vainio, H., Eds., International Agency for Research on Cancer, Lyon, 1984, 339-349.
50. **Forni, A.,** Cytogenetic methods for assessing human exposure to genotoxic chemicals, in *Occupational and Environmental Chemical Hazards — Chemical and Biochemical Indices for Monitoring Toxicity,* Foa, V., Emmet, E. A., Maroni, M., and Colombi, A., Eds., Ellis Horwood, Chichester, England, 1987, 403-410.
51. **Brøgger, A., Hagmar, L., Hansteen, I.-L, Heim, S., Hogstedt, B., Knudsen, L., Lambert, B., Linnainmaa, K., Mitelman, F., Nordenson, I., Reuterwall, C., Salomaa, S., Skerfving, S., and Sorsa, M.,** An inter-Nordic prospective study on cytogenetic endpoints and cancer risk, *Cancer Genet. Cytogenet.,* 44, 000, 1990.

52. **Heim, S., and Mitelman, F.,** *Cancer Cytogenetics,* Alan R. Liss, New York, 1987.
53. **Croce, C. M.,** Role of chromosome translocations in human neoplasia, *Cell,* 49, 155, 1987.
54. **Saiki, R. K., Scharf, S., Faloona, F., Mullis, K. B., Horn, G. T., Erlich, H. A., and Arnheim, N.,** Enzymatic amplification of beta-globin genomic sequences and restriction site analysis for diagnosis of sickle cell anemia, *Science,* 230, 1350, 1985.
55. **Cariello, N. F., Scott, J.K., Kat, A. G., Thilly, W. G., and Keohavong, P.,** Resolution of a missense mutant in human genomic DNA by denaturing gradient gel electrophoresis and direct sequencing using in vitro DNA amplification: HPRT, *Am. J. Hum. Genet.,* 42, 726, 1988.
56. **Messing, K., and Bradley, W. E. C.,** *In vivo* mutant frequency rises among breast cancer patients after exposure to high doses of gamma radiation, *Mutation Res.,* 152, 107, 1985.
57. **Chrysostomou, A., Seshandri, R. S., and Morely, A. A.,** Mutation frequency in nurses and pharmacists exposed to cytotoxic drugs, *Aust. N.Z. J. Med.,* 14, 831, 1984.
58. **Vogelstein, B., Fearon, E. R., Hamilton, S. R., Kern, S. E., Presinger, A. C., Leppert, M., Nakamura, Y., White, R., Smits, A. M. M., and Bos, J. L.,** Genetic alterations during colorectal-tumor development, *N. Engl. J. Med.,* 319, 525, 1988.
59. **Harris, C. C.,** Chemical and physical carcinogenesis: advances and perspectives for the 1990's, *Cancer Res.,* 51, 5023s, 1991.
60. **Kern, S., Kinzler, K., Bruskin, A., Jarosz, D., Friedman, P., Prives, C., and Vogelstein, B.,** Identification of p53 as a sequence-specific DNA-binding protein, *Science,* 252, 1708, 1991.
61. **Hollstein, M., Sidransky, D., Vogelstein, B., and Harris, C. C.,** p53 Mutations in human cancers, *Science,* 253, 49, 1991.
62. **Kinzler, K. W., Nilbert, M., Su, L.-K., Vogelstein, B., Bryan, T., Levy, D., Smith, K., Preisinger, A., Hedge, P., McKechnie, D., Finniear, R., Markham, A., Groffen, J., Boguski, M., Altschul, S., Horii, A., Ando, H., Miyoshi, Y., Miki, Y., Nishisho, I., and Nakamura, Y.,** Identification of FAP locus genes from chromosome 5q21, *Science,* 253, 661, 1991.
63. **Nishisho, I., Nakamura, Y., Miyoshi, Y., Miki, Y., Ando, H., Horii, A., Koyama, K., Utsunomiya, J., Baba, S., Hedge, P., Markham, A., Krush, A., Petersen, G., Hamilton, S., Nilbert, M., Levy, D., Bryan, T., Preisinger, A., Smith, K., Su, L. K., Kinzler, K., and Vogelstein, B.,** Mutations of chromosome 5q21 genes in FAP and colorectal cancer patients, *Science,* 253, 665, 1991.
64. **Langlois, R. G., Bigbee, W. L., and Jensen, R. H.,** Measurements of the frequency of human erythrocytes with gene expression loss phenotypes at the glycophorin A locus, *Hum. Genet.,* 74, 353, 1986.
65. **Brandt-Rauf, P. W.,** New markers for monitoring occupational cancer: the example of oncogene proteins, *J. Occup. Med.,* 5, 399, 1988.
66. **Brandt-Rauf, P. W., and Niman, H. L.,** Serum screening for oncogene proteins in workers exposed to PCBs, *Br. J. Ind. Med.,* 45, 689, 1988.
67. **Brandt-Rauf, P., Smith, S., Perera, F., Niman, H. L., Yohannan, W., Hemminki, K., and Santella, R. M.,** Serum oncogene proteins in foundry workers, *J. Soc. Occup. Med.,* 40, 11, 1990.
68. **Brandt-Rauf, P.,** personal communication.
69. **Breslow, J. L.,** Human apolipoprotein molecular biology and genetic variation, *Annu. Rev. Biochem.,* 54, 699, 1985.

70. **Brown, M. S., and Goldstein, J. L.,** A receptor-mediated pathway for cholesterol homeostasis, *Science,* 232, 34, 1986.
71. **Brown, M. S., and Goldstein, J. L.,** Plasma lipoproteins: teaching old dogmas new tricks, *Nature,* 330, 113, 1987.
72. **Hobbs, H. H., Brown, M. S., Russel, D. W., et al.,** Deletion in the gene for the low-density lipoprotein receptor in a majority of French Canadians with familial hypercholesterolemia, *N. Engl. J. Med.,* 317, 734, 1987.
73. **Aalto-Setala, K.,** The Finnish type of the LDL receptor gene mutation: molecular characterization of the deleted gene and the corresponding mRNA, *FEBS Lett.,* 234, 411, 1988.
74. **Berg, K.,** DNA polymorphism at the apolipoprotein B locus is associated with lipoprotein level, *Clin. Genet.,* 30, 515, 1986.
75. **Breslow, J. L.,** Apolipoprotein genetic variation and human disease, *Physiol. Rev.,* 85-132, 1988.
76. **Mahley, R. W.,** Apolipoprotein E: cholesterol transport protein with expanding role in cell biology, *Science,* 240, 622, 1988.
77. **Scott, J.,** The molecular and cell biology of apolipoprotein-B, *Mol. Biol. Med.,* 6, 65, 1989.
78. **Russel, D. W., Esser, V., and Hobbs, H. H.,** Molecular basis of familial hypercholesterolemia, *Arteriosclerosis,* 9(Suppl. 1), 8, 1989.
79. **Goldstein, J. L., and Brown, M. S.,** Regulation of the mevalonate pathway, *Nature,* 343, 425, 1990.
80. **Talmud, P. J., Lloyd, J. K., Muller, D. P. R., Collins, D. R., Scott, J., and Humphries, S.,** Genetic evidence from two families that the apolipoprotein B gene is not involved in abetalipoproteinemia, *J. Clin. Invest.,* 82, 1803, 1988.
81. **Innerarity, T. L., Wesigraber, K. H., Arnold, K. S., Mahley, R. W., Krauss, R. M., Vega, G. L., and Grundy, S. M.,** Familial defective apolipoprotein B-100: low density lipoproteins with abnormal receptor binding, *Proc. Natl. Acad. Sci. U.S.A.,* 84, 6919, 1987.
82. **Wiklund, O., Angelin, B., Olofsson, S. O., Eriksson, M., Fager, G., Berglund, L., and Bondjers, G.,** Apolipoprotein (a) and ischaemic heart disease in familial hypercholesterolaemia, *Lancet,* 335, 1360, 1990.
83. **Davignon, J., Gregg, R. E., and Sing, C. F.,** Apolipoprotein E polymorphism and atherosclerosis, *Arteriosclerosis,* 8, 1, 1988.
84. Lowering blood cholesterol to prevent heart disease, *J.A.M.A.,* 253, 2080, 1985.
85. **Tyroler, H. A.,** Lowering plasma cholesterol levels decreases risk of coronary heart disease: an overview of clinical trials, in *Hypercholesterolemia,* Steinberg, D., and Olefsky, J. M., Eds., Churchill Livingstone, New York, 1987, 99-116.
86. **Steinberg, D., Parthasarathy, S., Carew, T., Khoo, J., and Witztum, J.,** Beyond cholesterol. Modifications of low-density lipoprotein that increase its atherogenicity, *N. Engl. J. Med.,* 320, 915, 1989.
87. **Keating, M., Atkinson, D., Dunn, C., Timothy, K., Vincent, G., and Leppert, M.,** Linkage of a cardiac arrhythmia, the long QT syndrome, and the harvey *ras*-1 gene, *Science,* 252, 704, 1991.
88. **Iwakura, Y., Tosu, M., Yoshida, E., Takiguchi, M., Sato, K., Kitajima, I., Nishioka, K., Yamamoto, K., Takeda, T., Hatanaka, M., Yamamoto, H., and Sekiguchi, T.,** Induction of inflammatory arthropathy resembling rheumatoid arthritis in mice transgenic for HTLV-I, *Science,* 253, 1026, 1991.

89. **Paliard, C., West, S., Lafferty, J., Clements, J., Kappler, J., Marrack, P., and Kotzin, B.,** Evidence for the effects of a superantigen in rheumatoid arthritis, *Science,* 253, 325, 1991.
90. **Prockop, D. J., and Kivirikko, K. I.,** Heritable disease of collagen, *N. Engl. J. Med.,* 311, 376, 1984.
91. **Prockop, D. J.,** Mutations that alter the primary structure of type I collagen. The perils of a system for generating large structures by the principle of nucleated growth, *J. Biol. Chem.,* 265, 15349, 1990.
92. **Kuivaniemi, H., Tromp, G., and Prockop, D. J.,** Mutations in collagen genes: causes of rare and some common diseases in humans, *FASEB J.,* 5, 2052, 1991.
93. **Edfors-Lubs, M.,** Allergy in 7000 twin pairs, *Acta Allergol.,* 26, 249, 1971.
94. **Marsh, D. G., Hsu, S. H., Hussain, R., Meyers, D. A., Freidhoff, L. R., and Bias, W. B.,** Genetics of human immune response to allergens, *J. Allergy Clin. Immunol.,* 65, 322, 1980.
95. **Kay, A. B., Durham, S. R., Gin, W., Moqbel, R., MacDonald, A. J., Walsh, G. M., Shaw, R. J., Cromwell, O., Mackay, J., and Carroll, M.,** Inflammatory cells in early, late-phase and chronic asthma, *Prog. Resp. Res.,* 19, 211, 223, 1985.
96. **Cookson, W., Sharp, P., Faux, J., and Hopkin, J.,** Linkage between immunoglobulin E responses underlying asthma and rhinitis and chromosome 11q, *Lancet,* i, 1292, 1989.
97. **Valenta, R., Duchene, M., Pettenburger, K., Sillaber, C., Valent, P., Bettelheim, P., Breitenbach, M., Rumpold, H., Kraft, D., and Scheiner, O.,** Identification of profilin as a novel pollen allergen: IgE autoreactivity in sensitized individuals, *Nature,* 253, 557, 1991.
98. **Bjorksten, F., Suoniemi, I., and Koski, V.,** Neonatal birch pollen contact and subsequent allergy to birch pollen, *Clin. Allergy,* 10, 585, 1980.
99. Epidemiological studies of childhood insulin-dependent diabetes. Review, *Acta Paediatr. Scand.,* 80, 583, 1991.
100. **Todd, J. A., Bell, J., and Devitt, H. O.,** HLA-DQ gene contributes to susceptibility and resistance to insulin-dependent diabetes mellitus, *Nature,* 329, 599, 1987.
101. **Landin-Olsson, M., Karlsson, A., Dahlquist, G., Blom, L., Lernmark, Å., and Sundkvist, G.,** Islet all and other organ-specific auto-antibodies in all children developing type 1 diabetes mellitus in Sweden during one year and in matched control children, *Diabetologia,* 32, 387, 1989.
102. **Baekkestov, S., Aanstoot, H.-J., Christgau, S., Reetz, A., Solimena, M., Cascalho, M., Folli, F., RIchter-Olesen, H., and De-Camilli, P.,** Identification of the 64K autoantigen in insulin-dependent diabetes as the GABA-synthesizing enzyme glutamic acid decarboxylase, *Nature,* 347, 151, 1990.
103. **Dahlquist, G., Blom, L., Persson, L.-Å., Sundstrom, A., and Wall, S.,** Dietary factors and the risk of developing insulin-dependent diabetes in childhood, *Br. Med. J.,* 300, 1302, 1990.

8

Molecular Epidemiologic Approaches in Environmental Carcinogenesis

Regina M. Santella and Frederica P. Perera

Division of Environmental Science, School of Public Health, Columbia University, New York, NY

INTRODUCTION

The term molecular or biochemical epidemiology has been applied to methods that combine epidemiology with laboratory assays. These methods study the molecular biology of cancer and use human materials. As applied to environmental carcinogenesis, the ultimate goal of these studies is to better define exposure to carcinogens as well as to more accurately determine individual risk for cancer development. Identification of individual differences in susceptibility to disease is an important part of this risk determination.

A much utilized diagram of the sequence of events in the continuum from initial exposure to cancer development is given in Figure 1. Internal dose refers to the measurement of the amount of a carcinogen or its metabolites present in cells, tissues, or body fluids. Biologically effective dose refers to the binding of the carcinogen to DNA, while markers of biological response measure results of exposure. Influencing the process at all steps are individual susceptibility factors. The number of methods useful for molecular epidemiology has increased

0-87371-631-0/94/$0.00+$.50

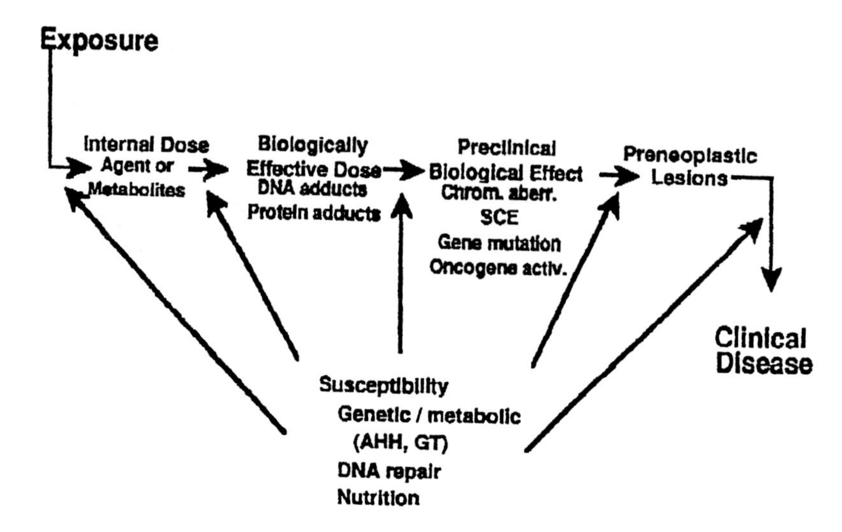

Figure 1. The relationship of biological markers to exposure and disease.

dramatically over the past few years as have reports of their application to environmental or occupational exposures. The various markers will be discussed in some detail and examples given of their application. This summary is not intended to be a comprehensive review of all studies.

INTERNAL DOSE

The first step in the process described in Figure 1 is the absorption and distribution of the chemical in the body. Examples of internal dose include measurement of cotinine in serum or urine resulting from cigarette smoke exposure, urinary levels of 1-hydroxypyrene resulting from polycyclic aromatic hydrocarbon (PAH) exposure or aflatoxin M_1 levels in urine from dietary exposure. These assays for specific chemicals utilize analytical chemical or immunologic methods for quantitation. An alternate method which is not chemical specific is the Ames *Salmonella typhimurium* mutagenesis assay to measure total mutagenicity of urine. Markers of internal dose take into account individual differences in absorption of the compound and demonstrate that exposure has resulted in actual increases in levels of the compound in the body. Examples of studies on internal dose are given in Table 1.

BIOLOGICALLY EFFECTIVE DOSE

DNA Adducts

The biologically effective dose reflects the amount of compound that has actually reacted with critical cellular targets. Since DNA is considered the

Table 1
Internal Dose

Compound analyzed[a]	Exposure source	Biologic sample	Population	Ref.
Aflatoxin M_1	Diet	Urine	China	1
Benzene, toluene	Cigarette smoke	Blood	Smokers, nonsmokers	2
CFA	Occupational exposure	Urine	Workers	3
Glu-P-1, Glu-P-2	Diet	Plasma	Uremic patients and control	4
1-Hydroxypyrene	Coal tar products	Urine	Workers, smokers	5,6
3-Hydroxy-BP	Coal tar products	Urine	Coal tar-treated patients	7
MeIQx	Diet	Urine	Fried beef consumers	8
Mutagens	Cigarette smoke, various occupational exposures	Urine	Smokers, workers	Reviewed in 9,10
Nitrosamino acids	N-nitroso compounds in diet	Urine	Chinese residing in areas of low and high cancer risk	11
N-nitrosoproline	Cigarette smoke	Urine	Smokers, nonsmokers	12
		unexposed		13
Trp-P-1, Trp-P-2	Diet	Plasma	Volunteers	14
Trp-P-1, Trp-P-2	Diet	Bile	Patients	15

[a] CFA, 3-chloro-4-fluoroaniline; Glu-P-1, 2-amino-6-methyldipyrido[1,2-a:3′,2′-d] imidazole; Glu-P-2, 2-aminodipyrido[1,2-a:3′,2′-d] imidazole; 3-Hydroxy-BP, 3 hydroxybenzoapyrene; MeIQx, 2-amino-3,8-dimethylimidazo[4,5-f]quinoxa line; Trp-P-1, 3-amino-1,4-dimethyl-5H-pyrido [4,3-b] indole; Trp-P-2, 3-amino-1-methyl-5H-pyrido[4,3-b] indole.

primary target for chemical carcinogens, measurement of carcinogen-DNA adducts in the target tissue provides a more relevant marker of exposure than measurement of the carcinogen itself. It takes into account individual differences in absorption, distribution, and metabolism of the chemical as well as repair of adducts once formed. Unfortunately, for many studies in humans, the target tissue is not readily accessible and thus surrogate tissues are sometimes utilized (e.g., peripheral blood cells and placenta). The relationship of adducts in more readily sampled tissue to that in the target tissue has not been well characterized in humans. In addition, the recency of exposure which can be measured is not well established. Limited studies of white blood cell DNA adducts suggest that adducts are rapidly reduced after exposure ends.[16] Studies in lung tissue of exsmokers indicated a decrease in cigarette smoking-related adducts within 2 years of smoking cessation.[17-19]

Quantitation of DNA adducts in humans by immunoassays, fluorescence spectroscopy, [^{32}P] postlabeling, and gas chromatography/mass spectroscopy have been reviewed recently.[20] Table 2 lists examples of DNA adducts which have been measured in various occupational or environmental settings.

These methods can quantitate adducts at the femto (10^{-15}) or atto (10^{-18}) mole level, which means detection of one adduct/10^{8-10} nucleotides. Immunoassays utilize either polyclonal or monoclonal antisera generated against specific adducts. Enzyme-linked immunosorbent assays (ELISA) are used most frequently because they do not require radiolabeled substrate, have high sensitivity, and can easily be applied to large numbers of samples. There may be significant cross-reactivity of both types of antisera with structurally related adducts. For example, antisera generated against benzo[a]pyrene diol epoxide modified DNA (BPDE-I-DNA) cross-reacts with DNAs modified by several other PAH diol epoxides.[59,60] Humans are exposed to BP as part of a complex mixture of PAH and the precise adducts present in each sample are unknown. Thus, absolute quantitation is impossible and only a semiquantitative measure of antigenicity of the sample can be determined. However, since multiple PAHs are carcinogenic, such measurements are relevant to ultimate risk for cancer development. Cross-reactivity problems can be eliminated by chromatographic separation of adducts followed by quantitation in the immunoassay. For example, HPLC separation of adducts followed by ELISA has been used to monitor the multiple adducts of *cis*-platinum.[47]

Antisera against BPDE-I-DNA have been used in a number of studies to investigate adduct formation in human populations. Adducts have been detected[61,62] in lung tissue of cancer patients at levels up to $1/10^6$. Adducts have also been investigated in placental and white blood cell DNA of smokers and nonsmokers. No significant difference was observed in placenta,[63] and there have been conflicting results in white blood cells.[64,65] Feeding studies have also suggested that diet may be an important source of PAH exposure[66] (See Chapter 9). In contrast to these studies, dramatic increases in adduct levels have been observed in workers occupationally exposed to high levels of PAHs. These have included foundry, aluminum plant, and coke oven workers.[65,67-70] PAH-DNA adducts have been linked to environmental exposure in individuals living in a highly polluted region of Poland.[71]

Table 2
Biologically Effective Dose — DNA Adducts

Compound analyzed[a]	Exposure source	Biologic sample	Population	Ref.
4-ABP-DNA	Cigarette smoke and occupation	Lung tissue	Volunteers	21
Acrolein-DNA	Cyclophosphamide Chemotherapy	WBC	Chemotherapy patients	22
AFB_1 guanine	Diet	Urine	China and Africa	23,24
AFB_1-DNA	Diet	Liver tissue	Taiwan	25,26
Cisplatinum-DNA,	Cisplatin Chemotherapy	WBC	Chemotherapy patients	27–31
0^4-Ethylthymine	Methylating agents	Liver tissue	Liver cancer patients	32
3-Methyladenine	Methylating agents	Urine	Volunteers	33
O^6-Methyldeoxy-guanosine	Nitrosamines in diet, smoking Chemotherapy	Esophageal and stomach mucosa, placenta, WBC	Chinese and European cancer patients, smokers	34 35–38
8-MOP-DNA	8-Methoxypsoralen chemotherapy	Skin	Psoriasis patients	39
5-OH-methyl uridine	Diet	WBC	Low fat and normal diet	40
PAH-DNA	PAH in cigarette smoke, in workplace	WBC, lung tissue, placenta	Lung cancer patients, smokers, workers	Reviewed in 20,41
Spectrum of DNA adducts	Betel and tobacco chewing, smoking, industrial exposures, wood smoke	Placenta, lung tissue, oral mucosa, WBC, bone marrow, colonic mucosa	Smokers, workers, volunteers	Reviewed in 20,41
Thymine glycol	Agents that cause oxidative damage to DNA	Urine	Volunteers	42

[a] 4-ABP, 4-aminobiphenyl; AFB_1, aflatoxin B_1; PAH, polycyclic aromatic hydrocarbons; WBC, white blood cells; 5-OH-methyl uridine; 5-hydroxymethyluridine.

Alkylated bases in DNA such as O^6-methyldeoxyguanosine (O^6-me-dGuo) or O^4-ethylthymine (O^4-et-dThy) have been measured in various tissues. Elevated levels of O^6-me-dGuo were seen in esophageal cancer patients compared to controls from Europe,[31] but no difference was seen in placenta of smokers compared to nonsmokers.[35] O^4-et-dThy was elevated in liver DNA of cancer patients.[72]

Another application of antisera to carcinogen adducts is in the isolation of adducts by affinity chromatography prior to quantitation by alternate chemical methods. For example, BPDE-I-DNA adducts have been measured by synchronous fluorescence spectroscopy or GC/MS in placental samples after affinity purification.[73]

Antisera to DNA adducts can also be used in immunohistochemical studies to localize adduct formation to specific cell and tissue types. This approach has been used to demonstrate adduct formation in skin biopsies from coal tar-treated psoriasis patients[74] and aflatoxin-B_1-DNA adducts in liver tissue of hepatocellular cancer patients.[26] The advantage of this approach is that much smaller amounts of material are required than for the ELISA. Thus, the method should also be applicable to biopsy samples.

The [^{32}P] postlabeling method for DNA adduct detection involves the enzymatic incorporation of high specific activity [^{32}P] into nucleotides.[75,76] DNA samples are first digested to the 3′-monophosphates, followed by labeling at the 5′ position with T4 polynucleotide kinase and γ-[^{32}P] ATP. Thin layer chromatography separates normal nucleotides from adducted ones, and further chromatography fingerprints adducts. The advantages of the method are its lack of specificity allowing multiple adducts to be detected, the small amounts of DNA required (1 to 50 μg), and the high sensitivity (detection of 1 adduct/10^{8-10} nucleotides). Absolute quantitation is a problem with unknown adducts since the efficiency of all the steps cannot be determined. Some adducts may not be digested or labeled efficiently, leading to an underestimation of actual adduct levels. In addition, once an adduct has been detected, it is usually not possible to characterize it because of the small amount available. Two procedures have been developed to increase the sensitivity of detection: butanol extraction of adducts[77] or nuclease P_1 digestion of normal nucleotides.[78] The methodology has been reviewed recently.[79]

Postlabeling has been used to detect cigarette smoking-related adducts in placenta, lung, cervix, etc.[17,18,63,80] In many tissues from smokers, a diagonal zone of radioactivity was found on the chromatography plate, suggesting a large number of adducts. Studies on white blood cells of smokers and nonsmokers have generally detected no difference.[81,82] Several studies on mononuclear cells (lymphocytes plus monocytes) or monocytes, however, have been able to demonstrate elevated levels of adducts in smokers compared to nonsmokers,[16,83,84] but others have not.[85,86] In contrast to smoking exposure, as with the immunoassays, studies of worker populations have demonstrated dramatic differences in adduct levels of white blood cells (see Chapter 7).

Since several carcinogens, including aflatoxin B_1 and BP, are fluorescent, spectrofluorescence methods have been developed for their quantitation. AFB_1-guanine adducts, excised from DNA, were measured in urine samples from China and Africa after C18 Sep Pak or immunoaffinity isolation followed by HPLC with fluorescence detection.[24,87,88] Quantitation of BP adducts has been carried out by release of BP tetrol from DNA by mild acid hydrolysis followed by synchronous fluorescence spectroscopy (SFS).[57,89] Detectable adducts were seen in coke oven workers[69,70] and aluminum plant workers.[90] Samples have been purified by immunoaffinity chromatography and HPLC before SFS to further confirm the presence of BP adducts.[89]

Finally, GC/MS techniques with femtogram sensitivity have been developed for several alkylated bases, including O^4-et-dThy, 7-methylguanosine, 3-methyladenine, N^2,3-ethenoguanine, 8-hydroxyadenine, and 8-hydroxyguanosine.[91-94] BP adducts in DNA and protein samples have also been monitored by GC/MS following derivitization of the released BP tetrols.[57]

Protein Adducts

Since chemical carcinogens bind to proteins as well as DNA, determination of protein adducts is being used as an alternate marker of exposure.[95] However, the relationship between protein adducts and DNA adducts may vary for each compound of interest and must be determined. Because of the large amounts in blood, both hemoglobin and albumin have been used for determination of the biologically effective dose. Both proteins will provide information on only relatively recent exposure since the lifespan of the red blood cell is 4 months and the half-life of albumin is 21 days. Table 3 provides examples of measurement of carcinogen protein adducts. Ethylene oxide hemoglobin adducts resulting from occupational exposure were the first protein adducts to be measured by GC/MS.[52] More recently, the development of an Edman degradation GC/MS method for quantitation of the N-terminal hydroxyethylvaline has simplified monitoring ethylene oxide exposure.[96] Elevated adduct levels have been seen in hospital sterilization workers as well as smokers.[54,97] Immunologic methods have also been developed for measurement of the valine adduct.[98]

4-Aminobiphenyl-hemoglobin adducts have been measured by release of the parent carcinogen from the protein, followed by extraction, derivatization, and GC/MS analysis.[43] Elevated levels of adducts were seen in smokers (154 pg/g Hb) compared to nonsmokers (28 pg/g),[43] and the persistence of adducts following smoking cessation has also been monitored.[45] Comparison of adducts in controls and blond and black tobacco smokers demonstrated significant differences between the three groups.[44] These differences were approximately proportional to the relative risk of each group. Studies in maternal and fetal blood demonstrated increased adduct levels in smokers compared to nonsmokers.[99] Maternal levels were about twofold higher than fetal blood levels.

Hemoglobin adducts of the tobacco-specific nitrosamines have also been measured by GC/MS methods.[55] 4-Hydroxy-1-(3-pyridyl)-1-butanone (HPB), is cleaved from the protein followed by extraction and derivitization. Snuff dippers had the highest adduct levels measured (mean, 517 fmol HPB/g Hb) compared to smokers (79.6 fmol/g) and nonsmokers (29.3 fmol/g).

Table 3
Biologically Effective Dose — Protein Adducts

Compound analyzed[a]	Exposure source	Biologic sample	Population	Ref.
4-ABP-Hb	Cigarette smoke	RBC	Smokers, nonsmokers	43-45
AFB_1-albumin	Diet	Plasma	China	46,47
Alkylated Hb	Propylene oxide	RBC	Workers	48
BP & CHRY-Hb	Cigarette smoke	RBC	Smokers	49
Cisplatinum-protein	Cisplatin chemotherapy	Plasma, RBC	Chemotherapy patients	50
Hydroxyethylhistidine: Hydroxyethylvaline	Ethylene oxide	RBC	Workers, smokers, volunteers	51-54
NNK,NNN-Hb	Cigarette smoke	RBC	Smokers	55
PAH-albumin	PAH in workplace, in cigarette smoke	Plasma	Workers, smokers	56-58

[a] 4-ABP, 4-aminobiphenyl; AFB_1, aflatoxin B_1; BP, benzo(a)pyrene; CHRY, chrysene; Hb-hemoglobin; NNK, 4-methylnitrosamino-1-3-pyridyl-1-butanone; NNN, N^1-nitrosonornicotine; PAH, polycyclic aromatic hydrocarbons; RBC, red blood cells.

Several methods have been developed for quantitation of BP protein adducts. Immunoassays which recognize multiple PAH adducts have been used to quantitate albumin adducts in foundry workers and roofers. In one assay, albumin was acid treated to release BP tetrols, followed by quantitation by ELISA;[56] while in the other, it was enzyme digested prior to ELISA.[58] Mean protein adduct concentrations for nonsmoking (0.240 fmol/μg) and smoking (0.280) foundry workers were significantly higher than mean values for nonsmoking (0.0723) and smoking (0.142) controls.[56]. Roofers also had elevated levels of adducts (5.19 fmol/μg) compared to controls (3.28).[58] The absolute difference in adduct levels between these two reports is probably related to differences in methodology, including sample workup and extraction. Released tetrols have also been measured after immunoaffinity chromatography by fluorescence and GC/MS analysis.[57,100]

The studies to date demonstrate the feasibility of measuring adducts in humans, but their ultimate use for risk assessment remains to be determined. Background levels of adducts have also been measured in populations, as well as large inter-individual variations. These individual differences in adducts with similar exposures demonstrate the advantage of biomonitoring over environmental monitoring.

BIOLOGICAL RESPONSE

The next step in the continuum comprises markers that indicate a biological response resulting from the exposure either at the target or an analogous site which is known to be pathogenically linked to cancer. A wide variety of biomarkers are in this category, including somatic mutations, DNA single-strand breaks, and the various cytogenetic assays, including sister chromatid exchange (SCE), micronuclei (MN), and chromosomal aberrations (CA). Table 4 provides examples of exposures and populations monitored for these endpoints. These markers are not chemical or exposure specific. Thus, extensive information on other lifestyle and environmental factors (e.g., cigarette smoking habits, age, diet, viral infection, and gender) that could affect the assay must be determined. Mutation assays are reviewed in Chapter 10 and are not discussed here.

Measurement of CAs in peripheral lymphocytes has been used extensively as a sensitive monitor of radiation exposure but it is less sensitive for detecting exposure to chemicals.[122] Ethylene oxide exposure has been monitored in several studies; two[101,102] demonstrated increased aberrations, but another did not[52] possibly due to the low exposure of the worker population followed.

MN reflect the consequences of lagging chromosomes or chromosome fragments. They consist of small amounts of DNA in the cytoplasm which are not incorporated into daughter nuclei during mitosis. MN are easy to score but reflect only a small proportion of the induced CAs. MN can be assayed in peripheral blood lymphocytes which are stimulated to divide *in vitro*. This cell division is necessary for the induction of MN. A major advantage of assaying for MN is that it can also be carried out on exfoliated cells such as oral mucosa cells, which can

Table 4
Early Biologic Effect or Response

Compound analyzed[a]	Exposure source	Biologic sample	Population	Ref.
Chromosomal aberrations	Occupational exposure, radiation	WBC	Workers	52,101,102
DNA hyperploidy	Aromatic amines	Bladder and lung cells	Workers	103
GPA mutation	Chemotherapeutic agents, radiation	RBC	Patients, Japanese atom bomb survivors	104-107
HPRT mutation	Chemotherapeutic agents, radiation	WBC	Workers, patients	108-111
Micronuclei	Organic solvents, heavy metals, cigarette smoke, betel quid	WBC, oral mucosa	Workers	112,113
Mutation in tumor suppressor genes	AFB_1	Tumor tissue	Patients	114,115
Oncogene activation	PAH, cigarette smoke	Serum	Patients, workers	116,117
Single-strand breaks	Styrene	WBC	Workers	118
Sister chromatid exchange	Occupational exposure, radiation	WBC	Workers	119,120
Unscheduled DNA synthesis	Propylene oxide	WBC	Workers	121

[a] AFB_1, aflatoxin B_1; GPA, glycophorin A; HPRT, hypoxanthine guanine phosphoribosyl transferase; PAH, polycyclic aromatic hydrocarbons; RBC, red blood cells; WBC, white blood cells.

be sampled noninvasively. Cigarette smoking has been evaluated, as well as dietary intervention with β-carotene in oral cells of betel nut chewers.[123]

SCE is considered a more sensitive, rapid, and simple cytogenetic endpoint than CA for evaluating exposure to genotoxic agents. Human peripheral blood lymphocytes are normally used because of ease of access. The induced lesions which lead to the formation of SCE may persist in these cells for days or months. Increased levels of SCEs have been demonstrated with exposure to cigarette smoke, occupational exposures, and diet and drug use.[124] The number of high frequency cells (HFCs) which display a number of SCEs per cell, which is higher than the 95th percentile of the distribution of SCEs per cell in the controls, can also be estimated. Under certain conditions, HFCs are more sensitive criteria for genotoxic exposure than mean values of SCE per cell.[125]

Two studies have investigated multiple cytogenetic markers, including CA, MN, and SCE in ethylene oxide-exposed workers.[97,126] Since hydroxyethyl valine hemoglobin adducts were also measured in both studies and HPRT mutations in one and DNA single-strand breaks and repair in the other, the relative sensitivity of the various methods could be established. The most sensitive indicator of recent exposure, by a factor of 10 to 100, was measurement of hydroxyethyl valine adducts. Its major drawback was that it can only measure exposure during the few months prior to blood sampling because of the lifespan of the red blood cell. In contrast, HPRT mutations, which were elevated in exposed workers compared to controls, may measure exposure occurring years earlier. Both studies demonstrated a good correlation between SCE and adducts which were elevated in workers compared to controls.

BIOMARKERS OF ALTERED STRUCTURE/FUNCTION

Biomarkers of altered structure/function include assays to measure genetic changes in oncogenes or tumor suppressor genes. Examples include measurement of *ras* mutations, *myc* amplifications, and mutations in p53. Recent studies on mutations in the p53 tumor supressor gene in liver tumors have suggested a mechanism for determining the causative agent in tumor development.[114,115] These studies were on tumors collected from different regions of the world with high exposure to aflatoxin B_1. The detection of specific mutations similar to those known to be produced by aflatoxin in *in vitro* studies suggests, but does not prove, that aflatoxin may be the causative agent. Measurement of serum levels of oncogene proteins by Western blot analysis has also been suggested as indicative of increased risk for development of disease.[116]

MARKERS OF SUSCEPTIBILITY

At all stages between initial exposure and cancer development, individual susceptibility factors may influence response. These factors can be environmental or genetic. Environmental factors can include nutritional status (including fat, fiber, alcohol, vitamins, and minerals) or exposure to other environmental agents. Genetic factors (Table 5) include differences in ability to metabolize carcinogens to their active form, inactivate reactive metabolites, and repair DNA after damage has occurred. Since most carcinogens require metabolic activation before binding to DNA, individuals with elevated metabolic capacity may be at increased risk for cancer development. One example is the relationship of cytochrome P450 levels to lung cancer risk. CYP2D6 metabolizes debrisoquine, a β-adrenergic receptor blocking agent used to treat hypertension. A urine test can determine metabolic capacity based upon excretion of debrisoquine and its 4-hydroxy metabolite. "Extensive metabolizers" have 10 to 200 times higher rates of metabolism than "poor metabolizers", who constitute 5 to 10% of the Caucasian population in the U.S. While poor metabolizers are at increased risk of adverse drug reactions, extensive metabolizers are at four- to sixfold increased risk of lung cancer.[127,128] The recent demonstration of the ability of CYP2D6 to metabolize the tobacco-specific nitrosamine, NNK, suggests a mechanism for the association of extensive metabolizers of debrisoquine with elevated lung cancer risk.[129] Polymerase chain reaction (PCR) techniques have been developed for genotyping individuals and should facilitate studies of this risk factor.[130]

A second metabolizer phenotype related to increased cancer risk is ability to N-acetylate aromatic amines. First identified in isoniazid-treated tuberculous patients, "slow" acetylators more slowly inactivated the drug than "fast" acetylators. This bimodality is based on a homozygous recessive mutation carried by approximately half the population. Aromatic amines are activated by N-hydroxylation, primarily in the liver. N-acetylation is a competing detoxification reaction catalyzed by N-acetyltransferase. This enzyme is noninducable and under autosomal dominant genetic control. "Slow" acetylators are homozygous for the slow acetylator gene, and "fast" acetylators are either heterozygous or homozygous for the fast gene.

Acetylation phenotype can be measured by administration of several different compounds, including sulfamethazine (an antibiotic) and caffeine, given as a measured dose of coffee.[131] The molar ratio of acetylated and nonacetylated metabolites in urine or blood are then determined.[132] The identification of the primary mutations in two alleles of the gene for the isozyme NAT2, which account for 90% of slow acetylator phenotypes in European populations, has allowed the development of a simple DNA amplification assay to genotype individuals.[131]

"Slow" acetylators are at increased risk of bladder cancer, especially those occupationally exposed.[133-136] In a study of blond and black tobacco smokers, "slow" acetylators had higher levels of 4-ABP-Hb adducts for the same type and quantity of cigarettes smoked than "fast" acetylators.[137,138] The Barsch study also measured the rate of N-hydroxylation of caffeine; those with a fast N-oxidation and slow N-acetylation phenotype had the highest hemoglobin adduct levels. Thus, the determination of both phenotypes may provide a better prediction of risk.

In contrast to the protective effect of the fast acetylator phenotype in bladder cancer, increased risk has been observed in two studies of colon cancer with odds

Table 5
Genetic Factors Influencing Susceptibility

Gene[a]	Population	Ref.
CYP 2D6	Lung cancer cases and controls	127,128,151
CYP 1A1	Lung cancer cases and controls	152
CYP 2E1	Lung cancer cases and controls	153
GSTM 1	Lung, stomach, colon cancer	143-145
NAT	Smokers colon	133,137 136,140,142
DNA repair genes		Reviewed in 147
Tumor suppressor genes, Rb, p53	Retinoblastoma, Li-Fraumeni syndrome	Reviewed in 149

[a] CYP, cytochrome P450; GST, glutathione-s-transferase; NAT, N-acetyltransferase; Rb, retinoblastoma.

ratios of 2.5 to 3.8.[139,140] While it is assumed that N-acetylation is a detoxification step, a potential activation role exists involving the formation of N-acetoxy arylamines, either by O-acetylation of N-OH arylamines or by N,O-acetyltransfer of arylhydroxamic acids.[141] However, a lack of association of acetylator phenotype with colon cancer has also been reported.[142] Studies of an association with lung, breast, and lymphoid cancers have also been negative.[142]

Genetic susceptibility may also be related to inability to inactivate metabolites of carcinogens. Conjugation of metabolites with glucuronide, glutathione, or sulfate to produce hydrophilic products for excretion is a pathway of inactivation. Glutathione S-transferases are a family of proteins of which the μ form is present in only about 50% of individuals. Enzyme activity has been measured in lymphocytes using *trans*-stilbene oxide as a substrate.[143] Studies in lung cancer patients and controls matched for age and smoking history demonstrated that controls had an increased likelihood of having activity (58%) compared to patients (36%).[144] A threefold greater risk for developing adenocarcinoma in individuals lacking in μ activity was also recently reported.[145] PCR techniques are available for genotyping this deletion, allowing large numbers of samples to be analyzed.[146]

Genetic susceptibitily to skin cancer after exposure to UV radiation has been known for many years and provided some of the first evidence linking DNA damage to carcinogenesis. Patients lack specific repair capacity inherited in an autosomal recessive mode. They are at elevated risk for the development of skin cancer in exposed areas.[147]

Genetic predisposition to cancer induction is also related to inherited mutations in tumor suppressor genes which regulate cell growth and terminal differentiation.[148] They are inherited in a recessive form requiring the loss of both copies for the phenotype to be expressed. Retinoblastoma and Li-Fraumeni syndrome are two examples of diseases involving mutations of tumor suppressor genes. In the familial form of retinoblastoma, patients inherit one defective allele of the Rb gene and the other is lost through later somatic mutation. In the spontaneous form, both copies of the gene are lost through gene deletions or mutations. In Li-Fraumeni syndrome patients with inherited mutations in p53, a phosphoprotein controlling cell proliferation, are at up to 1000-fold increased risk for cancers at multiple sites.[149]

APPLICATIONS

A major application of biomarkers may be in prevention studies. The use of intermediate markers such as adducts, SCEs, or gene mutations provide a method for assessing the efficacy of the intervention at early time points. For example, dietary intervention with micronutrient supplements could be monitored by determination of DNA adducts, SCE, etc. A recent study has demonstrated differences in levels of the oxidized base 5-hydroxymethyuracil in white blood cells of women on nonintervention and low-fat diets.[37] Women on the low-fat diet had about threefold lower levels of DNA damage than those on the nonintervention diet. This type of approach could be utilized to determine the efficacy of various dietary intervention trials.

Multiple genetic factors can influence DNA and protein adducts as well as biological response in the form of mutations or SCE. Thus, future studies should investigate as many of these factors as possible. For example, both activation and detoxifying phenotypes could be studied in the same individual and correlated with adducts or biological response. The development of methods to determine individual DNA repair capacity is also an important goal of future research and should provide further insight into individual risk for cancer development.

Finally, the ultimate goal of any biomarker is to predict disease outcome. This is not a simple task and requires the validation of biomarkers and application to transitional epidemiologic studies which try to bridge the gap between population-based epidemiology and laboratory experiments.[150] These transitional studies could address inter- and intraindividual variability or feasibility in the field such as marker stability in stored samples. They are essential before application of biomarkers to the prospective studies, which will be required for determination of their value in prediction of disease outcome.

ACKNOWLEDGMENTS

Research by the authors was supported by NIH ES05249, ES05116, CA21111, an award from the Lucille P. Markey Charitable Trust, and the American Cancer Society.

REFERENCES

1. **Zhu, J., Zhang, L., Hu, X., Xiao, Y., Chen, J., Xu, Y., Chu, J. F., and Chu, F. S.,** Correlation of dietary aflatoxin B1 levels with excretion of aflatoxin M1 in human urine, *Cancer Res.,* 47, 1848, 1987.
2. **Hajimiragha, H., Ewers, U., Brockhaus, A., and Boettger, A.,** Levels of benzene and other volatile aromatic compounds in the blood of non-smokers and smokers, *Int. Arch. Occup. Environ. Health,* 61, 513, 1989.
3. **Eadsforth, C. V., Coveney, P. C., and Sjoe, W. H. A.,** An improved analytical method, based on HPLC with electrochemical detection, for monitoring exposure to 3-chloro-4-fluoroaniline, *J. Anal. Toxicol.,* 12, 330, 1988.
4. **Manabe, S., Yanagisawa, H., Ishikawa, S., Kitagawa, Y., Kanai, Y., and Wada, O.,** Accumulation of 2-amino-6-methyldipyridol[1,2-a:3′,2′-d]i midazole and 2-aminodipyrido[1,2-a:3′,2′-d]imidazole, carcinogenic glutamic acid pyrolysis products, in plasma of patients with uremia, *Cancer Res.,* 47, 6150, 1987.
5. **Jongeneelen, F. J., Anzion, R. B. M., Theuws, J. L. G., and Bos, R. P.,** Urinary 1-hydroxypyrene levels in workers handling petroleum coke, *J. Toxicol. Environ. Health,* 26, 133, 1989.
6. **Tolos, W. P., Shaw, P. B., Lowry, L. K., MacKenzie, B. A., Deng, J.-F., and Markel, H. L.,** 1-Pyrenol: a biomarker for occupational exposure to polycyclic aromatic hydrocarbons, *Appl. Occup. Environ. Hyg.,* 5, 303, 1990.
7. **Jongeneelen, F. J., Bos, R. P., Anzion, R. B. M., Theuws, J. L. G., and Henderson, P. T.,** Biological monitoring of polycyclic aromatic hydrocarbons metabolites in urine, *J. Scand. Wk. Environ. Health,* 12, 137,1986.
8. **Murray, S., and Gooderham, N. J.,** Detection and measurement of MeIQx in human urine after digestion of a cooked meat meal, *Carcinogenesis,* 10, 763, 1989.
9. **Everson, R. B.,** Detection of occupational and environmental exposures by bacterial mutagenesis assays of human body fluids, *J. Occup. Med.,* 28, 647, 1986.
10. **Vainio, H., Sorsa, M., and Falck, K.,** Bacterial urinary assay in monitoring exposure to mutagens and carcinogens, in *IARC Scientific Publication #59,* Berlin, A., Draper, M., Hemminki, K., and Vainio, H., Eds., 1984.
11. **Lu, S.-H., Ohshima, H., Fu, H.-M., Tian, Y., Li, F.-M., Blettner, M., Wahrendorf, J., and Bartsch, H.,** Urinary excretion of N-nitrosamino acids and nitrate by inhabitants of high- and low-risk areas for esophageal cancer in Northern China: endogenous formation of nitrosoproline and its inhibition by vitamin C, *Cancer Res.,* 46, 1485, 1986.
12. **Garland, W. A., Kuenzing, W., Rubio, F., Kornychuk, H., Norkus, E. P., and Conney, A. H.,** Urinary excretion of nitrosodimethylamine and nitrosoproline in humans: interindividual differences and the effect of administered ascorbic acid and tocopherol, *Cancer Res.,* 46, 5392, 1986.
13. **Hoffmann, D., and Hecht, S. S.,** Nicotine-derived N-nitroamines and tobacco-related cancer: current status and future direction, *Cancer Res.,* 45, 935, 1985.
14. **Manabe, S., and Wada, O.,** Analysis of human plasma as an exposure level monitor for carcinogenic tryptophan pyrolysis products, *Mutation Res.,* 209, 33, 1988.
15. **Manabe, S., and Wada, O.,** Identification of carcinogenic tryptophan pyrolysis products in human bile by high-performance liquid chromatography, *Environ. Mol. Mutation,* 15, 229, 1990.

16. **Holz, O., Krause, T., Scherer, G., Schmidt-Preub, U., and Rudiger, H. W.,** ^{32}P-postlabeling analysis of DNA adducts in monocytes of smokers and passive smokers, *Int. Arch. Occup. Environ. Health,* 62, 299, 1990.
17. **Phillips, D. H., Hewer, A., Martin, C. N., Garner, R. C., and King, M. M.,** Correlation of DNA adduct levels in human lung with cigarette smoking, *Nature,* 336, 790, 1988.
18. **Randerath, E., Miller, R. H., Mittal, D., Avitts, T. A., Dunsford, H. A., and Randerath, K.,** Covalent DNA damage in tissues of cigarette smokers as determined by ^{32}P-postlabeling assay, *J. Natl. Cancer Inst.,* 81, 41, 1989.
19. **Cuzick, J., Routledge, M. N., Jenkins, D., and Garner, R. C.,** DNA adducts in different tissues of smokers and non-smokers, *Int. J. Cancer,* 45, 673, 1990.
20. **Santella, R. M.,** DNA adducts in humans as biomarkers of exposure to environmental and occupational carcinogens, *Environ. Carc. Rev.,* C9, 57, 1991.
21. **Wilson, V. L., Weston, A., Manchester, D. K., Trivers, G. E., Roberts, D. W., Kadlubar, F. F., Wild, C. P., Montesano, R., Willey, J. C., Mann, D. L., and Harris, C. C.,** Alkyl and aryl carcinogen adducts detected in human peripheral lung, *Carcinogensis,* 10, 2149, 1989.
22. **McDiarmid, M. A., Iype, P. T., Kolodner, K., Jacobson-Kram, D., and Strickland, P. T.,** Evidence for acrolein-modified DNA in peripheral blood leukocytes of cancer patients treated with cyclophosphamide, *Mutation Res.,* 248, 93, 1991.
23. **Groopman, J. D., Donahue, J. P. R., Zhu, J., Chen, J., and Wogan, G. N.,** Aflatoxin metabolism in humans: detection of metabolites and nucleic acid adducts in urine by affinity chromatography, *Proc. Natl. Acad. Sci. U.S.A.,* 82, 6492, 1985.
24. **Autrup, H., Bradley, K. A., Shumsuddin, A. K. M., Wakhisi, J., and Wasunna, Q.,** Detection of putative adduct with fluorecence characteristics identical to 2,3-dehydro-2-(7′-guanyl)-3-hydroxyaflatoxin B1 in human urine collected in Murang's District, Kenya, *Carcinogenesis,* 4, 1193, 1983.
25. **Hsieh, L. L., Hsu, S. W., Chen, D. S., and Santella, R. M.,** Immunological detection of aflatoxin B1-DNA adducts formed *in vivo, Cancer Res.,* 48, 6328, 1988.
26. **Zhang, Y.-J., Chen, C.-J., Lee, C.-S., Haghighi, B., Yang, G.-Y., Wang, L.-W., Feitelson, M., and Santella, R.,** Aflatoxin B1-DNA adducts and hepatitis B virus antigens in hepatocellular carcinoma and non-tumorous liver tissue, *Carcinogenesis,* 12, 2247, 1991.
27. **Reed, E., Sauerhoff, S., and Poirier, M. C.,** Quantitation of platinum-DNA binding after therapeutic levels of drug exposure-a novel use of graphite furnace spectrometry, *Atomic Spectrosc.,* 9, 93, 1988.
28. **Reed, E., Yuspa, S. H., Zwelling, L. A., Ozols, R. F., and Poirier, M. C.,** Quantiation of *cis*-diamminedichloroplatinum II (cisplatin)-DNA-intrastrand adducts in testicular and ovarian cancer patients receiving cisplatin chemotherapy, *J. Clin. Invest.,* 77, 545, 1986.
29. **Fichtinger-Schepman, A. M. J., Van Oosterom, A. T., Lohman, P. H. M., and Berends, F.,** *cis*-Diamminedichloroplatinum(II)-induced DNA adducts in peripheral leukocytes from seven cancer patients: quantitative immunochemical detection of adduct induction and removal after a single dose of *cis*-diamminedichloroplatinum(II), *Cancer Res.,* 47, 3000, 1987.
30. **Fichtinger-Schepman, A. M. J., van der Velde-Visser, S. D., van Dijk-Knijnenburg, H. C. M., van Oosterom, A. T., Baan, R. A., and Berends, F.,** Kinetics of the formation and removal of cisplatin-DNA adducts in blood cells and tumor tissue of cancer patients receiving chemotherapy: comparison with *in vitro* adduct formation, *Cancer Res.,* 50, 7887, 1990.

31. **Parker, R. J., Gill, I., Tarone, R., Vionnet, J. A., Grunberg, S., Muggia, F. M., and Reed, E.,** Platinum-DNA damage in leukocyte DNA of patients receiving carboplatin and cisplatin chemotherapy, measured by atomic absorptin spectrometry, *Carcinogenesis,* 12, 1253, 1991.
32. **Huh, N. H., Satoh, M. S., Shiga, J., Rajewsky, M. F., and Kuroki, T.,** Immunoanalytical detection of O^4-ethylthymine in liver DNA of individuals with or without malignant tumors, *Cancer Res.,* 49, 93, 1989.
33. **Prevost, V., Shuker, D. E. G., Bartsch, H., Pastorelli, R., Stillwell, W. G., Trudel, L. J., and Tannenbaum, S. R.,** The determination of urinary 3-methyladenine by immunoaffinity chromatography-monoclonal antibody-based ELISA: use in human biomonitoring studies, *Carcinogenesis,* 11, 1747, 1990.
34. **Umbenhauer, D., Wild, C. P., Montesano, R., Saffhill, R., Boyle, J. M., Huh, N., Kirstein, U., Thomale, J., Rajewsky, M. F., and Lu, S. H.,** O^6-methyldeoxyguanosine in oesophageal DNA among individuals at high risk of oesophageal cancer, *Int. J. Cancer,* 36, 661, 1985.
35. **Wild, C. P., Umbenhauer, D., Chapot, B., and Montesano, R.,** Monitoring of individual human exposure to aflatoxins (AF) and N-nitrosamines (NNO) by immunoassays, *J. Cell. Biochem.,* 30, 171, 1986.
36. **Hall, N., Badawi, A. F., O'Connor, P. J., and Saffhill, R.,** The detection of alkylation damage in the DNA of human gastro-intestinal tissues, *Br. J. Cancer,* 59, 1991.
37. **Souliotis, V. L., Kaila, S., Boussiotis, V. A., Pangalis, G. A., and Kyrtopoulos, S. A.,** Accumulation of O^6-methylguanine in human blood leukocyte DNA during exposure to procarbazine and its relationships with dose and repair, *Cancer Res.,* 50, 2759, 1990.
38. **Foiles, P. G., Miglietta, L. M., Akerkar, S. A., Everson, R. B., and Hecht, S. S.,** Detection of O^6-methyldeoxyguanosine in human placental DNA, *Cancer Res.,* 48, 4184, 1988.
39. **Santella, R. M., Yang, X. Y., DeLeo, V. A., and Gasparro, F. P.,** Detection and quantification of 8-methoxypsoralen-DNA adducts, in *Methods for Detecting DNA Damaging Agents in Humans: Applications in Cancer Epidemiology and Prevention,* Bartsch, H., Hemminki, K., and O'Neil, I. K., Eds., IARC, Lyon, 1988.
40. **Djuric, Z., Heilbrun, L. K., Reading, B. A., Boomer, A., Valeriote, F. A., and Martino, S.,** Effects of a low-fat diet on levels of oxidative damage to DNA in human peripheral nucleated blood cells, *J. Natl. Cancer Inst.,* 83, 766, 1991.
41. **Santella, R. M.,** Application of new techniques for the detection of carcinogen adducts to human population monitoring, *Mutation Res.,* 205, 271, 1988.
42. **Cathcart, R., Schweirs, E., Saul, R. L., and Ames, B. N.,** Thymine glycol and thymidine glycol in human and rat urine: a possible assay for oxidative DNA damage, *Proc. Natl. Acad. Sci. U.S.A.,* 81, 5633, 1984.
43. **Bryant, M. S., Skipper, P. L., and Tannenbaum, S. R.,** Hemoglobin adducts of 4-aminobiphenyl in smokers and nonsmokers, *Cancer Res.,* 47, 602, 1987.
44. **Bryant, M. S., Vineis, P., Skipper, P. L., and Tannebaum, S. R.,** Hemoglobin adducts of aromatic amines: associations with smoking status and type of tobacco, *Proc. Natl. Acad. Sci. U.S.A.,* 85, 9788, 1988.
45. **Maclure, M., Bryant, M. S., Skipper, P. L., and Tannenbaum, S. R.,** Decline of the hemoglobin adduct of 4-aminobiphenyl during withdrawal from smoking, *Cancer Res.,* 50, 181, 1990.
46. **Wild, C. P., Jiang, Y. Z., Sabbioni, G., Chapot, B., and Montesano, R.,** Evaluation of methods for quantitation of aflatoxin-albumim and their application to human exposure assessment, *Cancer Res.,* 50, 245, 1990.

47. **Gan, L. S., Skipper, P. L., Peng, X., Groopman, J. D., Chen, J. S., Wogan, G. N., and Tannebaum, S. R.,** Serum albumin adducts in the molecular epidemiology of aflatoxin carcinogenesis: correlation with aflatoxin B1 intake and urinary excretion of aflatoxin M1, *Carcinogensis,* 9, 1323, 1988.
48. **Osterman-Golkar, S., Bailey, E., Farmer, P. B., Gorf, S. M., and Lamb, J. H.,** Monitoring exposure to propylene oxide through the determination of hemoglobin alkylation, *Environ. Health,* 10, 99, 1984.
49. **Day, B. W., Skipper, P. L., Wishnok, J. S., Coghlin, J., Hammond, S. K., Gann, P., and Tannenbaum, S. R.,** Identificaiton of an *in vivo* chrysene diol epoxide adduct in human hemoglobin, *Chem. Res. Toxicol.,* 3, 340, 1990.
50. **Mustonen, R., Hemminki, K., Alhonen, A., Hietanen, P., and Kiilunen, M.,** Determination of of *cis*-platin in blood compartments of cancer patients, in *Methods for Detecting DNA Damaging Agents in Humans: Applications in Cancer Epidemiology and Prevention,* Bartsch, H., Hemminiki, K., and O'Neill, I. K., Eds., IARC, Lyon, 1988.
51. **Calleman, C. J., Ehrenberg, L., Jansson, B., Osterman-Golker, S., Segerback, D., Svensson, K., and Wachtneister, C. A.,** Monitoring and risk assessment by means of alkyl groups in hemoglobin in persons occupationally exposed to ethylene oxide, *J. Environ. Pathol. Toxicol.,* 2, 427, 1978.
52. **Van Sittert, N. J., Jong, G. D., Garner, R. C., Davies, R., Dean, B. J., Wren, L. J., and Wright, A. S.,** Cytogenicity, immunological, and haematological effects in workers in an ethylene oxide manufacturing plant, *Br. J. Ind. Med.,* 42, 19, 1985.
53. **Farmer, P. B., Bailey, E., Gorf, S. M., Tornqvist, M., Osterman-Golkar, S., Kautiainen, and Lewis-Enright, D. P.,** Monitoring human exposure to ethylene oxide by the determination of haemoglobin adducts using gas chromatography-mass spectrometry, *Carcinogenesis,* 7, 1, 1986.
54. **Tornqvist, M., Osterman-Golkar, S., Kautiainen, S., Jensen, S., Farmer, P. B., and Ehrenberg, L.,** Tissue doses of ethylene oxide in cigarette smokers determined from adduct levels in hemoglobin, *Carcinogenesis,* 7, 1519, 1986.
55. **Carmella, S. G., Kagan, S. S., Kagan, M., Foiles, P. G., Palladino, G., Quart, A. M., Quart, E., and Hecht, S. S.,** Mass spectrometric analysis of tobacco-specific nitrosamine hemoglobin adducts in snuff dippers, smokers, and nonsmokers, *Cancer Res.,* 50, 5438, 1990.
56. **Sherson, D., Sabro, P., Sigsgaard, T., Johansen, F., and Autrup, H.,** Biological monitoring of foundry workers exposed to polycyclic aromatic hydrocarbons, *Br. J. Ind. Med.,* 47, 448, 1990.
57. **Weston, A., Rowe, M., Manchester, D. K., Farmer, P. B., Mann, D. L., and Harris, C. C.,** Fluorescence and mass spectral evidence for the formation of benzo(a)pyrene anti-diol-epoxide-DNA and -hemoglobin adducts in humans, *Carcinogenesis,* 10, 251, 1989.
58. **Lee, B. M., Baoyun, Y., Herbert, R., Hemminki, K., Perera, F. P., and Santella, R. M.,** Immunologic measurement of polycyclic aromatic hydrocarbon-albumin adducts in foundry workers and roofers, *Scand. J. Work Environ. Health,* 17, 190, 1991.
59. **Santella, R. M., Gasparro, F. P., and Hsieh, L. L.,** Quantitation of carcinogen-DNA adducts with monoclonal antibodies, *Prog. Exp. Tumor Res.,* 31, 63, 1987.
60. **Weston, A., Trivers, G., Vahakangas, K., Newman, M., and Rowe, M.,** Detection of carcinogen-DNA adducts in human cells and antibodies to these adducts in human sera, *Prog. Exp. Tumor Res.,* 31, 76, 1987.

61. **Perera, F. P., Poirier, M. C., Yuspa, S. H., Nakayama, J., Jaretzki, A., Curnen, M. M., Knowles, D. M., and Weinstein, I. B.,** A pilot project in molecular cancer epidemiology: determination of benzo[a]pyrene-DNA adducts in animal and human tissues by immunoassays, *Carcinogenesis,* 3, 1405, 1982.
62. **vanSchooten, F. J., Hillebrand, M. J. X., vanLeeuwen, F. E., Lutgerink, J. T., vanZandwijk, N., Jansen, H. M., and Kriek, E.,** Polycyclic aromatic hydrocarbon-DNA adducts in lung tissue from lung cancer patients, *Carcinogenesis,* 11, 1677, 1990.
63. **Everson, R. B., Randerath, E., Santella, R. M., Cefalo, R. C., Avitts, T. A., and Randerath, K.,** Detection of smoking-related covalent DNA adducts in human placenta, *Science,* 231, 54, 1986.
64. **Perera, F. P., Santella, R. M., Brenner, D., Poirier, M. C., Munshi, A. A., Fischman, H. K., and VanRyzin, J.,** DNA adducts, protein adducts and SCE in cigarette smokers and nonsmokers, *J. Natl. Cancer Inst.,* 79, 449, 1987.
65. **vanSchooten, F., vanLeeuwen, F. E., Hillebrand, M. J. X., deRijke, M. E., Hart, A. A. M., vanVeen, H. G., Oosterink, S., and Kriek, E.,** Determination of benzo(a)pyrene diol epoxide-DNA adducts in white blood cell DNA from coke-oven workers: the impact of smoking, *J. Natl. Cancer Inst.,* 82, 927, 1990.
66. **Rothman, N., Poirier, M. C., Baser, M. E., Hansen, J. A., Gentile, C., Bowman, E. D., and Strickland, P. T.,** Formation of polycyclic aromatic hydrocarbon-DNA adducts in peripheral white blood cells during consumption of charcoal-broiled beef, *Carcinogenesis,* 11, 1241, 1990.
67. **Perera, F. P., Hemminki, K., Young, T. L., Santella, R. M., Brenner, D., and Kelly, G.,** Detection of polycyclic aromatic hydrocarbon-DNA adducts in white blood cells of foundry workers, *Cancer Res.,* 48, 2288, 1988.
68. **Shamsuddin, A. K. M., Sinopoli, N. T., Hemminki, K., Boesch, R. B., and Harris, C. C.,** Detection of benzo[a]pyrene: DNA adducts in human white blood cells, *Cancer Res.,* 45, 66, 1985.
69. **Harris, C. C., Vahakangas, K., Newman, J. M., Trivers, G. E., Shamsuddin, A., Sinopoli, N., Mann, D. L., and Wright, W. E.,** Detection of benzo[a]pyrene diol epoxide-DNA adducts in peripheral blood lymphocytes and antibodies to the adducts in serum from coke oven workers, *Proc. Natl. Acad. Sci. U.S.A.,* 82, 6672, 1985.
70. **Haugen, A., Becher, G., Benestad, C., Vahakangas, K., Trivers, G. E., Newman, M. J., and Harris, C. C.,** Determination of polycyclic aromatic hydrocarbons in the urine, benzo[a]pyrene diol epoxide-DNA adducts in lymphocyte DNA, and antibodies to the adducts in sera from coke oven workers exposed to measured amounts of polycyclic aromatic hydrocarbons, *Cancer Res.,* 46, 4178, 1986.
71. **Hemminki, K., Grzybowska, E., Chorazy, M., Twardowska-Saucha, K., Sroczynski, J. W., Putnam, K. L., Randerath, K., Phillips, D. H., Hewer, A., Santella, R. M., Young, T. L., and Perera, F. P.,** DNA adducts in humans environmentally exposed to aromatic compounds in an industrial area of Polai..., *Carcinogenesis,* 11, 1229, 1990.
72. **Huitfeldt, H. S., Spangler, E. F., Baron, J., and Poirier, M. C.,** Microfluorometric determination of DNA adducts in immunofluorescent-stained liver tissue from rats fed 2-acetylaminofluorene, *Cancer Res.,* 47, 2098, 1987.
73. **Manchester, D. K., Wilson, V. L., Hsu, I.-C., Choi, J.-S., Parker, N. B., Mann, D. L., Weston, A., and Harris, C. C.,** Synchronous fluorescence spectroscopic, immunoaffinity chromatographic and ^{32}P-postlabeling analysis of human placental DNA known to contain benzo[a]pyrene diol epoxide adducts, *Carcinogenesis,* 11, 553, 1990.

74. **Zhang, Y. J., Li, Y., DeLeo, V. A., and Santella, R. M.,** Detection of DNA adducts in skin biopsies of coal tar-treated psoriasis patients: immunofluorescence and ^{32}P postlabeling, *Skin Pharm.,* 3, 171, 1990.
75. **Randerath, D., Reddy, M. V., and Gupta, R. C.,** [^{32}P]-labeling test for DNA damage, *Proc. Natl. Acad. Sci. U.S.A.,* 78, 6126, 1981.
76. **Gupta, R. C., Reddy, M. V., and Randerath, K.,** [^{32}P]-postlabeling analysis of nonradioactive aromatic carcinogen-DNA adducts, *Carcinogenesis,* 3, 1081, 1982.
77. **Gupta, R. C.,** Enhanced sensitivity of ^{32}P-postlabeling analysis of aromatic carcinogen: DNA adducts, *Cancer Res.,* 45, 5656, 1985.
78. **Reddy, M. V., and Randerath, K.,** Nuclease P1-mediated enhancement of sensitivity of [^{32}P]-postlabeling test for structurally diverse DNA adducts, *Carcinogenesis,* 7, 1543, 1986.
79. **Watson, W. P.,** Post-radiolabelling for detecting DNA damage, *Mutagenesis,* 2, 319, 1987.
80. **Phillips, D. H., Hewer, A., Malcolm, A. D. B., Ward, P., and Coleman, D. V.,** Smoking and DNA damage in cervical cells, *Lancet,* 335, 417, 1990.
81. **Phillips, D. H., Hewer, A., and Grover, P. L.,** Aromatic DNA adducts in human bone marrow and peripheral blood leukocytes, *Carcinogenesis,* 7, 2071, 1986.
82. **Savela, K., Hemminki, K., Hewer, A., Phillips, D. H., Putman, K. L., and Randerath, K.,** Interlaboratory comparison of the ^{32}P-postlabeling assay for aromatic DNA adducts in white blood cells of iron foundry workers, *Mutation Res.,* 224, 485, 1989.
83. **Savela, K., and Hemminki, K.,** DNA adducts in lymphocytes and granuloctyes of smokers and nonsmokers detected by the ^{32}P-postlabeling assay, *Carcinogenesis,* 12, 503, 1991.
84. **Schoket, B., Phillips, D. H., Hewer, A., and Vincze, I.,** ^{32}P-postlabeling detection of aromatic DNA adducts in peripheral blood lymphocytes from aluminum production plant workers, *Mutation Res.,* 260, 89, 1991.
85. **Phillips, D. H., Schoket, B., Hewer, A., Bailey, E., Kostic, S., and Vincze, I.,** Influence of cigarette smoking on the levels of DNA adducts in human bronchial epithelium and white blood cells, *Int. J. Cancer,* 46, 569, 1990.
86. **Jahnke, G. D., Thompson, C. L., Walker, M. P., Gallagher, J. E., Lucier, G. W., and DiAugustine, R. P.,** Multiple DNA adducts in lymphocytes of smokers and nonsmokers determined by ^{32}P-postlabeling analysis, *Carcinogenesis,* 11, 205, 1990.
87. **Autrup, H., Seremet, T., Wakhisi, J., and Wasunna, A.,** Aflatoxin exposure measured by urinary excretion of aflatoxin guanine adduct and hepatitis B virus infection in areas with different liver cancer incidence in Kenya, *Cancer Res.,* 47, 3430, 1987.
88. **Groopman, J. D., and Donahue, K. F.,** Aflatoxin, a human carcinogen: determination in foods and biological samples by monoclonal antibody affinity chromatography, *J. Assoc. Off. Anal. Chem.,* 71, 861, 1988.
89. **Weston, A., Manchester, D. K., Poirier, M. C., Choi, J.-S., Trivers, G. E., Mann, D. L., and Harris, C. C.,** Derivative fluorescence spectral analysis of polycyclic aromatic hydrocarbon-DNA adducts in human placenta, *Chem. Res. Toxicol.,* 2, 104, 1989.
90. **Vahakangas, K., Trivers, G., Rowe, M., and Harris, C. C.,** Benzo[a]pyrene diolepoxide-DNA adducts detected by synchronous fluorescence spectrophotometry, *Environ. Health Persp.,* 62, 101, 1985.

91. **Adams, J., David, M., and Giese, R. W.,** Pentafluorobenzylation of O^4-ethylthymidine and analogues by phase-transfer catalysis for determination by gas chromatography with electron capture detection, *Anal. Chem.,* 58, 345, 1986.
92. **Stillwell, W. G., Glogowski, J., Xu, H.-X., Wishnok, J. S., Zavala, D., Montes, G., Correa, P., and Tannenbaum, S. R.,** Urinary excretion of nitrate, N-nitrosoproline, 3-methyladenine, and 7-methylguanine in a Columbian population at high risk for stomach cancer, *Cancer Res.*, 51, 190, 1991.
93. **Fedtke, N., Boucheron, J. A., Turner, M. J., Jr., and Swenberg, J. A.,** Vinyl chloride-induced DNA adducts. I. Quantitative determination of N-2,3-ethenoguanine based on electrophore labeling, *Carcinogenesis,* 11, 1279, 1990.
94. **Shuker, D. E. G., Bailey, E., Parry, A., Lamb, J., and Farmer, P. B.,** The determination of urinary 3-methyladenine in humans as a potential monitor of exposure to methylating agents, *Carcinogenesis,* 8, 959, 1987.
95. **Skipper, P. L., and Tannenbaum, S. R.,** Protein adducts in the molecular dosimetry of chemical carcinogens, *Carcinogenesis,* 11, 507, 1990.
96. **Tornqvist, M., Mowrer, J., Jensen, S., and Ehrenberg, L.,** Monitoring of environmental cancer initiators through hemoglobin adducts by a modified Edman degration method, *Anal. Biochem.,* 154, 255, 1986.
97. **Mayer, J., Warburton, D., Jeffrey, A. M., Pero, R., Walles, S., Andrews, L., Toor, M., Latriano, L., Wazneh, L., Tang, D., Tsai, W. Y., Kuroda, M., and Perera, F. P.,** Biologic markers in ethylene oxide-exposed workers and controls, *Mutation Res.,* 248, 163, 1991.
98. **Wraith, M. J., Watson, W. P., Eadsforth, C. V., van Sittert, N. J., and Wright, A. S.,** An immunoassay for monitoring human exposure to ethylene oxide, in *Methods for Detecting DNA Damaging Agents in Humans: Applications in Cancer Epidemiology and Prevention,* Bartsch, H., Hemminki, K., and O'Neill, I. K., Eds., IARC Publications, Lyon, 1988.
99. **Coghlin, J., Gann, P. H., Hammond, S. K., Skipper, P. L., Taghizadeh, K., Paul, M., and Tannenbaum, S. R.,** 4-Aminobiphenyl hemoglobin adducts in fetuses exposed to the tobacco smoke carcinogen in utero, *J. Natl. Cancer Inst.,* 83, 274, 1991.
100. **Day, B. W., Naylor, S., Gan, L. S., Sahali, Y., Nguyen, T. T., Skipper, P. L., Wishnok, J. S., and Tannenbaum, S. R.,** Molecular dosimetry of polycyclic aromatic hydrocarbon epoxides and diol epoxides via hemoglobin adducts, *Cancer Res.,* 50, 4611, 1990.
101. **Sarto, F., Cominato, I., Pinto, A. M., Brovedani, P. G., Faccioli, C. M., Bianchi, V., and Levis, A. G.,** Cytogenetic damage in workers exposed to ethylene oxide, *Mutation Res.,* 138, 185, 1984.
102. **Galloway, S. M., Berry, P. K., Nichols, W. W., Wolman, S. R., Soper, K. A., Stolley, P. D., and Archer, P.,** Chromosome aberrations in individuals occupationally exposed to ethylene oxide, and in a large control population, *Mutation Res.,* 107, 55, 1986.
103. **Hemstreet, G. P., Schulte, P. A., Ringen, K., Stringer, W., and Altekruse, E. B.,** DNA hyperploidy as a marker for biological response to bladder carcinogen exposure, *Intl. J. Cancer,* in press.
104. **Langlois, R. G., Bigbee, W. L., Kgoizumi, S., Nakamura, N., Bean, M. A., Akiyama, M., and Jensen, R. H.,** Evidence for increased somatic cell mutations at the glycophorin A locus in atom bomb survivors, *Science,* 236, 445, 1987.

105. **Jensen, R. H., Langlois, R. G., and Bigbee, W. L.,** Determination of somatic mutations in human erythrocytes by flow cytometry, in *Genetic Toxicology of Environmental Chemicals, Part B. Genetic Effects and Applied Mutagenesis,* Ramel, C., Lambert, B., and Magnusson, J., Eds., Alan R. Liss, New York, 1986.
106. **Bigbee, W. L., Wyrobek, A. J., Langlois, R. G., Jensen, R. H., and Everson, B. J.,** The effect of chemotherapy on the *in vivo* frequency of glycophorin A 'null' variant erythrocytes, *Mutation Res.,* 240, 165, 1990.
107. **Kyoizumi, S., Nakamura, N., Hakoda, M., Awa, A. A., Bean, M. A., Jensen, R. H., and Akiyama, M.,** Detection of somatic mutations at the glycophorin A locus in erythrocytes of atomic bomb survivors using a single beam flow sorter, *Cancer Res.,* 49, 581, 1989.
108. **O'Neill, J. P., McGinniss, M. J., Berman, J. K., Sullivan, L. M., Nicklas, J. A., and Albertini, R. J.,** Refinement of at T-lympocyte cloning assay to quantify the *in vivo* thioguanine-resistant mutant frequency in humans, *Mutagenesis,* 2, 87, 1987.
109. **Messing, K., Seifert, A. M., and Bradley, W. E. C.,** *In vivo* mutant frequency of technicians professionally exposed to ionizing radiation, in *Monitoring of Occupational Genotoxicants,* Sorsa, M., and Norppa, H., Eds., Alan R. Liss, New York, 1986.
110. **McGinniss, M. J., Falta, M. T., Sullivan, L. M., and Albertini, R. J.,** *In vivo* HPRT mutant frequencies in T-cells of normal human newborns, *Mutation Res.,* 240, 117, 1990.
111. **Ostrosky-Wegman, P., Montero, R., Palao, A., Cortinas de Nava, C., Hurtado, F., and Albertini, R. J.,** 6-Thioguanine-resistant T-lymphocyte autoradiographic assay. Determination of variation frequencies in individuals suspected of radiation exposure, *Mutation Res.,* 232, 49, 1990.
112. **Hogstedt, B., Akesson, B., Axell, K., Gullberg, B., Mitelman, F., Pero, R. W., Skerving, S., and Welinder, H.,** Increased frequency of lymphocyte micronuclei in workers producing reinforced polyester resin with low exposure to styrene, *Scand. J. Work Environ. Health,* 49, 271, 1983.
113. **Stich, H. F., and Dunn, B. P.,** DNA adducts, micronuclei and leukoplakias as intermediate endpoints in interventions trails, in *Methods for Detecting DNA Damaging Agents in Humans: Applications in Cancer Epidemiology and Prevention,* Bartsch, H., Hemminki, K., and and O'Neil, I. K., Eds., IARC Publications, Lyon, 1988.
114. **Hsu, I. C., Metcalf, R. A., Sun, T., Welsh, J. A., Wang, N. J., and Harris, C. C.,** Mutational hotspot in the p53 gene in human hepatocellular carcinomas, *Nature,* 350, 427, 1991.
115. **Bressac, B., Kew, M., Wands, J., and Ozturk, M.,** Selective G to T mutations of p53 gene in hepatocellular carcinoma from southern Africa, *Nature,* 350, 429, 1991.
116. **Brandt-Rauf, P. W., and Niman, H. L.,** Serum screening for oncogene proteins in workers exposed to PCBs, *Br. J. Ind. Med.,* 45, 689, 1988.
117. **Brandt-Rauf, P. W.,** Oncogene proteins as biomarkers in the molecular epidemiology of occupational carcinogenesis, *Int. Arch. Occup. Environ. Health,* 63, 1, 1991.
118. **Walles, S. A. S., Norppa, H., Osterman-Golkar, S., and Maki-Paakkanen, J.,** Singel-strand breaks in DNA of peripheral lymphocytes of styrene-exposed workers, in *Methods for Detecting DNA Damaging Agents in Humans: Applications in Cancer Epidemiology and Prevention,* Bartsch, H. Hemminki, K., and O'Neil, I. K.,Eds., IARC, Lyon, 1988.
119. **Carrano, A. V., and Moore, D. H.,** The rationale and methodology for quantifying sister chromatid exchange in humans, in *Mutagenicity: New Horizons in Genetic Toxicology,* Heddle, J. A., Ed., Academic Press, New York, 1982.

120. **Wilcosky, T. C., and Rynard, M. R.,** Sister chromatid exchange, in *Biological Markers in Epidemiology,* Hulka, B. B. S.. Wilcosky, T. C., and Griffith, J. D., Eds., Oxford University Press, New York, 1990.
121. **Pero, R. W., Bryngelesson, T., Widergren, B., Hogstedt, B., and Welinder, H.,** A reduced capacity for unscheduled DNA synthesis in lymphocytes from individuals exposed to propylene oxide and ethylene oxide, *Mutation Res.,* 104, 193, 1982.
122. **Schwartz, G. G.,** Chromosome Aberrations, in *Biological Markers in Epidemiology,* Hulka, B. S., Wilcosky, T. C., and Griffith, J. D., Eds., Oxford University Press, New York, 1990.
123. **Vine, M. F.,** Micronuclei, in *Biological Markers in Epidemiology,* Hulka, B. X., Wilcosky, T. C., and Griffith, J. D., Eds., Oxford University Press, New York, 1990.
124. **Das, B. C.,** Factors that influence formation of sister chromatid exchanges in human blood lymphocytes, *Crit. Rev. Toxicol.,* 19, 43, 1988.
125. **Moore, D. H., II, and Carrano, A. V.,** Statistical analysis of high SCE frequency cells in human lymphocytes, in *Sister Chromatid Exchanges. Part A. The Nature of SCEs,* Tice, R. R., and Hollaender, A., Eds., Plenum, New York, London, 1984, 469.
126. **Tates, A. D., Grummt, T., Tornqvist, M., Farmer, P. B., van Dam, F. J., van Mossel, H., Schoemaker, H. M., Osterman-Golkar, S., Uebel, C. H., Tang, Y. S., Zwinderman, A. H., Natarajan, A. T., and Ehrenberg, L.,** Biological and chemical monitoring of occupational exposure to ethylene oxide, *Mutation Res.,* 250, 483, 1991.
127. **Ayesh, R., Idle, J. R., Ritchie, J. C., Crothers, M. J., and Hetzel, M. R.,** Metabolic oxidation phenotypes as markers for susceptibility to lung cancer, *Nature,* 312, 169, 1984.
128. **Caporaso, N., Hayes, R. B., Dosemeci, M., Hoover, R., Ayesh, R., Hetzel, M., and Idle, J.,** Lung cancer risk, occupational exposure, and the debrisoquine metabolic phenotype, *Cancer Res.,* 49, 3675, 1989.
129. **Crespi, C. L., Penman, B. W., Gelboin, H. V., and Gonzalez, F. J.,** A tobacco smoke-derived nitrosamine, 4-(methylnitrosamino)-1-(3pyridyl)-1-butanone, is activated by multiple human cytochrome P450s including the polymorphic human cytochrome P4502D6, *Carcinogenesis,* 12, 1197, 1991.
130. **Heim, M., and Meyer, U. A.,** Genotyping of poor metabolizers of debrisoquine by allele-specific PCR amplification, *Lancet,* 336, 529, 1990.
131. **Blum, M., Demierre, A., Grant, D. M., Heim, M., and Meyer, U. A.,** Molecular mechanism of slow acetylation of drugs and carcinogens in humans, *Proc. Natl. Acad. Sci. U.S.A.,* 88, 5237, 1991.
132. **Weber, W. W., and Hein, D. W.,** N-acetylation pharmacogenetics, *Pharmacol. Rev.,* 37, 25, 1985.
133. **Cartwright, R. A., Glashan, R. W., Rogers, J. J., and Al, E. T.,** The role of N-acetyltransferase in bladder carcinogenesis: a pharmacogenetic epidemiological approach to bladder cancer, *Lancet,* 2, 842, 1982.
134. **Evans, D. A. P., Eze, L. C., and Whibley, E. J.,** The association of the slow acetylator phenotype with bladder cancer, *J. Med. Genet.,* 20, 330, 1983.
135. **Karkaya, A. E., Cok, L., Sardas, S., Gogus, O., and Sardar, O. S.,** N-acetyltransferase phenotype of patients with bladder cancer, *Hum. Toxicol.,* 5, 333, 1986.
136. **Mommsen, S., and Aagaard, J.,** Susceptibility in urinary bladder cancer: acetyltransferase phenotypes and related risk factors, *Cancer Lett.,* 32, 199, 1986.
137. **Vineis, P., Caporaso, N., Tannenbaum, S. R., Skipper, P. L., Glogowski, J., Bartsch, H., Coda, M., and Talaska, F.,** Acetylation phenotype, carcinogen-hemoglobin adducts, and cigarette smoking, *Cancer Res.,* 50, 3002, 1990.

138. **Barsch, H., Caporaso, N., Coda, M., Kadlubar, F., Malaveille, C., Skipper, P., Talasaka, G., Tannenbaum, S. R., and Vineis, P.,** Carcinogen hemoglobin adducts, urinary mutagenicity, and metabolic phenotype in active and passive cigarette smokers, *J. Natl. Cancer Inst.,* 82, 1826, 1990.
139. **Ilett, K. F., David, B. M., Dethon, P., Castleden, W. M., and Kwa, R.,** Acetylation phenotype in colorectal carcinoma, *Cancer Res.,* 47, 1466, 1987.
140. **Wohlleb, J. C., Hunter, C. F., Blass, B., Kadlubar, F. F., Chu, D. Z. J., and Lang, N. P.** Aromatic amine acetyltransferase as a marker for colorectal cancer: environmental and demographic associations, *Int. J. Cancer,* 46, 22, 1990.
141. **Kirlin, W. G., Ogolla, F., Andrews, A. F., Trinidad, A., Ferguson, R. J., Yerokun, T., Mpezo, M., and Hein, D. W.,** Acetylator genotype-dependent expression of arylamine N-acetyltransferanse in human colon cytosol from non-cancer and colorectal cancer patients, *Cancer Res.,* 51, 549, 1991.
142. **Ladero, J. M., Gonzalez, J. F., Benitex, J., Vargas, E., Fernandex, M. J., Baki, W., and Diaz-Rubio, M.,** Acetylator polymorphism in human colorectal carcinoma, *Cancer Res.,* 51, 2098, 1991.
143. **Seidegard, J., Pero, R. W., Miller, D. G., and Beattie, E. J.,** A glutathione transferase in human leukocytes as a marker for the susceptibility to lung cancer, *Carcinogenesis,* 7, 751, 1986.
144. **Seidegard, J., Pero, R. W., Markowitz, M. M., Roush, G., Miller, D. G., and Beattie, E. J.,** Isoenzyme(s) of glutathione transferase (class Mu) as a marker for the susceptibility to lung cancer: a follow up study, *Carcinogenesis,* 11, 33, 1990.
145. **Strange, R. C., Matharoo, B., Faulder, G. C., Jones, P., Cotton, W., Elder, J. B., and Deakin, M.,** The human glutathione S-transferases: a case-control study of the incidence of the GST1 0 phenotype in patients with adenocarcinoma, *Carcinogenesis,* 12, 25, 1991.
146. **Seidegard, J., Vorachek, W. R., Pero, R. W., and Person, W. R.,** Hereditary differences in the expression of the human glutathione transferase active on *trans*-stilbene oxide are due to a gene deletion, *Proc. Natl. Acad. Sci. U.S.A.,* 85, 7293, 1988.
147. **Friedberg, E. C.,** *DNA Repair,* W. H. Freeman, New York, 1985.
148. **Marshall, C. J.,** Tumor suppressor genes, *Cell,* 64, 313, 1991.
149. **Li, F. P.,** Familial cancer syndromes and clusters, *Curr. Probl. Cancer,* 49, 75, 1990.
150. **Hulka, B. S.,** Epidemiological studies using biological markers: issues for epidemiologists, *Cancer Epi. Biomarkers Prev.,* 1, 13, 1991.
151. **Caporaso, N. E., Tucker, M. A., Hoover, R. N., Hayes, R. B., Pickle, L. W., Issaq, H. J., and Muschik, G. M.,** Lung cancer and the debrisoquine metabolic phenotype, *J. Natl. Cancer Inst.,* 82, 1264, 1990.
152. **Kawajiri, K., Nakachi, K., Imai, K., Yoshii, A., Shinoda, N., and Watanabe, J.,** Identification of genetically high risk individuals to lung cancer by DNA polymorphisms of the cytochrome P450IA1 gene, *FEBS Lett.,* 263,131, 1990.
153. **Uematsu, F., Kikuchi, H., Motomiya, M., Abe, T., Sagami, I., Ohmachi, T., Wakui, A., Kanamaru, R., and Watanabe, M.,** Association between restriction fragment length polymorphism of the human cytochrome p450IIe1 gene and susceptibility to lung cancer, *Jpn. J. Cancer Res.,* 82, 254, 1991.

9

Molecular Dosimetry of Ingested Carcinogens

Paul T. Strickland and John D. Groopman

Department of Environmental Health Sciences, Johns Hopkins School of Hygiene and Public Health, Baltimore, MD

INTRODUCTION

The major routes of human exposure to environmental carcinogens are through inhalation, ingestion, and percutaneous absorption. Food-borne carcinogens constitute the primary source of ingested carcinogens. This is consistent with epidemiological analyses indicating that a large percentage (20 to 50%) of all human cancer is due to dietary causes.[1,2] The incidence of cancers of specific organs varies considerably in different countries. For example, the incidence of colorectal and breast cancer is up to tenfold higher in Western countries than in Far Eastern and developing countries.[2,3] Substantial increases in the risk of these and other cancers are observed among populations migrating from low- to high-risk areas, suggesting that international differences in cancer incidence can be attributed primarily to environmental or lifestyle factors rather than genetic factors.[4-6] Indeed, Doll and Peto have suggested[2] that up to 20% of breast and 90% of colon cancer in the U.S. may be avoidable by altering the diet. In some countries (or regions therein), specific dietary carcinogens have been associated with high incidences of particular cancer types. For example, aflatoxin in the diet is related

0-87371-631-0/94/$0.00+$.50

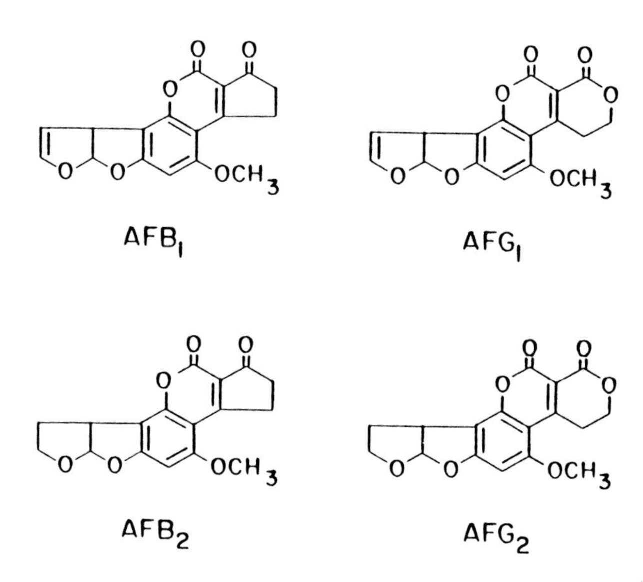

Figure 1. Structures of aflatoxins.

to high liver cancer rates in Sub-Saharan Africa and certain regions in China. This chapter will examine the use of molecular biomarkers in studies of several classes of carcinogens to which humans are exposed through the ingestion of food.

AFLATOXIN AND LIVER CANCER

Sources and Levels

Virtually all human populations are exposed to aflatoxins by the consumption of foodstuffs that have been directly contaminated by the fungal strains *Aspergillus flavus* and *A. parasiticus* during growth, harvest, or storage. Worldwide variation in exposure to aflatoxins is at least 1000-fold and the level of contamination is directly related to the technologies used to protect the grain from mold contamination and growth. These molds produce aflatoxin B_1 (AFB_1), aflatoxin B_2 (AFB_2), aflatoxin G_1 (AFG_1), and aflatoxin G_2 (AFG_2) (Figure 1); and there is about a 100-fold range in toxicologic potency among these compounds, with AFB_1 being the most potent agent. A comprehensive list of the grains and foodstuffs that have been found to be contaminated with aflatoxins can be found in Busby and Wogan,[7] and includes corn, peanuts, milo, sorghum, copra, and rice. While contamination by the molds may be universal, there are significant differences in levels within a localized area. The levels or final concentrations of aflatoxins in the grain product can vary from less than 1 μg/kg (1 ppb) to greater than 12,000 μg/kg (12 ppm). This problem is compounded by the unequal distribution of the mycotoxin within a lot of grain. For example, in many peanut lots, only one peanut in 10,000 may contain aflatoxin, but the level within a single peanut may be up to several hundred micrograms.[8] Thus, contamination of an entire shipment will occur once it has been blended, ground, and processed. It is for these reasons that the estimation of human exposure to aflatoxins from foodstuff sampling data is difficult.

Human epidemiology and experimental animal data have provided the statistical association and biological information necessary to suggest that aflatoxins are risk factors for human liver cancer.[7,9] The degree that aflatoxins contribute to this disease will be influenced by a number of human health factors, including hepatitis B virus infection, nutritional status, age, as well as the extent of aflatoxin exposure.[7] Since liver cancer causes at least 200,000 deaths worldwide per year, prevention measures must be developed to reduce the incidence of this largely fatal disease. Preventive strategies will be facilitated by the identification of individuals at high risk. It is the goal of molecular dosimetry to develop reliable biomarkers to identify people at high risk for carcinogen exposure and consequent adverse health effects. Methods have been developed to detect the major aflatoxin-DNA adduct, aflatoxin-N^7-guanine (AFB-N^7-gua), in urine and thus allow the dose-response characteristics of this biomarker to be examined in humans.[10,11] In addition, complementary approaches have been validated for the measurement of aflatoxin bound to peripheral blood albumin.[12,13] The AFB-N^7-Gua adduct has been monitored in the urine of rats whose risk for developing liver cancer has been modulated with dietary chemoprotective agents such that independent groups of animals receiving the same dose of AFB_1 were at either high or low risk for tumorigenesis.[14,15] These complementary human and experimental data have been used to evaluate the DNA and protein adducts as molecular biomarkers for determining both aflatoxin exposure and risk of genetic damage in target organs.

Metabolism, DNA, and Protein Adduct Formation

Aflatoxins, like most chemical carcinogens, require metabolic activation to their ultimate carcinogenic forms, the 8,9-epoxides, in order to produce covalent macromolecular adducts. Shimada and Guengerich[16] reported an *in vitro* study with human liver, indicating that the major cytochrome P450 involved in the bioactivation of AFB_1 to its genotoxic epoxide derivative is cytochrome CYP3A4. This is a major cytochrome P450 in human liver in terms of amount and catalytic activity and had previously been characterized as the protein that also catalyzes the oxidation of nifedipine and other dihydropyridines, quinidine, macrolide antibiotics, various steroids, and other compounds. In addition, cytochrome CYP3A4 is involved in the activation of AFB_1, benzo[a]pyrene diol-epoxide, and several other carcinogens (reviewed in Guengerich et al.[17]). Evidence was obtained using activation of AFB_1 as monitored by umuC gene expression response in *Salmonella typhimurium* TA1535/pSK1002, enzyme reconstitution, immunochemical inhibition, correlation of response with levels of CYP3A4 (also known as CYPNF), and nifedipine oxidase activity in different liver samples. Liver samples with increasing levels of P450NF also produced higher amounts of AFB-N^7-Gua formed in DNA *in vitro*. Thus, it appears that in human liver a major inducible form of cytochrome P-450 is responsible for the principal metabolism of AFB_1 to its ultimate carcinogenic form.

The major AFB_1-DNA adduct was previously identified by Essigmann et al.[18] as AFB-N^7-Gua, and its presence *in vivo* was also established.[19] The binding of AFB_1 residues to DNA *in vivo* was essentially a linear function of dose.[20] A number of other DNA components were isolated[21,22] from nucleic acid hydrolysates activated *in vivo* and *in vitro* with AFB_1. These adducts were identified as 8,9-dihydro-8-(N^5-formyl-2,5,6-triamino-4-oxopyrimidin-N^5-yl)-9-hydroxyAFB_1 (AFB-FAPyr) a formamido-pyrimidine derivative of AFB-N^7-Gua which contained an opened imidazole ring. At the present time, it appears that between 95 and 98% of the aflatoxin residues bound to DNA have been accounted for by these chemical structures.

The investigation of the interactions and biological consequences of AFB_1-DNA binding has been extensively studied. In recent years, with the heightened interest to develop molecular dosimetry methods for human populations, the structural analysis of the aflatoxin-protein adducts has been completed. Albumin is the only protein in serum which binds AFB_1 to any significant extent in monkeys and rats;[23,24] 1 to 3% of a single dose is bound to serum albumin after 24 hr in rats. Further, a constant relationship between AFB_1 bound to plasma albumin and liver DNA has been observed in Wistar rats following single (3.5 to 200 μg/kg) and multiple doses (3.5 μg/kg) of AFB_1.[24] This suggested that albumin-bound aflatoxin might be a particularly useful surrogate marker reflecting DNA damage.

Sabbioni et al.[12] have elucidated the structure of the major aflatoxin albumin adduct found *in vivo*. Their data suggested that the lysine adduct formed in serum albumin from the 8,9-epoxide proceeded through the conversion of the epoxide to a dihydrodiol and sequential oxidation to the dialdehyde and condensation with the epsilon amino group of lysine. This adduct is a Schiff base which undergoes Amadori rearrangement to an alpha-amino ketone. This protein adduct is a completely modified aflatoxin structure retaining only the coumarin and cyclopentenone rings of the parent compound. Thus, with the structural knowledge of both the major DNA and serum albumin adducts of aflatoxin known, precise experiments could now be done in animals and exposed human populations to characterize dose-response relationships.

Molecular Dosimetry in Animals

There have been numerous animal studies in many different species conducted over the years to examine the relation between AFB_1 dose and the formation of DNA and protein adducts. A comprehensive review of these investigations is beyond the scope of this chapter; however, there are relatively few studies examining the relation between chronic exposure to AFB_1 and macromolecular adduct formation.[24-27] These studies form an important basis for future human analyses and the reader is urged to examine these efforts. The following describes some recent research to develop molecular dosimetry methods for AFB-N^7-Gua excretion in urine and serum albumin adduct formation.

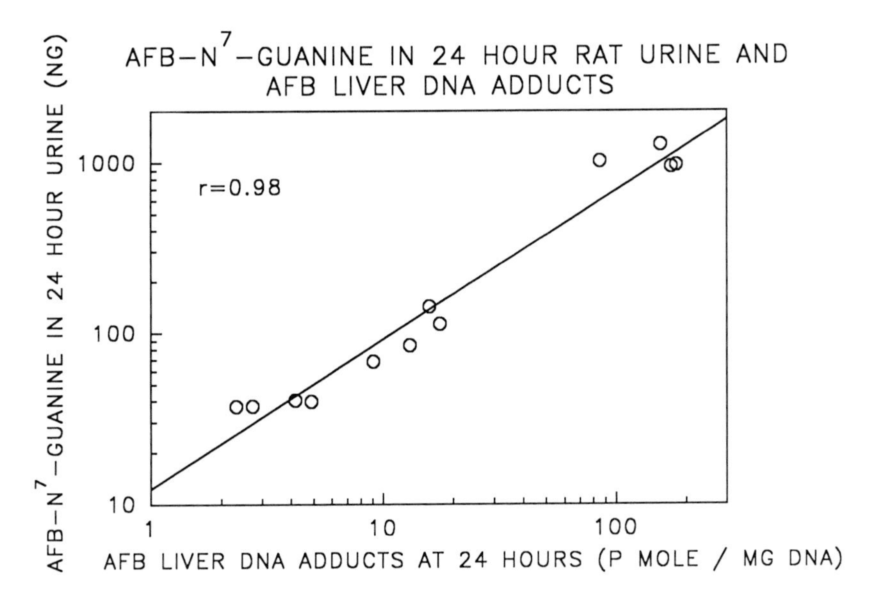

Figure 2. Aflatoxin-DNA adduct excretion in urine.

Initial studies were done in rats following a single exposure to AFB_1 to characterize excretion kinetics of specific metabolites. Urines were collected over 24 hr from 12 male F344 rats dosed orally at levels ranging from 0.030 to 1.00 mg AFB_1/kg body weight.[28] Aliquots of each urine were preparatively purified by immunoaffinity chromatography and individual metabolites were quantified by HPLC analysis. AFB-N^7-Gua, AFQ_1, AFP_1, AFM_1, and AFB_1 accounted for 7.5, 3.0, 31.5, 2.2, and 0.3% of the total aflatoxins injected on the HPLC, respectively. The relationship between AFB_1 dose and the excretion of the AFB-N^7-Gua adduct over the 24 hr following exposure was then determined with a correlation coefficient of 0.99. This analysis demonstrates an excellent linear correspondence between oral dose and excretion of a biologically relevant metabolite in urine. In contrast, other oxidative metabolites, such as AFP_1, revealed no linear excretion characteristics. Finally, at 24 hr postdosing, the residual level of aflatoxin liver DNA adducts was determined and compared to AFB-N^7-Gua excretion in urine (Figure 2). The correlation coefficient was 0.98, supporting the concept that measurement of the AFB-N^7-Gua adduct in urine reflects DNA damage in the primary target organ.

The risk for aflatoxin hepatocarcinogenesis can be modified in animals by using a number of chemoprotective agents, including phenolic antioxidants, ethoxyquin, and dithiolethiones. A particularly effective cancer chemoprotective agent for aflatoxin carcinogenesis is the substituted dithiolethione, oltipraz.[14] Male F344 rats were fed a purified diet supplemented with 0.075% w/w oltipraz for 4 weeks. In this study, rats received 10 intragastric doses of AFB_1 (25 µg/rat/day; days 8 to 12 and 15 to 19) and this 10-dose exposure to AFB_1 produced an 11% incidence of hepatocellular carcinoma at 23 months, with an additional 9%

of the rats exhibiting hyperplastic nodules in their livers. By contrast, feeding rats a diet supplemented with 0.075% oltipraz for a 4-week period surrounding the time of AFB_1 exposure afforded complete protection against AFB_1-induced hepatocellular carcinomas and hyperplastic nodules. None of these lesions were observed in the oltipraz-fed, AFB_1-treated animals. Further, no tumors were at secondary, extrahepatic sites for AFB_1 carcinogenesis such as the colon and kidney. Thus, the chemoprotection model affords an experimental system to ask basic questions about the relationships between levels of AFB-N^7-gua in urine for high- and low-risk animals.

Structure-activity studies with dithiolethione indicated that the cancer chemoprotective activity of oltipraz is exclusively embodied in the 1,2-dithiole-3-thione nucleus of the molecule. Recently completed experiments indicated that the unsubstituted congener of oltipraz, 1,2-dithiole-3-thione, is also an effective inhibitor of AFB_1-induced tumorigenesis, as determined by analyses for preneoplastic foci expressing either GGT+ or GST-π.[29] These data justified an examination of the impact of chemoprotection by 1,2-dithiole-3-thione on the molecular dosimetry of AFB-N^7-Gua in urine and to compare the modulation of this biomarker with levels of aflatoxin-DNA adduct in the liver. The effects of 1,2-dithiole-3-thione on the kinetics of hepatic aflatoxin-DNA adduct formation and removal in rats receiving 250 μg AFB_1/kg by gavage on each of days 0 to 4 and 7 to 11 are shown in Figure 3A. Maximal levels of carcinogen binding were achieved following the third dose in the control group and declined thereafter despite continued exposure to aflatoxin. This diminution of binding, particularly during the second dosing cycle, has been observed previously,[24-26] and may be a consequence of the induction of glutathione S-transferases and/or other enzymes involved in AFB_1 detoxication following chronic exposure to aflatoxin.[29] Inclusion of 0.03% 1,2-dithiole-3-thione in the diet, beginning 1 week prior to dosing with AFB_1, resulted in substantially lower levels of hepatic aflatoxin-DNA adducts throughout the exposure period. Binding was reduced by 76% over the initial 18-day period.

The levels of total aflatoxin equivalents in 24-hr urine samples collected over the 2-week exposure period were determined. There are no remarkable differences in the levels of aflatoxin metabolites in rats fed the control AIN-76A diet compared to those fed the 1,2-dithiole-3-thione-supplemented diet. Urinary aflatoxin levels rise rapidly following dosing with AFB_1 and drop equally quickly following cessation of dosing, reflecting the overall short *in vivo* half-life of this carcinogen. The lack of an effect by 1,2-dithiole-3-thione is not surprising, given that exposures to AFB_1 were identical in both dietary groups. However, a distinctly different pattern emerged when the urines were subjected to sequential monoclonal antibody immunoaffinity chromatography and HPLC. Shown in Figure 3B are the levels of AFB-N^7-Gua in serial 24-hr urine samples collected from rats undergoing a chemoprotective intervention with 0.03% 1,2-dithiole-3-thione. The highest level of AFB-N^7-Gua excretion occurred on day 2 in both groups following the third dosage of AFB_1. This outcome is identical to that observed with hepatic levels of aflatoxin-DNA adducts and with the serum albumin adduct formation

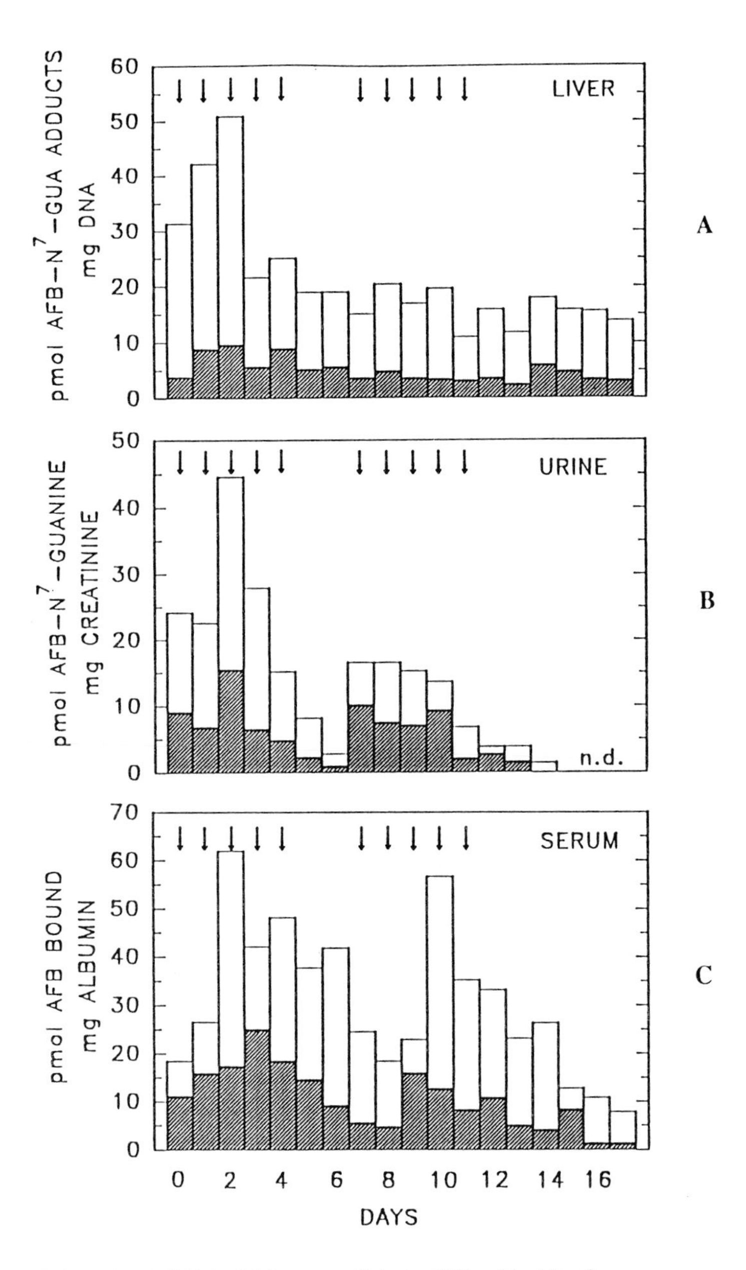

Figure 3. Effect of 1,2-dithiole-3-thione on aflatoxin-DNA adduct levels.

(Figure 3C). Over the 15-day collection period in which AFB-N^7-Gua adducts were detectable in the urine, feeding of 1,2-dithiole-3-thione produced an overall reduction of 62% in the elimination of this aflatoxin-DNA adduct excision product, mirroring the data on overall levels of hepatic aflatoxin-DNA adducts. The amount of AFB-N^7-Gua in urine represents only 1% of the total aflatoxin metabolites in urine, which suggests why the dramatic differences seen between treatment groups in AFB-N^7-Gua levels are not reflected in the levels of total urinary aflatoxin metabolites. Thus, these data indicate that the excreted DNA adduct in urine and the formation of the serum albumin adduct accurately reflects the amount of genotoxic damage at the target organ site in the liver. In addition, these data indicate that the measurement of these adducts may reflect risk for disease development in the animal.

Molecular Dosimetry in Humans

Liver cancer is the third leading cause of cancer related deaths in China, and there are regions of the country where this disease is very prevalent. The Guanxgi Autonomous Region is among the areas of highest incidence of liver cancer. The following study was done to examine biomarkers for aflatoxin exposures. The dietary intake of aflatoxins were monitored for 1 week in a study group consisting of 30 males and 12 females, ages ranging from 25 to 64 years, living in Fusui county, Guangxi Autonomous Region, P.R.C.[11] The vast majority of AFB_1 exposure was from contaminated corn. The average male intake of AFB_1 was 48.4 μg per day, giving a total mean exposure over the study period of 276.8 μg. The average female daily intake was 92.4 μg per day, resulting in a total average exposure over the 7-day period of 542.6 μg AFB_1. The maximum intake for the male and female subjects over the 1-week collection was 963.9 and 1035 μg, respectively, and the minimum exposure for male and female subjects was 56.7 and 90 μg AFB_1, respectively. Total 24-hr urine samples were collected starting on the 4th day as consecutive 12-hr fractions.

Total aflatoxin metabolites in the urine samples were measured by competitive radioimmunoassay using a monoclonal antibody that recognizes AFB_1, aflatoxin P1 (AFP_1), and aflatoxin M_1 (AFM_1) with equal affinity and cross-reacts with AFB-N^7-Gua with five- to tenfold less affinity. The relation between AFB_1 intake per day and total aflatoxin metabolite excretion per day revealed a correlation coefficient of 0.26 with a statistical significance level of 0.10. Thus, total metabolites in urine, as measured by this monoclonal antibody, does not provide data to indicate that total metabolites were an appropriate dosimeter measurement for exposure status.

These data prompted an analysis of the urine samples by combined preparative monoclonal antibody affinity chromatography/analytical HPLC to determine levels of individual aflatoxin metabolites and nearly 550 individual analytical HPLCs were performed. AFB-N^7-Gua, AFM_1, AFP_1, and AFB_1 were the aflatoxins most commonly detected and quantified in the urine samples. The linear regression analyses for the urinary levels of each of these individual aflatoxins was compared to aflatoxin intake, and the correlation coefficients for AFB-N^7-Gua, AFM1, AFP_1, and AFB_1 were 0.65, 0.55, 0.02, and –0.10, respectively. Further, it was noted that AFP_1 contributes substantially to the overall levels of aflatoxins in the urine samples as detected by competitive RIA, thereby masking the association

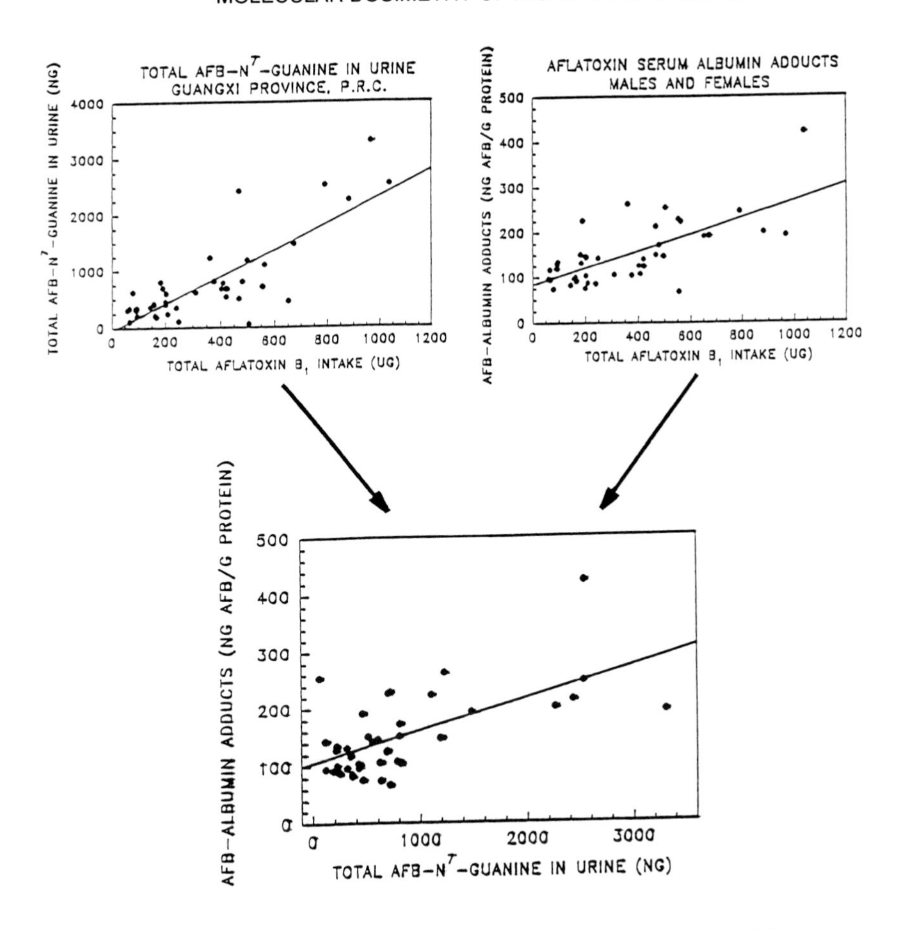

Figure 4. Correlation between urinary DNA adducts and serum albumin adducts.

between exposure and minor metabolites. However, the resolution of the total aflatoxin metabolite content in the urine permits the association of AFB-N^7-Gua and AFM$_1$ excretion as biomarkers of exposure to be revealed.

One objective of this study was to determine the number of samples required from an individual and the time frame for sample collection necessary to validate a biomarker as reflecting a biologically effective dose of AFB$_1$ in people. There was particular interest in characterizing the molecular dosimetry of AFB-N^7-Gua because of the putative relationship of this metabolite with the cancer initiation process. Figure 4 shows total AFB-N^7-Gua excretion in the urine of the male and female subjects over the complete urine collection period plotted against the total AFB$_1$ exposure in the diet for each of the subjects. This analysis smooths the day-to-day variations in both intake and excretion of AFB-N^7-gua and reveals a correlation coefficient of 0.80 and a p value of <0.0000001. This analysis clearly demonstrates that a summation of excretion and exposure status provides a stronger association between exposure and a molecular dosimetry marker than was seen in prior statistical analyses and supports the concept that quantitation of the AFB-N^7-Gua adduct in urine is a reliable biomarker for AFB_1 exposures.

In the same subjects described above, Gan et al.[30] examined serum albumin adduct formation. Serum albumin was isolated from blood by affinity chromatography on Reactive Blue 2-Sepharose and subjected to enzymatic proteolysis using Pronase. Immunoreactive products were purified by immunoaffinity chromatography and quantified by competitive radioimmunoassay. A highly significant correlation of adduct level with intake (r = 0.69, $p < 0.000001$) was observed (Figure 4). From the slope of the regression line for adduct level as a function of intake, it was calculated that 1.4 to 2.3% of ingested AFB_1 becomes covalently bound to serum albumin, a value very similar to that observed when rats are administered AFB_1. When levels of DNA adducts in urine and serum albumin adducts are compared (Figure 4), a statistically significant relationship is seen with a correlation coefficient of 0.73. Thus, both of these markers appear to be valid in human monitoring studies.

Similar studies of DNA adduct excretion and serum albumin adduct formation have been done in The Gambia. These studies reveal similar relationships between dietary exposure and adduct formation.[31,32] The correlation coefficients for the DNA adduct excretion and serum albumin adduct formation were 0.82 and 0.55, respectively. Thus, in two populations at high risk for liver cancer, living in very different geographic regions, and having different dietary sources of aflatoxins, these dosimetry markers appear to be rigorous for exposure determination.

The investigations described above revealed that both the DNA and albumin adducts of aflatoxin are useful molecular dosimetry markers. However, from a practical perspective, the measurement and quantification of the serum albumin adduct offers an approach that can be used to screen very large numbers of people. These methods have been extensively validated in experimental and human sample analysis and the technique is described in detail in Wild et al.[13] In this study, three complementary approaches to the quantitation of adducts are described. They are an ELISA performed directly on intact albumin (direct ELISA), an ELISA performed on an albumin hydrolysate (hydrolysis ELISA), and high-performance liquid chromatographic fluorescence detection of the aflatoxin-lysine adduct after albumin hydrolysis and immunoaffinity purification; the detection limits of approximately 100, 5.0, and 5.0 pg AF/mg human albumin, respectively, were determined for the three methods.

Using the hydrolysis ELISA method, an extensive geographic study was carried out on hundreds of samples.[33] Measurements of aflatoxin bound to serum albumin in children and adults from various African countries show that between 12 and 100% contain aflatoxin-albumin adducts, with levels up to 350 pg AFB_1-lysine equivalent/mg albumin. In Thailand, lower levels and prevalence of this adduct were observed, while no positive sera were detected from France or Poland. Data are presented showing that exposure to this carcinogen can occur throughout life and significant seasonal variation, consistent with high and low aflatoxin levels in foods, was observed. Taken together, the ELISA approach serves as an important screening tool for epidemiologic studies.

Interest in aflatoxin as a risk factor for human liver cancer has been heightened by recent reports that showed that one half of the human liver tumors examined from China and Southern Africa had a hotspot guanine-to-thymine transversion mutation in the tumor suppressor gene p53 at codon 249. Given previous experimental data showing that aflatoxin causes this type of mutation, the suggestion was raised that AFB_1 could be an etiological agent for these tumors.[34,35] In addition, an extensive analysis of this specific mutation in p53 found in liver tumors from diverse geographic regions shows a strong association between putative exposure to this mycotoxin and the formation of this particular mutation.[36] Also, recent work has shown that aflatoxin-epoxide can bind to this particular codon 249 of p53 in a plasmid, providing further indirect evidence for a putative role of aflatoxin exposure in p53 mutagenesis.[37] The biological importance of mutant p53 is suggested for a wide number of human tumors,[38] and the molecular dosimetry methods described may become useful in epidemiological investigations to examine the role of this gene in liver cancer.

Taken together, the development of the molecular dosimetry markers for aflatoxin exposure can be used in a variety of epidemiological investigations. These studies include cross-sectional, case-control, and cohort or prospective investigations. A recent cross-sectional study by Campbell et al.[39] revealed no association between composite aflatoxin metabolites in human urine and liver cancer disease rates. These findings are consistent with the animal and human data described previously that suggest that the composite measurement in urine is not a good marker for exposure and risk from aflatoxins.

A case-control study of potential risk factors for hepatocellular carcinoma was reported for northeast Thailand.[40,41] In this study, 65 cases from 3 hospitals, with matched controls, were examined. Infection with hepatitis-B virus was the major risk factor discerned and chronic carriers of hepatitis surface antigen had an estimated relative risk of 15.2. No increase in risk was found with recent aflatoxin intake, as estimated by consumption of possibly contaminated foods, or by measuring aflatoxin-albumin adducts in serum. Regular use of alcohol (two or more glasses of spirits per week) was associated with a nonsignificant elevation in risk (odds ratio = 3.4, 95% C.I. = 0.8–14.6), but the number of regular drinkers in the population was small. Thus, in this case-control study, the contribution of recent exposure to aflatoxin and liver cancer was not significant.

In most instances, the most rigorous test of an association between an agent and disease outcome is found in prospective epidemiological studies, where healthy people are followed up until the disease is diagnosed. Recent data analyzed from a prospective nested case control study begun in Shanghai in 1986 to examine the relation between markers for aflatoxin and hepatitis B virus and the development of liver cancer has been published.[42] Over a $3^1/_2$-year period, 18,244 urine samples were collected from healthy males between the ages of 45 and 64; 22 of these individuals subsequently developed liver cancer and their urine samples were each age-matched with 5 to 10 controls and analyzed for aflatoxin biomarkers and hepatitis B virus (HBV) surface antigen status. The data revealed a highly significant increase in the relative risk (RR = 4.9) for those liver cancer cases where

AFB-N^7-Gua was detected. There were also elevated risks for other aflatoxin urinary markers. The relative risk for people who tested positive for the HBV surface antigen was also about five, but individuals with both urinary aflatoxins and positive HBV surface antigen status had a relative risk for developing liver cancer of over 60. These results show for the first time a chemical carcinogen-specific biomarker relationship with cancer risk, and also for the first time demonstrates a synergistic interaction between two major risk factors for liver cancer, hepatitis B virus and AFB_1.

CARCINOGENS IN COOKED MEATS AND COLON CANCER

Etiology of Colon Cancer

Colon cancer is strongly associated with consumption of red meat and animal fat. This association has been observed in a number of international correlative studies and case-control studies.[43-47] Consumption of fried foods and barbecued/smoked meats has also been associated with increased risk for cancer at specific colorectal subsites in young men.[48] A prospective study of colon cancer risk among women demonstrated an association with consumption of red meat and animal fat, but not vegetable fat, independent of total energy intake.[49] The increased risk associated with animal fat was due primarily to meat intake as opposed to dairy product intake. The relative risk of colon cancer in women who consumed beef, pork, or lamb daily was 2.5 (95% confidence interval, 1.2–5.0) compared with those who consumed these meats less than once a month.[49] Other studies have shown associations between pancreatic or bladder cancer and consumption of fried or grilled meats.[50,51]

Despite these observations, the specific chemical etiology of colon and other diet-associated cancers remains unclear.[6,52] Several hypotheses have been proposed to explain the strong association of colon cancer risk with red meat and animal fat. It is known that diets high in fat increase the incidence of chemically induced colon cancers in rats and also increase the excretion of primary bile acids which are converted to secondary bile acids by bacterial metabolism.[53] Some secondary bile acids (e.g., deoxycholic and lithocholic acid) are colon tumor promotors in the rat.[53,54] Thus, it has been hypothesized that increased animal fat in the human diet leads to promotion of colon tumors by secondary bile acids. An alternative (or complementary) hypothesis is that carcinogens in cooked meats contribute to colon carcinogenesis. The cooking of meat produces at least two major classes of carcinogens, polycyclic aromatic hydrocarbons (PAHs) and heterocyclic aromatic amines (HAs), which induce genotoxic damage or cancer in the gastrointestinal tract of animals receiving these compounds orally.[55-58] Cooked meats are one of the major sources of animal fat in Western diets, and it has been suggested that cooking-induced carcinogens in meat, rather than animal fat, play a causative role in colon cancer.[59,60]

Polycyclic Aromatic Hydrocarbons in Food

PAHs are produced by incomplete combustion or pyrolysis of organic materials, and humans are exposed to these compounds from a variety of sources.[61,62] Many different PAHs have been reported in foods, including benz[a]anthracene, benzo[a]pyrene (BP), benzo[b]fluoranthene, chrysene, dibenz[a,h]anthracene, 7,12-dimethylbenz[a]anthracene, and 3-methylcholanthrene. Some of the highest concentrations of PAHs in food are found in broiled, barbecued, or smoked meats and fish (2 to 200 μg/kg).[63-65] In the U.S., diet constitutes a major source of exposure to PAHs.[66-69] Mean daily dietary intake of BP estimated in two studies[66,70,71] was 0.12 and 0.08 μg/person, ranging from 0.001 to 1.15 μg/person. Mean daily consumption of BP in other countries is comparable (0.08 to 0.25 μg/person), with maximum individual exposures an order of magnitude higher.[64,72,73]

Experimental studies demonstrate that cultured human colon tissue metabolizes BP to the proximate carcinogen 7,8-dihydro-7,8-dihydroxy-BP and that BP metabolites bind to DNA in cultured colon.[74] Oral administration of BP to rodents produces BP-DNA adducts in liver, stomach, colon, and intestine[56] and cancers of the esophagus, forestomach, intestine, lungs, and mammary gland.[65] Several biochemical pathways are involved in the metabolism of PAHs and BP in particular.[75,76] The initial step in BP metabolism (Figure 5) involves the epoxidation of an aromatic double bond by one of the cytochrome P450 monooxygenases (CYP1A1). The epoxide-BP intermediates may form phenols or glutathione conjugates or be further oxidized by epoxide hydrolase to form dihydrodiol-BP.[75] These metabolites can undergo a second oxidation step, resulting in the highly reactive 7,8-dihydrodiol-9,10-epoxide-BP (BPDE). A number of studies indicate a role for aryl hydrocarbon hydroxylase (CYP1A1) in the oxygenation of PAHs in animal models,[77] and recent experiments using human liver tissue show that nifedipine oxidase (CYP3A3/4) activates 7,8-dihydro-7,8-dihydroxy-BP to a genotoxic species.[78,79] The reactive species BPDE can form covalent adducts with DNA and protein or can hydrolyze in the presence of water to 7,8,9,10-tetrahydroxy-tetrol-BP (Figure 5).

Molecular Dosimetry of Ingested PAHs

PAH-DNA adducts have been detected in peripheral white blood cells (WBCs) from individuals occupationally exposed to high levels of airborne PAHs such as coke oven workers, foundry workers, and firefighters.[80-84] In addition, several epidemiological and experimental studies have demonstrated a consistent association between recent consumption of char-broiled (CB) foods and PAH-DNA adduct levels in peripheral WBCs.

Analysis of PAH-DNA adducts in peripheral WBCs from urban firefighters[80,81] suggested that adduct levels increased with exposure to several potential sources of PAH, including firefighting and broiled or smoked food consumption. Peripheral blood samples (40 ml) were collected from 43 urban

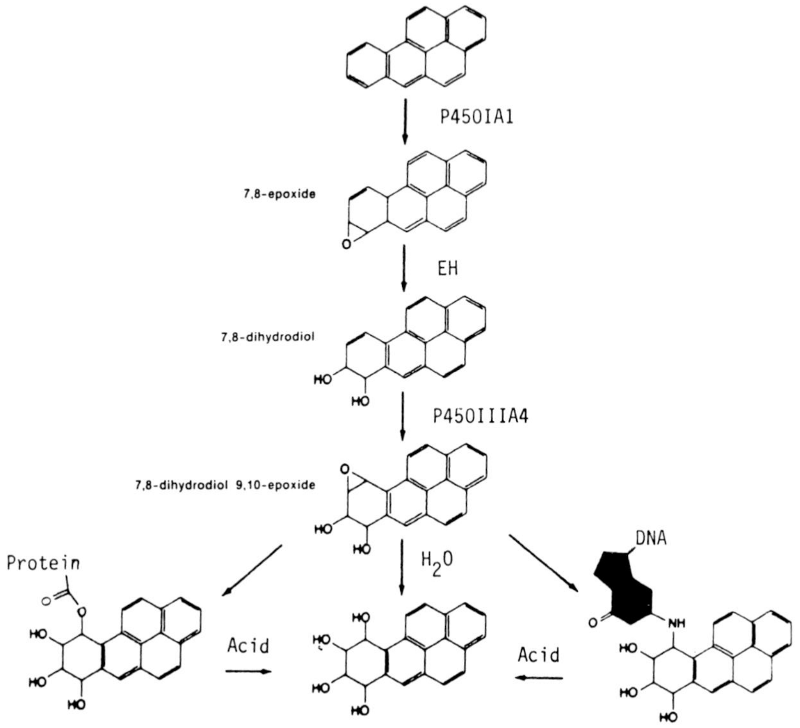

Figure 5. Metabolic activation of benzo[a]pyrene and formation of DNA and protein adducts. EH, epoxide hydrolase.

firefighters and 38 controls; DNA was extracted from the WBC fraction (buffy coat) and analyzed for PAH-DNA content by ELISA.[80,85] PAH-DNA adducts were detectable (≥0.3 fmol adduct/µg DNA) in 35% of the individuals tested. The study population was dichotomized into those with detectable adduct levels (28/81) and those with nondetectable adduct levels (53/81) for analysis. Smoking increased the unadjusted risk for detectable PAH-DNA adducts (odds ratio = 2.4; 95% confidence interval, 0.96 to 6.26).[80] Caucasian firefighters[64,81] exhibited a higher risk for the presence of detectable PAH-DNA adducts than Caucasian controls after adjustment for CB food consumption (odds ratio = 3.4; 95% confidence interval, 1.1 to 10.5; $p = 0.04$). Individual PAH-DNA adduct levels were also associated with frequency of consumption of CB foods in the previous month by linear regression ($p = 0.006$). When analysis was restricted to those with detectable adduct levels, linear regression indicated that consumption of CB foods 3 or more times per month was positively associated with increased PAH-DNA adduct levels ($p = 0.04$) after adjustment for firefighting. Mean adduct levels in the Caucasian participants (22 detectable per 66 tested) increased with exposure to one or more of these three sources of PAHs (Table 1). Thus, dietary PAHs appeared to contribute to the overall PAH-DNA adduct load in this study.

Table 1
Mean Levels of PAH-DNA Adducts

Number of exposure sources[a]	Mean adduct levels all samples[b] (n = 66)	(fmol/µg DNA) ± SD detectable samples (n = 22)
0	0.22 ± 0.16	0.50 ± 0.15
1	0.48 ± 0.53	1.13 ± 0.40
2	0.73 ± 0.11	1.75 ± 0.90
3	1.36 ± 2.48	2.97 ± 2.79

Source: Strickland et al.[86] With permission.

[a] Exposure sources are firefighting, smoking, and CB food consumption.
[b] Setting undetectable (<0.3 fmol/µg DNA) = 0.15 fmol/µg DNA. Linear regression analysis with number of exposure sources as independent variable: $p = 0.023$ (all samples); $p = 0.025$ (detectable samples).

Dietary exposure to PAH was examined in more detail in a controlled CB beef feeding experiment in which each individual served as his own control. This study was conducted in order to test the hypothesis that dietary sources of PAH can significantly contribute to PAH-DNA adduct load.[87] Four healthy Caucasian nonsmoking male volunteers, ages 33 to 39, adhered to a diet free of CB foods for 1 month. During this period, three blood samples were collected for baseline measurements. Volunteers then consumed CB ground beef daily for 7 days (approximately 280 g cooked weight per person per day). Blood samples were collected twice during this week of feeding, and three additional times in the 3 weeks after feeding. During the 3 weeks following feeding, volunteers refrained from eating any CB foods. DNA was extracted from the WBC fraction and PAH-DNA adduct content was analyzed by ELISA using a fluorescent enzyme substrate, 4-methylumbelliferyl phosphate.[85] This modification reduced the lower limit of detection to 0.04 fmole adduct/µg DNA.

Baseline levels of PAH-DNA adducts prior to CB beef consumption were between nondetectable and 0.19 fmol/µg DNA for all individuals. During the week of CB beef consumption, two individuals exhibited an increase in PAH-DNA adduct level.[87] One volunteer exhibited a sixfold increase after 1 day of CB beef and declined to baseline levels 4 days after cessation of CB beef consumption. Another volunteer exhibited a threefold increase after 4 days of CB beef consumption and declined to baseline levels 1 day after cessation of CB beef consumption. PAH-DNA adduct level in the other two volunteers did not increase at any time during the study.

These results suggest that individuals may respond differently to ingested PAHs. Although the mechanism for such a variable response is unclear, it may be due to interindividual differences in constitutive or induced physiological parameters such as digestion, absorption, metabolism, excretion, or DNA adduct repair.[88-91] The rapid decline in adducts in the two responding individuals may be due to the relatively short halftime (<24 hr) of the majority of WBCs in the peripheral blood. Only 15 to 30% of WBCs (primarily lymphocytes) remain in the peripheral blood for long periods (weeks to months). These findings indicate that dietary sources of PAH can contribute to PAH-DNA adduct load in peripheral blood cells and should be considered when using PAH-DNA adducts as a marker of exposure to PAH from occupational or environmental sources. This is particularly important when assessing exposure in individuals with no known occupational exposure, where diet may be the predominant source of PAH exposure.

In a subsequent study of combustion product exposure in wildland (forest) firefighters, a very strong association between PAH-DNA adduct levels and frequency of recent CB food consumption was observed.[92-93] Blood samples were collected from 47 California wildland firefighters early and late in an active forest fire season. A questionnaire requesting information on demographics, recent consumption of CB food, and firefighting work practices was administered. During the subsequent 8-week study period which ended late in the fire season, each participant prospectively recorded daily hours of firefighting activity. DNA extracted from WBCs was analyzed for PAH-DNA adduct content by ELISA as described above in the feeding study. The lower limit of detection was 0.04 fmol/μg DNA; 60% of samples assayed had detectable adducts. Nondetectable samples were assigned a value of half the detection limit (0.02 fmol/μg DNA).

Mean adduct level in the study population did not change between early (0.11 ± 0.02 fmol/μg DNA) and late (0.10 ± 0.01 fmol/μg DNA) fire season.[93] Furthermore, firefighting activity was not associated with PAH-DNA adduct level in either early or late season or in a combined analysis.[92] Thus, potential exposure to PAHs during wildland firefighting did not appear to contribute significantly to the PAH-DNA adduct load in this population. In contrast, the number of times that individuals reported eating CB food in the previous 2 weeks was positively associated with PAH-DNA adduct levels, both in the early and late fire season and in a combined analysis.[92,93]

Adduct level was significantly higher in individuals who had consumed CB food in the previous week[93] and there was an inverse relationship between individual adduct levels and days since CB food was last consumed (Rothman et al., in preparation). These results are consistent with the cross-sectional study of urban firefighters and the controlled feeding study in human volunteers described above. In particular, the importance of recent consumption of CB food observed in this study is consistent with the rapid kinetics of adduct loss observed in the feeding study.

Although PAHs are ubiquitous in the diet, CB food has been shown to contain relatively high PAH levels compared to other commonly consumed foods.[64] The wildland firefighter data suggest that CB food represented one of the major sources of dietary PAH intake in the population, half of whom had consumed CB food at least once in the previous 6 days. Those individuals who consumed CB food more than twice in the previous 2 weeks had a fourfold increased risk of elevated (>0.2 fmol/μg DNA) adduct levels compared to individuals consuming CB food 2 or less times ($p = 0.009$).[93]

Taken together, these studies indicate that dietary PAHs can affect PAH-DNA adduct levels in peripheral WBCs. The data also suggest that interindividual differences exist in response to ingested PAHs. The quantitation of metabolites of

Table 2
Amount of BP and PhIP in Cooked Meats

	Amount (μg/kg cooked meat)		
Sample	**BP**	**PhIP**	**Ref.**
Broiled beef	2.6/11.0	15.7	72, 103, 104
Fried beef	<1	0.6/15.0	72, 105
Broiled chicken	3.6/4.0 (2-5)	38.1	72, 73, 103
Fried codfish	?	69.2	72

PAHs excreted in the urine[94-96] should provide additional insight into the molecular basis for this biological variability.

Heterocyclic Amines in Cooked Meats

In addition to PAHs, highly mutagenic HAs are formed during broiling or frying of meat and fish due to pyrolysis of amino acids and proteins.[97] More than a dozen HAs have been identified[57,98] and the most common forms are quinolines, quinoxalines, pyridines, and carbolines. Examples of these include 2-amino-3-methylimidazo(4,5-f)quinoline (IQ); 2-amino-3,8-dimethylimidazo(4,5-f)quinoxaline (8-MeIQx); 2-amino-1-methyl-6-phenylimidazo(4,5-b)pyridine (PhIP); and 3-amino-1,4-dimethyl-5H-pyrido(4,3-b)indole (Trp-P-1). Some HAs are known to be N-hydroxylated by CYP1A and further activated to mutagenic forms by acetyltransferases.[79,99,100] These compounds are highly mutagenic in bacteria and moderately carcinogenic in animals causing colon cancer, mammary cancer, liver cancer, and lymphoma.[58,101,102]

PhIP is the most common HA in broiled meats occurring at levels comparable to or greater than those of BP (Table 2). Male rats fed PhIP develop colon adenocarcinomas, whereas female rats develop mammary adenocarcinomas.[58] PhIP is activated to the mutagenic metabolite 2-hydroxyamino-PhIP by CYP1A1, but is metabolized to an inactive form, 4′-hydroxy-PhIP, by CYP1A2.[106-108] These metabolites can subsequently form conjugates with glucuronide, glutathione, and sulfate (Figure 6).[109,110]

Molecular Dosimetry of Ingested HAs

Mutagenic activity can be detected in the urine of individuals ingesting fried or well-done meats.[111,112] However, the identification of cooking-associated mutagens and their metabolites in human urine has only recently become possible and pilot studies are now under way.

Urinary 8-MeIQx has been measured by gas chromatographic/mass spectrometry using negative-ion chemical ionization.[113] Ingestion of fried ground beef (approx. 320 g cooked weight) by six individuals resulted in an increase in

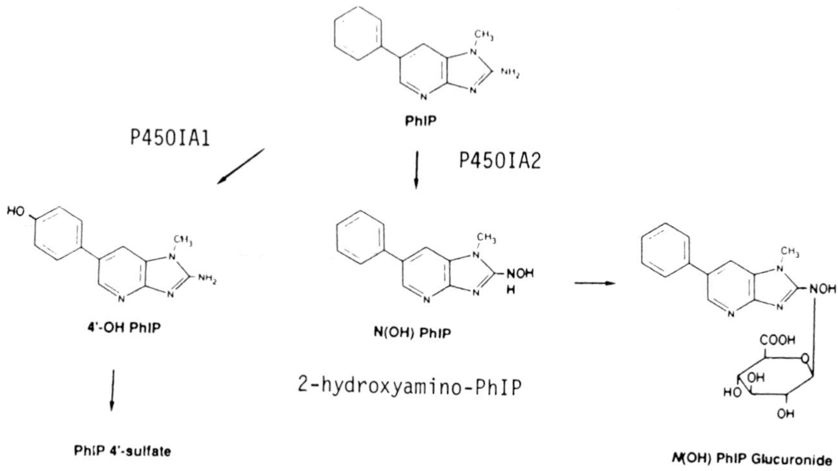

Figure 6. Metabolism of 2-amino-1-methyl-6-phenylimidazo(4,5-b)pyridine (PhIP). Activation of PhIP to mutagenic species, 2-hydroxyamino-PhIP, by CYPIA1.

concentration of urinary 8-MeIQx from undetectable (<5 pg/ml urine) to a mean of 14.7 pg/ml (range: 10.6 to 17.8) in the 12 hr immediately after ingestion. The mean urinary 8-MeIQx excretion in this 12-hr period was 12.4 ng (range: 9.1 to 15.5) which represented 1.8 to 4.9% of the 8-MeIQx ingested.[113] These results indicate that 8-MeIQx ingested during the consumption of cooked meat is absorbed and bioavailable. The proportion of 8-MeIQx measured in the urine is comparable to that reported in animal models administered this compound orally.[114]

In another study, four carcinogenic HAs were detected and quantitated in the urine of 10 healthy volunteers eating normal diets, but not in the urine of three patients receiving intravenous feeding.[115] Unmetabolized PhIP, 8-MeIQx, Trp-P-1, and 3-amino-1-methyl-5H-pyrido(4,3-b)indole (Trp-P-2) were measured by HPLC following partial purification by ion exchange chromatography of 24-hr urines (Table 3). These individuals appear to be continually exposed to these carcinogenic HAs through their normal diet. Since PhIP is generally more common in the diet than the other three HAs analyzed, these results suggest that the proportion of PhIP excreted unchanged in the urine is much smaller than that of the other HAs. Smoking did not elevate HA levels in the three study participants who were smokers.[115] In contrast, another study demonstrated that mutagens extracted from smoker's urine produce PhIP-like adducts when reacted with DNA *in vitro* and subjected to [^{32}P]-postlabeling analysis.[116]

Immunoaffinity chromatography has also been used to concentrate HAs and their metabolites in rat[117] and human[118] urine prior to their analysis by HPLC. The two monoclonal antibodies used in these studies recognize PhIP, N-substituted PhIP metabolites, and 4′-hydroxy-PhIP, but not metabolites with bulky conjugates at the 4′ or 5 position. In a controlled feeding study, a tenfold increase (180 ng vs

Table 3
Heterocyclic Amines in Human Urine

	Amount excreted in 24 hr (ng)[a]		
	Normal diet		Liquid diet (i.v.)
	5 male	5 female	(2 male + 1 female)
PhIP	0.30 (0.14–0.43)	0.77 (0.12–1.97)	<0.01
Trp–P–1	0.53 (0.04–1.43)	0.11 (0.06–0.26)	<0.01
Trp–P–2	0.28 (0.05–0.68)	0.07 (0.03–0.17)	<0.01
8–MeIQx	29 (16–44)	29 (11–47)	<1

Source: Adapted from Ushiyama et al.[115] With permission.

[a] Mean (range).

17 ng) in 24-hr urinary PhIP excretion was observed in one individual after switching from a 4-day vegetarian diet to a 4-day diet of fried hamburger.[118] In addition, several putative PhIP-conjugates and related compounds were detected. It was estimated that about 2.7% of the daily ingested PhIP was excreted as unmetabolized PhIP in the urine and a comparable percentage was excreted as putative PhIP-conjugates.[118] The levels of urinary PhIP measured in this study were significantly greater than those measured by ion exchange chromatography and HPLC in individuals consuming normal diets.[115]

METABOLIC PARAMETERS

A major area of related research is the study of individual metabolic phenotypes and their role in determining biomarker levels and human susceptibility to environmental carcinogens.[119] Several cytochrome P450 isozymes have been linked to increased cancer risk. For example, inducible CYP1A1 activity is higher in cultured lymphocytes from lung cancer cases than controls.[120] A genetic polymorphism in the CYP1A1 gene has also been tentatively associated with lung cancer risk in Japan.[121] Debrisoquine 4-hydroxylase (CYP2D6) is a polymorphic isozyme whose activity varies several thousand-fold among individuals. The high activity phenotype (extensive hydroxylator) is more common in patients with lung, liver, or bladder cancer than in controls without cancer.[122-125] An assay to assess hepatic CYP1A2 activity by caffiene demethylation in urine has also been developed[126-128] and is being applied in case-control studies of bladder and colon cancer.[129]

Hepatic arylamine N-acetyltransferase is associated with individual differences in susceptibility to both bladder and colorectal cancer in man.[130] A genetic polymorphism of this enzyme exists in man, resulting in slow- and rapid-acetylator phenotypes. Interestingly, the rapid-acetylator phenotype appears to protect aromatic amine-exposed individuals from bladder cancer;[130] however, the same rapid phenotype may enhance risk of colorectal cancer.[131] These results may be explained by the fact that N-acetylation of arylamines competes against the formation of reactive arylamine metabolites that reach the bladder; whereas, acetyltransferase in the colonic mucosa can O-acetylate N-hydroxy-arlyamine metabolites to highly reactive derivatives.[131]

Of particular interest with respect to the dietary carcinogens described in this chapter are those metabolic enzymes involved in the activation of aflatoxins, PAHs, and HAs to reactive species. As mentioned earlier, nifedipine oxidase (CYP3A4) appears to be important in the activation of 7,8-dihydroxy-BP to a genotoxic species. One of the substrates of CYP3A4 is the endogenous corticosteroid cortisol,[77] which is metabolized to 6-OH-cortisol. Measurement of cortisol and 6-OH-cortisol excreted in urine can be used to assess systemic CYP3A4 activity.[132-134]

Phenacetin deethylase (CYP1A2) catalyzes the N-oxidation of a number of HAs, including IQ and PhIP.[79-99] This enzyme is induced in humans by ingestion of CB beef.[135] In addition, caffeine 3-demethylation is selectively catalyzed by CYP1A2 and is the major pathway for caffeine metabolism in humans.[136,137] Thus, caffeine metabolites in urine can be used to determine the aromatic amine N-oxidation phenotype of individuals.[137,138] A recent modification of this method[129] should allow simultaneous determination of both N-oxidation and acetylation phenotypes.

Although no phenotyping assay has been developed to assess systemic CYP1A1 activity, a recent report[121] indicates that a rare polymorphism in the putative promotor region of the CYP1A1 gene may be associated with increased risk for lung cancer. A simple MspI restriction fragment length polymorphism (RFLP) assay was used to detect this polymorphism in DNA samples amplified by PCR.[121,139] If this polymorphism is indeed associated with cancer risk, or if a particular metabolic phenotype can be assigned to this polymorphism, it may prove useful in molecular epidemiological studies. Further discussion of the value of CYP1A1 phenotypes and genotypes as markers for environmental exposure and disease susceptibility may be found in Chapter 11.

SUMMARY AND FUTURE DIRECTIONS

A number of physiological steps are involved in the digestion, absorption, metabolism, and excretion of ingested carcinogens. Molecular biomonitoring of this process in humans is now plausible with current and developing methodology. The biomarkers currently being used to assess exposure and/or risk in these investigations are macromolecular adducts and carcinogen metabolites in peripheral blood and urine.

The major focus of molecular dosimetry/epidemiology in recent years has been the association of specific biomarkers with exposure to carcinogens. Thus, these biomarkers have been evaluated as markers of exposure or dose. A related, and potentially more important, evaluation is as markers of risk. Indeed, the early description of carcinogen-DNA adducts as markers of "biologically effective dose" implies the hope that these biomarkers indicate dose with biological effect (which in turn may have a higher degree of association with disease risk than would exposure or dose). The first rigorous test of DNA adducts as biomarkers of cancer risk in humans has now been reported.[42] Detectable levels of AFB-N^7-Gua, as well as several other aflatoxin B1 metabolites, in urine were shown to be predictive of liver cancer development. In addition, a strong interaction was observed between urinary aflatoxin B1 biomarkers and HBV, the other major risk factor for liver cancer.

While both epidemiological and experimental studies suggest that broiled foods and ingested PAHs and HAs may adversely affect health, the relative contribution of PAHs and HAs to carcinogenic risk in man has not been addressed. While most HAs are more mutagenic in bacterial test systems than PAHs, Sugimura et al. have noted[140] that both classes of compounds have roughly comparable carcinogenic potency in animal models. Thus, both classes may be potentially important in the etiology of diet-associated cancers. One approach to studying the genotoxic effects of ingested PAHs and HAs is to quantitate DNA adducts and metabolites formed in humans by representative members of these two classes of compounds.

In conclusion, molecular biomonitoring provides an opportunity to address certain issues of exposure, susceptibility, and risk in diet-associated human carcinogenesis, including (1) the kinetics of carcinogen adduct and metabolite formation and excretion, (2) the metabolic basis for interindividual differences in adduct and metabolite formation, (3) the biological significance of interindividual variabilty, and (4) identification of individuals at increased risk for cancer.

ACKNOWLEDGMENTS

The authors' research was supported in part by grants from the NIH (ES03819, ES06052, CA48409, CA01517) and the USDA Forest Service (INT89383).

REFERENCES

1. **Wynder E. L., and Gori G. B.,** Contribution of the environment to cancer incidence: an epidemiologic exercise, *J. Natl. Cancer Inst.,* 58, 825, 1977.
2. **Doll, R., and Peto, R.,** *The Causes of Cancer,* Oxford University Press, New York, 1981.
3. **Waterhouse, J., Muir, C., Shanmugaratnam, K., et al.,** *Cancer Incidence in Five Continents,* Vol. 4, IARC Sci. Publ. No. 42, Lyon, 1982.
4. **Haenszel, W., and Kurihara, M.,** Studies of Japanese migrants. I. Mortality from cancer and other diseases among Japanese in the U.S., *J. Natl. Cancer Inst.,* 40, 43, 1968.

5. **Staszewski, J., and Haenszel, W.,** Cancer mortality among the Polish-born in the U.S., *J. Natl. Cancer Inst.,* 35, 291, 1965.
6. **Willett, W.,** The search for the causes of breast and colon cancer, *Nature,* 338, 389, 1989.
7. **Busby, W. F., and Wogan, G. N.,** Aflatoxins, in *Chemical Carcinogens,* 2nd ed., Searle, C. E., Ed., American Chemical Society, Washington, D.C., 1985, 945-1136.
8. **Campbell, A. A., Whitaker, T. B., Pohland, A. E., Dickens, J. W., and Park, D. L.,** Sampling, sample preparation, and sampling plans for foodstuffs for mycotoxion analysis, *Pure Appl. Chem.,* 58, 305, 1986.
9. **Groopman, J. D., Sabbioni, G., and Wild, C. P.,** Molecular dosimetry of aflatoxin exposures, in *Molecular Dosimetry of Human Cancer: Epidemiological, Analytical and Social Considerations,* Groopman, J. D., and Skipper, P., Eds., CRC Press, Boca Raton, FL, 1991, 302-324.
10. **Groopman, J. D., Donahue, P. R., Zhu, J., Chen, J., and Wogan, G. N.,** Aflatoxin metabolism in humans: detection of metabolites and nucleic acid adducts in urine by affinity chromatography. *Proc. Natl. Acad Sci. U.S.A.,* 82, 6492, 1985.
11. **Groopman, J. D., Zhu, J., Donahue, P. R., Pikul, A., Zhang, L.-S., Chen, J.-S., and Wogan, G. N.,** Molecular dosimetry of urinary aflatoxin DNA adducts in people living in Guangxi Autonomous Region, People's Republic of China, *Cancer Res.,* 52, 45, 1992.
12. **Sabbioni, G., Skipper, P., Buchi, G., and Tannenbaum, S. R.,** Isolation and characterization of the major serum albumin adduct formed by aflatoxin B1 *in vivo* in rats, *Carcinogenesis,* 8, 819, 1987.
13. **Wild C. P., Jiang, Y. Z., Sabbioni, G., Chapot, B., and Montesano, R.,** Evaluation of methods for quantitation of aflatoxin-albumin adducts and their application to human exposure assessment, *Cancer Res.,* 50, 245, 1990.
14. **Roebuck, B. D., Liu, Y.-L., Rogers, A. E., Groopman, J. D., and Kensler, T. W.,** Protection against aflatoxin B1-induced hepatocarcinogenesis in F344 rats by 5-(2-pyrazinyl)-4-methyl-1,2-dithiol-3-thione (Oltipraz): predictive role for short-term molecular dosimetry, *Cancer Res.,* 51, 5501, 1991.
15. **Groopman, J. D., Egner, P., Love-Hunt, A., DeMatos, P., and Kensler, T. W.,** Molecular dosimetry of aflatoxin DNA and serum albumin adducts in chemoprotection studies using 1,2-dithiole-3-thione in rats. *Carcinogenesis,* 13, 101, 1992.
16. **Shimada, T., and Guengerich, F. P.,** Evidence for cytochrome P-450NF, the nifedipine oxidase, being the principal enzyme involved in the bioactivation of aflatoxins in human liver, *Proc. Natl. Acad. Sci. U.S.A.,* 86, 462, 1989.
17. **Guengerich, F. P., Shimada, T., Raney, K. D., Yun, C.-H., Meyer, D. J., Ketterer, B., Harris, T. M., Groopman, J. D., and Kadlubar, F. F.,** Elucidation of catalytic specificities of human cytochrome P-450 and glutathione S-transferase enzymes and relevance to molecular epidemiology studies, *Environ. Health Persp.,* in press.
18. **Essigmann, J. M., Croy, R. G., Nadzan, A. M., Busby, W. , Jr., Reinhold, V. N., Buchi, G., and Wogan, G. N.,** Structural identification of the major DNA adduct formed by aflatoxin B1 *in vitro, Proc. Natl. Acad. Sci. U.S.A.,* 74, 1870, 1977.
19. **Croy, R. G., Essigmann, J. M., Reinhold, V. N., and Wogan, G. N.,** Identification of the principle aflatoxin B1-DNA adduct formed *in vivo* in rat liver, *Proc. Natl. Acad. Sci. U.S.A.,* 75, 1745, 1978.
20. **Groopman, J. D., Busby, W. F., and Wogan, G. N.,** The nuclear distribution of aflatoxin B1 and its interaction with histones *in vivo, Cancer Res.,* 40, 4343, 1980.

21. **Lin, J. K., Miller, J. A., and Miller, E. C.,** 2,3-Dihydro-2-(guan-7-yl)-3-hydroxy-aflatoxin B1, a major acid hydrolysis product of aflatoxin B1-DNA or -ribosomal RNA adducts formed in hepatic microsome mediated reactions in rat liver *in vivo, Cancer Res.,* 37, 4430, 1977.
22. **Hertzog, P. J., Lindsay-Smith, J. R., and Garner, R. C.,** Production of monoclonal antibodies to guanine imidazole ring-opened aflatoxin B1-DNA, the persistent DNA adduct *in vivo, Carcinogenesis,* 3, 723, 1982.
23. **Dalezios, J. I.,** Aflatoxin P1: A New Metabolite of Aflatoxin B1. Its Isolation and Identification, Ph.D. thesis, Massachusetts Institute of Technology, Cambridge, MA, 1971.
24. **Wild, C. P., Garner, R. G., Montesano, R., and Tursi, F.,** Aflatoxin B1 binding to plasma albumin and liver DNA upon chronic administration to rats, *Carcinogenesis,* 7, 853, 1986.
25. **Croy, R. G., and Wogan, G. N.,** Temporal patterns of covalent DNA adducts in rat liver after single and multiple doses of aflatoxin B1, *Cancer Res.,* 41, 197, 1981.
26. **Kensler, T. W., Egner, P. A., Davidson, N. E., Roebuck, B. D., Pikul, A., and Groopman, J. D.,** Modulation of aflatoxin metabolism, aflatoxin-N7-guanine formation, and hepatic tumorigenesis in rats fed ethoxyquin: role of induction of glutathione S-transferase, *Cancer Res.,* 46, 3924, 1986.
27. **Goeger, D. E., Shelton, D., Hendricks, J. D., Pereira, C., and Bailey, G. S.,** Comparative effect of dietary butylated hydroxyanisole and β-naphthoflavone on aflatoxin B1 metabolism, DNA adduct formation and carcinogenesis in rainbow trout, *Carcinogenesis,* 9, 1793, 1988.
28. **Groopman, J. D., Hasler, J. A., Trudel, L. J., Pikul, A., Donahue, P. R., and Wogan, G. N.,** Molecular dosimetry in rat urine of aflatoxin-N7-guanine and other aflatoxin metabolites by multiple monoclonal antibody affinity chromatography and immunoaffinity/high performance liquid chromatography, *Cancer Res.,* 52, 267, 1992.
29. **Kensler, T. W., Groopman, J. D., Eaton, D. L., Curphey, T. J., and Roebuck, B. D.,** Potent inhibition of aflatoxin-induced hepatic tumorigenesis by the monofunctional enzyme inducer 1,2-dithiole-3-thione, *Carcinogenesis,* 12, 95, 1992.
30. **Gan, L.-S., Skipper, P. L., Peng, X.-C., Groopman, J. D., Chen, J. S., Wogan, G. N., and Tannenbaum, S. R.** Serum albumin adducts in the molecular epidemiology of aflatoxin carcinogenesis: correlation with aflatoxin B1 intake and urinary excretion of aflatoxin M1, *Carcinogenesis,* 9, 1323, 1988.
31. **Groopman, J. D., Hall, A., Whittle, H., Hudson, G., Wogan, G. N., Montesano, R., and Wild, C. P.,** Molecular dosimetry of aflatoxin-N7-guanine in human urine obtained in The Gambia, West Africa, *Cancer Epidemiol., Biomarkers Prevention,* 1, 221, 1992.
32. **Wild, C. P., Hudson, G., Sabbioni, G., Wogan, G. N., Whittle, H., Montesano, R., and Groopman, J. D.,** Correlation of dietary intake of aflatoxins with the level of albumin bound aflatoxin in peripheral blood in The Gambia, West Africa, *Cancer Epidemiol., Biomarkers Prevention,* 1, 229, 1992.
33. **Wild, C. P., Jiang, Y. Z., Allen, S. J., Jansen, L. A., Hall, A. J., and Montesano, R.,** Aflatoxin-albumin adducts in human sera from different regions of the world, *Carcinogenesis,* 11, 2271, 1990.
34. **Hsu, I. C., Metcalf, R. A., Sun, T., Wesh, J. A., Wang, N. J., and Harris, C. C.,** Mutational hotspot in the p53 gene in human hepatocellular carcinomas, *Nature,* 350, 427, 1991.

35. **Bressac, B., Kew, M., Wands, J., and Ozturk, M.,** Selective G to T mutations of p53 gene in hepatocellualr carcinoma from Southern Africa, *Nature,* 350, 429, 1991.
36. **Ozturk, M, et al.,** p53 Mutation in hepatocellular carcinoma after aflatoxin exposure, *Lancet,* 338, 1356, 1991.
37. **Puisieux, A., Lim, S., Groopman, J. D., and Ozturk, M.,** Selective targeting of p53 gene mutational hotspots in human cancers by etiologically defined carcinogens, *Cancer Res.,* 51, 6185, 1991.
38. **Hollstein, M., Sidransky, D., Vogelstein, B., and Harris, C. C.,** p53 Mutations in human cancers, *Science,* 253, 49, 1991.
39. **Campbell, T. C., Chen, J., Liu, C., Li, J., and Parpia, B.,** Nonassociation of aflatoxin with primary liver cancer in a cross-sectional ecological survey in the People's Republic of China, *Cancer Res.,* 50, 6881, 1990.
40. **Srivatanakul, P., Parkin, D. M., Khlat, M., Chenvidhya, D., Chotiwan, P., Insiripong, S., L'Abbe, K. A., and Wild, C. P.,** Liver cancer in Thailand. II. A case-control study of hepatocellular carcinoma, *Int. J. Cancer,* 48, 329, 1991.
41. **Srivatanakul, P., Parkin, D. M., Jiang, Y.-Z., Khlat, M., Koa-Ian, U.-T., Sontipong, S., and Wild, C. P.,** The role of infection by opisthorchis viverrini, hepatitis B virus, and aflatoxin exposure in the etiology of liver cancer in Thailand, *Cancer,* 68, 2411, 1991.
42. **Ross, R., Yuan, J.-M., Yu, M., Wogan, G. N., Qian, G.-S., Tu, J.-T., Groopman, J. D., Gao, Y.-T. and Henderson, B.E.,** Urinary aflatoxin biomarkers and risk of hepatocellular carcinoma, *Lancet,* 339, 943, 1992.
43. **Armstrong, B., and Doll, R.,** Environmental factors and cancer incidence and mortality in different countries with special reference to dietary practices, *Int. J. Cancer,* 15, 617-631, 1975.
44. **Rose, D. P., Boyar, A. P., and Wynder, E. L.,** International comparisons of mortality rates for cancer of the breast, ovary, prostate, and colon and per capita food consumption, *Cancer,* 58, 2363, 1986.
45. **Potter, J. D., and McMichael, A. J.,** Diet and cancer of the colon and rectum: a case-control study, *J. Natl. Cancer Inst.,* 76, 557, 1986.
46. **Graham S., Marshall, J., Haughey, B., et al.,** Dietary epidemiology of cancer of the colon in western New York, *Am. J. Epidemiol.,* 128, 490-503, 1988.
47. **Slattery, M. L., Schumacher, M. C., Smith, K. R., West, D. W., and Abd-Elghany, N.,** Physical activity, diet, and risk of colon cancer in Utah, *Am. J. Epidemiol.,* 128, 989, 1988.
48. **Peters, R. K., Garabrant, D. H., Yu, M. C., and Mack, T. M.,** A case-control study of occupational and dietary factors in colorectal cancer in young men by subsite, *Cancer Res.,* 49, 5459, 1989.
49. **Willett, W. C., Stampfer, M. J., Colditz, G. A., Rosner, B. A., and Speizer, F. E.,** Relation of meat, fat, and fiber intake to the risk of colon cancer in a prospective study among women, *N. Engl. J. Med.,* 323, 1664, 1990.
50. **Norell, S. E., Ahlbom, A., Erwald, R., et al.,** Diet and pancreatic cancer: a case-control study, *Am. J. Epidemiol.,* 124, 894, 1986.
51. **Steineck, G., Hagman, U., Gerhardsson, M., and Norell, S. E.,** Vitamin A supplements, fried foods, fat and urothelial cancer. A case-referent study in Stockholm in 1985-87, *Int. J. Cancer,* 45, 1006, 1990.
52. **Joossens, J. V., Hill, M. J., and Geboers, J., Eds.,** *Diet and Human Carcinogenesis,* Elsevier: Amsterdam, 1985.

53. **Narisawa, T., Magadia, N. E., Weisburger, J. H., and Wynder, E. L.,** Promoting effect of bile acids on colon carcinogenesis after intrarectal instillation of N-methyl-N′-nitro-N-nitrosoguanidine in rats, *J. Natl. Cancer Inst.,* 53, 1093, 1974.
54. **Reddy, B. S., Watanabe, K., Weisburger, J. H., and Wynder, E. L.,** Promoting effect of bile acids in colon carcinogenesis in germ-free and conventional F344 rats, *Cancer Res.,* 37, 3238, 1977.
55. **Wargovich, M. J., Goldberg, M. T., Newmark, H. L., and Bruce, W. R.,** Nuclear aberrations as a short-term test for genotoxicity to the colon: evaluation of nineteen agents in mice, *J. Natl. Cancer Inst.,* 71, 133, 1983.
56. **Dunn, B. P.,** Wide-range linear dose response curve for DNA binding of orally administered benzo(a)pyrene in mice, *Cancer Res.,* 43, 2654-2658, 1983.
57. **Sugimura, T., Sato, S., and Wakabayashi, K.,** Mutagens/carcinogens in pyrolysates of amino acids and proteins and in cooked foods: heterocyclic aromatic amines, in *Chemical Induction of Cancer: Structural Bases and Biological Mechanisms,* Woo, Y. T., et al., Eds., Academic Press, New York, 1988, 681-710.
58. **Ito, N., Hasegawa, R., Sano, M., Tamano, S., Esumi, H., Takayama, S., and Sugimura, T.,** A new colon and mammary carcinogen in cooked food, 2-amino-1-methyl-6-phenyl-imidazo(4,5-b)pyridine (PhIP), *Carcinogenesis,* 12, 1503, 1991.
59. **Bruce, W. R.,** Recent hypotheses for the origin of colon cancer, *Cancer Res.,* 47, 4237, 1987.
60. **Schiffman, M. H., and Felton, J. S.,** Fried foods and the risk of colon cancer, *Am. J. Epidemiol.,* 131, 376-378, 1990.
61. **Baum, E. J.,** Occurrence and surveillance of polycyclic aromatic hydrocarbons, in *Polycyclic Hydrocarbons and Cancer,* Vol. 1, Gelboin, H. V., and Ts'o, P.O.P., Eds., Academic Press, New York, 1978, 45-70.
62. **Sontag, J. M.,** *Carcinogens in Industry and the Environment,* Marcel Dekker, New York, 1981, 167-281, 467-475.
63. **Lijinsky, W., and Shubik, P.,** Benzo(a)pyrene and other polynuclear hydrocarbons in charcoal-broiled meat, *Science,* 145, 53, 1964.
64. **Fazio, T., and Howard, J. W.,** Polycyclic aromatic hydrocarbons in foods, in *Handbook of Poylycyclic Aromatic Hydrocarbons,* Vol. 1, Bjorseth, A., Ed., Marcel Dekker, New York, 1983, 461-505.
65. **Int. Agency Res. Cancer,** IARC monograph on the evaluation of the carcinogenic risk of chemicals to man: certain polycyclic aromatic hydrocarbons and heterocyclic components, *IARC,* 3, 91, 1973.
66. **Lioy, P. L., Waldman, J. M., Greenberg, A., Harkov, R., and Pietarinen, C.,** The total human environmental exposure study (THEES) to benzo(a)pyrene: comparison of the inhalation and food pathways, *Arch. Environ. Health,* 43, 304, 1988.
67. **Greenberg, A., Luo, S., Hsu, C. H., Creighton, P., Waldman, J. M., and Lioy, P. L.,** Benzo(a)pyrene in composite prepared meals: results from the THEES (total human exposure to environmental substances) study, *Polycyclic Aromatic Compds.,* 1, 221, 1990.
68. **Surgeon General,** *Smoking and Health: A Report of the Surgeon General,* Washington, D.C., U.S. Dept. Health Educ. Welfare [DHEW publication No.(PHS)79-50066], 1979.
69. **Perera, F., Santella, R., and Poirier, M.,** Biomonitoring of workers exposed to carcinogens: immunoassays to benzo(a)pyrene-DNA adducts as a prototype, *J. Occup. Med.,* 28, 1117, 1986.

70. **Waldman, J. M., Lioy, P. J., Greenberg, A., and Butler, J.,** Analysis of human exposure to B(a)P via inhalation and food ingestion in the total human exposure to environmental substances (THEES) study, *J. Toxicol. Environ. Health,* in press.
71. **Vaessen, H. A., Jekel, A. A., and Wilburs, A. A.,** Dietary intake of polycyclic aromatic hydrocarbons, *Toxicol. Environ. Chem.,* 16, 281, 1988.
72. **Wakabayashi, K., Ushiyama, H., Takahashi, M., Nukaya, H., Kim, S. B., Hirose, M., Ochiai, M., Sugimura, T., and Nagao, M.,** Exposure to heterocyclic amines (food pyrolysates). *Environ. Health Persp.,* 99, 129-133, 1993.
73. **Larson, B. K., Sahlberg, G. P., Eriksson, A. T., and Busk, L.A.,** Polycyclic aromatic hydrocarbons in grilled food. *J. Agri. Food Chem.,* 31, 867, 1983.
74. **Autrup, H., Harris, C. C., Trump, B. F., and Jeffrey, A. H.,** Metabolism of benzo(a)pyrene and identification of the major benzo(a)pyrene-DNA adducts in cultured human colon, *Cancer Res.,* 38, 3689, 1978.
75. **Gelboin, H. W.,** Benzo(a)pyrene metabolism, activation, and carcinogenesis: role and regulation of mixed-function oxidases and related enzymes, *Physiol. Rev.,* 60, 1107, 1980.
76. **Conney, A. H.,** Induction of microsomal enzymes by foreign chemicals and carcinogenesis by polycyclic aromatic hydrocarbons, *Cancer Res.,* 42, 4875, 1982.
77. **Guengerich, F. P.,** Characterization of human microsomal cytochrome P-450 enzymes, *Annu. Rev. Pharmacol. Toxicol.,* 29, 241, 1989.
78. **Shimada, T., Martin, M. V., Pruess-Schwartz, D., Marnett, L. J., and Guengerich, F. P.,** Role of individual human cytochrome P-450 enzymes in the bioactivation of benzo(a)-pyrene, 7,8-dihydroxy-7.8-dihyrobenzo(a)pyrene, and other dihydrodiol derivatives of polycyclic aromatic hydrocarbons, *Cancer Res.,* 49, 6304, 1989.
79. **McManus, M. E., Burgess, W. M., Veronese, M. E., Huggett, A., Quattrochi, L. C., Tukey, R. H.,** Metabolism of 2-acetylaminofluorene and benzo(a)pyrene and activation of food-derived heterocyclic amine mutagens by human cytochromes P-450, *Cancer Res.,* 50, 3367, 1990.
80. **Liou, S. H., Jacobson-Kram, D., Poirier, M. C., Nguyen, D., Strickland, P. T., and Tockman, M. S.,** Biological monitoring of firefighters: sister chromatid-exchange and polycyclic aromatic hydrocarbon-DNA adducts in periopheral blood cells, *Cancer Res.,* 49, 4929, 1989.
81. **Strickland, P. T., Rothman, N., Baser, M. E., and Poirier, M. C.,** Polycyclic aromatic hydrocarbon-DNA adduct load in peripheral blood cells: contribution of multiple exposure sources, in *Immunoassays for Trace Chemical Analysis,* Vanderlaan, M., et al., Eds, American Chemical Society, Washington D.C., 1990, 257-263.
82. **Harris, C. C., Vahakangas, K., Newman, M. J., Trivers, G. E., Shamsuddin, A. K. M., Sinopoli, N., Mann, D. L., and Wright, W. E.,** Detection of benzo(a)pyrene diol epoxide-DNA adducts in peripheral blood lymphocytes and antibodies to the adducts in serum from coke-oven workers, *Proc. Natl. Acad. Sci. U.S.A.,* 82, 6672, 1985.
83. **Perera, F. P., Hemminki, K., Young, T. L., Brenner, D., Kelly, G., and Santella, R. M.,** Detection of polycyclic aromatic hydrocarbon-DNA adducts in white blood cells of foundry workers, *Cancer Res.,* 48, 2288, 1988.
84. **Haugen, A., Becher, G., Bemestad, C., Vahakangas, K., Trivers, G. E., Newman, M. J., and Harris, C. C.,** Determination of polycyclic hydrocarbons in the urine, benzo(a)pyrene diol epoxide-DNA adducts in lymphocyte DNA, and antibodies to the adducts in sera from coke oven workers exposed to measured amounts of polycyclic aromatic hydrocarbons in the work atmosphere, *Cancer Res.,* 46, 4178, 1986.

85. **Santella, R. M, Weston, A., Perera, F. P., Trivers, G. T., Harris, C. C., Young, T. L., Nguyen, D., Lee, B. M., and Poirier, M. C.,** Interlaboratory comparison of antisera and immunoassays for benzopyrene-diolepoxide-modified DNA, *Carcinogenesis,* 9, 1265, 1988.
86. **Strickland, P. T., Liou, S. H., Poirier, M. C., Nguyen, D., and Tockman, M. S.,** Contribution of occupation, diet, and smoking to PAH-DNA adduct load in peripheral blood cells, in *Human Carcinogen Exposure: Biomonitoring and Risk Assessment,* Garner, R. C., Ed., IRL Oxford University Press, Oxford, 1991, 399-405.
87. **Rothman, N., Poirier, M. C., Baser, M. E., Hansen, J. A., Gentile, C., Bowman, E. D., and Strickland, P. T.,** Formation of polycyclic aromatic hydrocarbon-DNA adducts in peripheral white blood cells during consumption of charcoal-broiled beef, *Carcinogenesis,* 11, 1241, 1990.
88. **Harris, C. C.,** Interindividual variation among humans in carcinogen metabolism, DNA adduct formation and DNA repair, *Carcinogenesis,* 10, 1563, 1989.
89. **Conney, A. H., Pantuck, E. J., Hsiao, K. C., Kuntzman, R., Alvares, A. D., and Kappas, A.,** Regulation of drug metabolism in man by environmental chemicals and diet, *Fed. Proc.,* 36, 1647, 1977.
90. **Nowak, D., Schmidt-Preuss, U., Jorres, R., Liebke, F., and Rudiger, H. W.,** Formation of DNA adducts and water-soluble metabolites of benzo(a)pyrene in human monocytes is genetically controlled, *Int. J. Cancer,* 41, 169, 1989.
91. **Thompson, C. L., McCoy, Z., Lambert, J. M., Andries, M. J., and Lucier, G. W.,** Relationship among benzo(a)pyrene metabolism, benzo(a)pyrene-diolepoxide: DNA adduct formation, and sister chromatid exchanges in human lymphocytes from smokers and nonsmokers, *Cancer Res.,* 49, 6503, 1989.
92. **Rothman, N., Poirier, M. C., Correa-Villasenor, A., Ford, P. D., Hansen, J. A., O'Toole, T., and Strickland, P. T.,** Association of char-broiled food ingestion with polycyclic aromatic hydrocarbon-DNA adducts in white blood cells from wildland firefighters, *Proc. Am. Assoc. Cancer Res.,* 32, 96, 1991.
93. **Rothman, N., Poirier, M. C., Correa-Villasenor, A., Ford, P. D., Hansen, J. A., O'Toole, T., and Strickland, P. T.,** Association of PAH-DNA adducts in peripheral white blood cells with dietary exposure to PAHs, *Environ. Health Persp.,* 99, 265, 1993.
94. **Bowman, E. D., Carr, P., Strickland, P. T., Rothman, N., and Weston, A.,** Detection of metabolites of benzo(a)pyrene in human urine, *Proc. Am. Assoc. Cancer Res.,* 33, 103, 1992.
95. **Gomes, M., and Santella, R. M.,** Immunologic methods for the detection of benzo(a)-pyrene metabolites in urine, *Chem. Res. Toxicol.,* 3, 307, 1990.
96. **Jongeneelen, F. J., Bos, R. P., Anzion, R. B. M., Theuws, J. L. G., and Henderson, P. T.,** Biological monitoring of polycyclic aromatic hydrocarbons, *Scand. J. Work Environ. Health,* 12, 137, 1986.
97. **Sugimura, T., and Sato, S.,** Mutagens-carcinogens in foods, *Cancer Res.,* 43, 2415s, 1983.
98. **Felton, J. S., and Knize, M. G.,** Occurrence, identification, and bacterial mutagenicity of heterocyclic amines in cooked food, *Mutation Res.,* 259, 205, 1991.
99. **Shimada, T., Iwasaki M., Martin, M. V., and Guengerich, F. P.,** Human liver microsomal cytochrome P-450 enzymes involved in the bioactivation of procarcinogens detected by umu gene response in Salmonella typhimurium TA 1535/pSK1002, *Cancer Res.,* 49, 3218, 1989.

100. **Beland, F. A., and Kadlubar, F. F.,** Metabolic activation and DNA adducts of aromatic amines and nitroaromatic hydrocarbons, in *Chemical Carcinogenesis and Mutagenesis I,* Cooper, C. S., and Grover, P. L., Eds, Springer-Verlag, Berlin, 1990, 267-325.
101. **Adamson, R. H., Thorgeirsson, U. P., Snyderwine, E. G., Thorgeirsson, S. S., Reeves, J., Dalgard, D. W., Takayama, S., and Sugimura, T.,** Carcinogenicity of 2-amino-3-methylimidazo-(4,5-f)quinoline in nonhuman primates: induction of tumors in three macaques, *Jpn. J. Cancer Res.,* 81, 10, 1990.
102. **Esumi, H., Ohgaki, H., Kohzen, E., Takayama, S., and Sugimura, T.,** Induction of lymphoma in CDF1 mice by the food mutagen, 2-amino-1-methyl-6-phenylimidazo(4,5-b)-pyridine, *Jpn. J. Cancer Res.,* 80, 1176, 1989.
103. **Lijinsky, W.,** The formation and occurrence of polycyclic aromatic hydrocarbons associated with food, *Mutation Res.,* 259, 251, 1991.
104. **Greenberg, A., Hsu, C. H., Rothman, N., and Strickland, P. T.,** PAH profiles of char-broiled hamburgers: pyrene/B(a)P ratios and presence of reactive PAH, *Polycyclic Aromatic Compds.,* 3, 101-110, 1993.
105. **Felton, J. S., and Knize, M. G.,** Heterocyclic-amine mutagens/carcinogens in food, in *Chemical Carcinogenesis and Mutagenesis I,* Cooper, C. S., and Grover, P. L., Eds, Springer-Verlag, Berlin, 1990, 471-502.
106. **Holme, J. A., Wallin, H., Brunborg, G., Soderlund, E. J., Hongslo, J. K., and Alexander, J.,** Genotoxicity of the food mutagen 2-amino-1-methyl-6-phenylimidazo(4,5-b)pyridine (PhIP): formation of 2-hydroxyamino-PhIP, a directly acting genotoxic metabolite, *Carcinogenesis,* 10, 1389, 1989.
107. **Wallin, H., Mikalsen, A., Guengerich, F. P., Ingelman-Sundberg, M., Solberg, K. E., Rossland, O. J., and Alexander, J.,** Differential rates of metabolic activation and detoxication of the food mutagen 2-amino-1-methyl-6-phenylimidazo(4,5-b)pyridine by different cytochrome P450 enzymes, *Carcinogenesis,* 11, 489, 1990.
108. **Buonarati, M. H., and Felton, J.S.,** Activation of 2-amino-1-methyl-6-phenylimidazo(4,5-b)pyridine (PhIP) to mutagenic metabolites, *Carcinogenesis,* 11, 1133, 1990
109. **Alexander, J., Wallin, H., Holme, J. A., and Becher, G.,** 4-(2-Amino-1-methylimidazo-(4,5-b)pyrid-6-yl)phenyl sulfate — a major metabolite of the food mutagen 2-amino-1-methyl-6-phenylimidazo(4,5-b)pyridine (PhIP) in the rat, *Carcinogenesis,* 10, 1543, 1989.
110. **Alexander, J., Wallin, H., Rossland, O. J., Solberg, K. E., Holme, J. A., Becher, G., Andersson, R., and Grivas, S.,** Formation of a glutathione conjugate and a semistable transportable glucuronide conjugate of N2-oxidized species of 2-amino-1-methyl-6-phenylimidazo-(4,5-b)pyridine (PhIP) in rat liver, *Carcinogenesis,* 12, 2239, 1991.
111. **Doolittle, D. J., Rahn, C. A., Burger, G. T., Lee, C. K., Reed, B., Riccio, E., Howard, G., Passananti, G. T., Vesell, E. S., and Hayes, A. W.,** Effect of cooking methods on the mutagenicity of food and on urinary mutagenicity of human consumers, *Food Chem. Toxic.,* 27, 657, 1989.
112. **Baker, R., Arlauskas, A., Bonin, A., and Angus, D.,** Detection of mutagenic activity in human urine following fried pork or bacon meals, *Cancer Lett.,* 16, 81-89, 1982.
113. **Murray, S., Gooderham, N. J., Boobis, A. R., and Davies, D. S.,** Detection and measurement of MeIQx in human urine after ingestion of a cooked meat meal, *Carcinogenesis,* 10, 763, 1989.

114. **Gooderham, N. J., Watson, D., Rice, J. C., Murray, S., Taylor, G. W., and Davies, D. S.,** Metabolism of the mutagen MeIQx *in vivo:* metabolite screening by liquid chromatography-thermospray mass spectrometry, *Biochem. Biophys. Res. Commun.,* 148, 1377, 1987.
115. **Ushiyama, H., Wakabayashi, K., Hirose, M., Itoh, Sugimura, T., and Nagao, M.,** Presence of carcinogenic heterocyclic amines in urine of healthy volunteers eating normal diet, but not of inpatients receiving parenteral alimentation, *Carcinogenesis,* 12, 1417, 1991.
116. **Peluso, M., Castegnaro, M., Malaveille, C., Friesen, M., Garren, L., Hautefeuille, A., Vineis, P., Kadlubar, F., and Bartsch, H.,** ^{32}P-Postlabelling analysis of urinary mutagens from smokers of black tobacco implicates 2-amino-1-methyl-6-phenylimidazo(4,5-b)pyridine (PhIP) as a major DNA-damaging agent, *Carcinogenesis,* 12, 713, 1991.
117. **Vanderlaan, M., Alexander, J., Thomas, C., Djanegara, T., Hwang, M., Watkins, B. E., and Wallin, H.,** Immunochemical detection of rodent hepatic and urinary metabolites of cooking-induced food mutagens, *Carcinogenesis,* 12, 349, 1991.
118. **Vanderlaan, M., Hwang, M., Djanegara, T., Wallin, H., and Alexander, J.,** Detection of 2-amino-1-methyl-6-phenylimidazo(4,5-b)pyridine (PhIP), its metabolites, and related chemicals by immunoaffinity/HPLC/fluorescence, submitted.
119. **Kadlubar, F., O'Neill, I. K., and Bartsch, H.,** Biomonitoring and susceptibility markers in human cancer: applications in molecular epidemiology and risk assessment, *Environ. Health Persp.,* 98, 1992.
120. **Kouri, R. E., McKinney, C. E., Slomiany, D. J., Snodgrass, D. R., Wray, N. P., and McLemore, T. L.,** Positive correlation between high aryl hydrocarbon hydroxylase activity and primary lung cancer as analyzed in cryopreserved lymphocytes, *Cancer Res.,* 42, 5030, 1982.
121. **Kawajiri, K., Nakachi, K., Imai, K., Yoshii, A., Shinoda, N., and Watanabe, J.,** Identification of genetically high risk individuals to lung cancer by DNA polymorphisms of the cytochrome CYPIA1 gene, *FEBS Lett.,* 263, 131, 1990.
122. **Idle, J. R., Mahgoub, A., Sloan, T. P., Smith, R. L., Mbanefo, C. O., and Bababunmi, E. A.,** Some observations on the oxidation phenotype status of Nigerian patients presenting with cancer, *Cancer Lett.,* 11, 331, 1981.
123. **Ayesh, R., Idle, J. R., Ritchie, J. C., Crothers, M. J., and Hetzel, M. R.,** Metabolic oxidation phenotypes as markers for susceptibility to lung cancer, *Nature,* 312, 169, 1984.
124. **Kaisary, A., Smith, P., Jaczq, E., McAllister, C. B., Wilkinson, G. R., Ray, W. A., and Branch, R. A.,** Genetic predisposition to lung cancer: ability to hydroxylate debrisoquine and mephenytoin as risk factors, *Cancer Res.,* 47, 5488, 1987.
125. **Caporaso, N., Hayes, R. B., Dosemeci, M., Hoover, R., Ayesh, R., Hetzel, M., and Idle, J.,** Lung cancer risk, occupational exposure, and the debrisoquine metabolic phenotype, *Cancer Res.,* 49, 3675, 1989.
126. **Butler, M. A., Caporaso, N., Vineis, P., Lang, N. P., and Kadlubar, F. F.,** Characterization of the carcinogenic arylamine N-oxidation phenotype in humans by urinalysis of caffeine metabolites, *Proc. Am. Assoc. Cancer Res.,* 31, 231, 1990.
127. **Grant, D. M., Tang, B. K., and Kalow, W.,** A simple test for acetylator phenotype using caffeine, *Br. J. Clin. Pharm.,* 17, 459, 1984.

128. **Evans, W. E., Relling, M. V., Petro, W. P., Meyer, W. H., Mirro, J., Jr., and Crom, W. R.,** Dextromethorphan and caffeine as probes for simultaneous determination of debrisoquine-oxidation and N-acetylation phenotype in children, *Clin. Pharm. Ther.,* 45, 568, 1989.
129. **Kadlubar, F. F., Butler, M. A., Kaderlik, K., Chou, J., and Lang, N. P.,** Polymorphisms for aromatic amine metabolism in humans: relevance for human carcinogenesis, *Environ. Health Persp.,* 98, 69, 1992.
130. **Hein, D. W.,** Acetylator genotype and arylamine-induced carcinogenesis, *Biochim. Biophys. Acta,* 948, 37, 1988.
131. **Lang, N. P., Chu, D. Z. J., Hunter, C. F., Kendall, D. C., Flammang, T. J., and Kadlubar, F. F.,** Role of aromatic amine acetyltransferase in human colorectal cancer, *Arch. Surg.,* 121, 1259, 1986.
132. **Joellenbeck, L., Qian, Z., Zarba, A., and Groopman, J. D.,** Urinary 6b-hydroxycortisol/cortisol ratios measured by HPLC for use as a marker for the human cytochrome P-450 3A4, *Cancer Epidemiol. Biomarkers Prev.,* 1, 567, 1992.
133. **Olnhaus, E. E., and Park, B. K.,** Measurement of urinary 6-beta-hydroxycortisol excretion as an *in vivo* parameter in the clinical assessment of the microsomal enzyme-inducing capacity of antipyrine, phenobarbitone and rifampicin, *Eur. J. Clin. Pharm.,* 15, 139, 1979.
134. **Ged, C., Rouillon, J. M., Pichard, L., Combalbert, J., Bressot, N., Bories, P., Michel, H., Beaune, P., and Maurel, P.,** The increase in urinary excretion of 6B-hydroxy-cortisol as a marker of human hepatic cytochrome P4503A induction, *Br. J. Clin. Pharm.,* 28, 373, 1989.
135. **Pantuck, E. J., Hsiao, K. C., Conney, A. H., Garland, W. A., Kappas, A., Anderson, K. E., and Alvares, A. P.,** Effect of charcoal-broiled beef on phenacetin metabolism in man, *Science,* 194, 1055, 1976.
136. **Butler, M. A., Iwasaki, M., Guengerich, F. P., and Kadlubar, F. F.,** Human cytochrome P-450PA (P-450IA2), the phenacetin O-deethylase, is primarily responsible for the hepatic 3-demethylation of caffeine and N-oxidation of carcinogenic arylamines, *Proc. Natl. Acad. Sci. U.S.A.,* 86, 7696, 1989.
137. **Campbell, M. E., Spielberg, S. P., and Kalow, W. A.,** Urinary metabolite ratio that reflects systemic caffeine clearance, *Clin. Pharm. Ther.,* 42, 157, 1987.
138. **Kadlubar, F. F., Talaska, G., Butler, M. A., Teitel, C. H., Massengill, J. P., and Lang, N. P.,** Determination of N-oxidation phenotype in humans by analysis of caffeine urinary metabolites, in *Mutation and the Environment, Part B,* Mendelsohn, M. L., and Albertini, R. J., Eds., Wiley-Liss: New York, 1990, 107-114.
139. **Shields, P. G., Sugimura, H., Caporaso, N. E., Resau, J. H., Trump, B. F., Weston, A., and Harris, C. C.,** Restriction fragment length polymorphism analysis of CYP1A1 and relationship with lung cancer risk, *Environ. Health Persp.,* 98, 191, 1992.
140. **Ohgaki, H, Takayama, S, and Sugimura, T.,** Carcinogenicities of heterocyclic amines in cooked food, *Mutation Res.,* 259, 399, 1991.

10

Molecular Analysis of Mutations Induced *In Vivo* in Humans

J. Patrick O'Neill, Richard J. Albertini, and Janice A. Nicklas

Department of Medicine and Vermont Cancer Center Genetics Laboratory, University of Vermont, Burlington, VT

INTRODUCTION

It is clear that mutagenic agents exist in our environment and that humans are exposed to these agents. It is equally certain that somatic cell gene mutations occur *in vivo* in humans and that gene mutations underlie a variety of diseases, such as cancer and possibly aging. Disease relevance alone is motivation for understanding mechanisms by which mutations arise, especially *in vivo* in humans. In addition, exposures to environmental agents that result in increases in the frequencies of *in vivo* somatic mutations would be expected to give increases in the incidence of genotoxic diseases. Risk assessments for these diseases will be facilitated by knowledge of the environmentally induced mutagenic effects in humans, even if mutations studied are not in genes directly involved in the disease. Assays for mutations in "indicator" genes may allow assessments of adverse environmental effects in a general sense, and even characterize specific mutagenic events.

0-87371-631-0/94/$0.00+$.50

Agents are defined as mutagens primarily through *in vitro* assays in a variety of prokaryotic and eukaryotic cell systems. These assays assess the enzymatic metabolism of chemical agents, define the nature of DNA damage caused by these agents, and measure the capacity of cells to repair this damage. In addition, specific mutagenic events are often defined. However, while *in vitro* approaches demonstrate the mutagenic activity of a given agent in different cell systems, they do not provide information on the genetic risk of an individual's environmental exposure to that agent. Effective risk assessments require measurements of mutations occurring *in vivo* in exposed individuals. This goal necessitates the development of assays to quantify *in vivo* mutations in human somatic cells.

Exposure *in vivo* to a known mutagenic agent may result initially in DNA lesions in somatic cells which can be measured by a variety of methods for determining chemical adducts, DNA strand breaks, or cellular repair processes.[1] Such "processed" lesions can be measured as chromosome aberrations, sister chromatid exchanges, or micronuclei. These lesions may then be repaired or "processed" into stable sequences that alter the original informational content of the DNA. If the lesions are processed *in vivo* so as to result in damage which is not lethal but which disrupts or deletes a gene or its controlling element, a gene mutation can result. Alternatively, the lesion may itself be directly miscoding, or be processed into a miscoding sequence, to produce a mutagenic change. Thus, mutagenic events can range from single base changes to total gene deletions. Mutations can occur in any gene in a cell and be expressed as a functional alteration in that gene. All genes share this common pathway for somatic genetic damage and functional significance in that a heritable alteration in genetic information is the result, i.e., a somatic gene mutation. When these mutations occur in critical regions, they produce genotoxic disease.

Measures of somatic mutation in any gene are relevant for human population monitoring because all share this common functional pathway for processing of DNA lesions. For investigations of the mutation process itself and for monitoring the mutagenic effects of human exposures to environmental mutagens, recourse is made to somatic mutations in simple "housekeeping" or other genes which result in easily recognized cellular phenotypic changes. These mutations can serve as "recorders" of the mutagenic process. They can give measurements of the frequency of mutagenic events and, in some genes, even determine the specific type of mutagenic event. This chapter will describe an assay which allows both measurement and molecular analysis of mutations which occur *in vivo* in somatic cells. Specific results with humans exposed to ionizing irradiation will be presented as an example of the use of this assay in a prototype study for other environmental exposures.

ASSAYS FOR SOMATIC MUTATIONS IN HUMANS

There are at present four assays for detecting *in vivo* somatic mutations in humans. All measure altered cellular phenotypes in blood cells. Somatic mutations in the hemoglobin genes and in the glycophorin A gene are reflected as altered phenotypes in red blood cells.[2-8] Somatic mutations in the hypoxanthine-guanine phosphoribosyltransferase *(hprt)* gene and in one or more of the HLA genes are reflected as altered phenotypes in T-lymphocytes.[9-14] This chapter will describe the *hprt* assay alone. Comparisons with the other three assays have been presented previously.[15]

The *hprt* assay measures mutation in the gene which codes for the HPRT enzyme by selection for the phenotype of resistance to the purine analogue, 6-thioguanine (2-amino,6-mercaptopurine; TG). The HPRT enzyme mediates the phosphoribosylation of hypoxanthine and guanine for the recycling of these purine bases. It also phosphoribosylates purine analogues such as TG and converts TG into a cytotoxic nucleotide. The loss of HPRT enzyme can be selected by resistance to this analogue.

The 44 kb *hprt* gene is located on the X-chromosome (Xq26) and the entire gene region (57 kb) has been sequenced. The coding sequence of the gene contains 654 bases distributed in 9 exons.[16,17] The X-chromosome location of the gene renders it actually or functionally hemizygous in males or females, respectively. Thus, mutations in this gene can be measured in all humans without limitation related to heterozygosity as would be the case with an autosomal gene. There is a large background of information concerning *hprt* mutations induced *in vitro* in several mammalian cell systems.

There are two basic methods for determining the frequency of TG-resistant (TG^r) cells in human T-lymphocytes. One approach is a short-term DNA replication assay which measures the frequency of cells able to overcome TG inhibition of the first round of phytohemagglutinin (PHA) stimulated DNA synthesis *in vitro*. Peripheral blood T-lymphocytes in humans reside almost exclusively in the G_0 phase of the cell cycle *in vivo*. Incubation *in vitro* with PHA results in a mitogenic response, which can be measured as DNA replication through the incorporation of [^{3}H]-thymidine or 5-bromodeoxyuridine into the DNA. The frequency of these replicating cells can be determined by autoradiography or differential staining/fluorescence, respectively. The ratio of labeling index in the presence and absence of TG defines the *hprt* variant frequency.[9,18] While this approach does provide a measure of the *hprt* mutant frequency, a disadvantage is that mutant cells cannot be recovered for further study. However, the short-term nature of this assay, the possibility of automation, the ability to freeze samples for later analysis, and the relatively low cost make it useful for population studies. In order to actually analyze *hprt* mutations, a T-cell cloning assay has been developed.

THE *hprt* T-LYMPHOCYTE CLONING ASSAY

The T-lymphocyte cloning assay is based on the ability of human T-cells to initiate proliferation *in vitro* in response to PHA and to continue to proliferate in the presence of the lymphokine, interleukin-2 (IL-2). Only *hprt* mutant cells are able to proliferate in the presence of TG. The assay was first described in 1982 by

Albertini et al.,[10] and it has been modified and refined by several laboratories since then.[11-13]

This cloning assay generally employs the mononuclear cell (MNC) fraction of human peripheral blood. This fraction is isolated from heparinized whole blood by density sedimentation. The MNC fraction can be used immediately (fresh) or cryopreserved for future study. The ability to utilize cryopreserved cell samples allows the establishment of sample repositories which are particularly advantageous for population studies, especially for retrospective studies of disease state correlations. For cryopreservation, we routinely freeze cells at 10 to 20 × 10^6 cells per 1.0 or 1.8-ml ampule in medium RPMI1640 containing 20% serum and 8% dimethylsufoxide at 1°C per min and store under liquid nitrogen. Comparisons of fresh and cryopreserved aliquots of the same MNC sample have shown that each yields similar mutant frequency results.[19,20]

The MNC are cultured at limiting dilutions in the presence and absence of TG in round-bottom wells of 96 well microtiter dishes in medium containing IL-2 and feeder cells. Growing colonies are scored by use of an inverted phase contrast microscope and cloning efficiencies (C.E.) are calculated from the Poisson relationship $P_0 = e^{-x}$, where P_0 is the proportion of negative (without growing colonies) wells, and "x" is the average number of clonable cells per well; "x" divided by the number of cells added per well gives the C.E. The ratio of C.E. in the presence and absence of TG defines the TG^r mutant frequency.

Our most recent procedure for this cloning assay uses medium RPMI 1640, containing 20% nutrient medium HL-1 (Ventrex, Portland, ME), 5% defined, supplemented newborn bovine serum (NBS, heat inactivated at 56°C for 30 min; Sterile Systems, Logan, UT), PHA, and IL-2. The feeder cells are an *hprt*⁻ derivative of the human B-lymphoblastoid cell strain WIL-2 (designated TK6) which are irradiated with 90Gy and added at 0.5 to 1 × 10^4 cells per well. We have employed as sources of IL-2 both crude preparations of T-cell growth factor and purified, recombinant IL-2. The studies described here employed the supernatant from a lymphokine activated killer cell (LAK) therapy incubation of human MNCs with recombinant IL-2. (The optimal amount of supernatant was defined by a cell cloning assay.) Either fresh or thawed, washed cryopreserved MNC are incubated at 1 × 10^6 cells/ml at a density of 1.25 to 2.5 × 10^6 cells/cm^2 in a cell culture vessel in medium RPMI 1640 containing 20% HL-1, 5% NBS, and 1 μg/ml PHA (HA17 Wellcome Diagnostics) for 36 to 40 hr. This time is sufficient to achieve mitogen stimulation, but not cell division. The cells are sedimented, washed, and the cell number determined. They are then added to the 96 wells in 200 μl medium RPMI 1640 containing 20% HL-1, 5% NBS, 10 to 20% LAK supernatant, 0.125 μg/ml PHA, and irradiated feeder cells. By limiting dilution, cells are plated at 2 × 10^4 cells/well in the presence of 10 μ*M* TG, and at 1, 2, 5, and 10 cells per well in the absence of TG. The C.E. in the presence of TG divided by the mean C.E. in the absence of TG yields the *hprt* mutant frequency.

Growing colonies can be propagated *in vitro* for further study. The contents of wells containing colonies are transferred to larger wells containing the growth medium described above and 2.5 × 10^5 irradiated feeder cells/cm^2, usually into 1-

or 2-cm^2 wells initially, and subsequently into 4-cm^2 wells. Appropriate numbers of cells are then collected for the desired analysis.

The conditions for the cloning of T-cells both in the absence and presence of TG and for the further propagation of isolated colonies has been defined in detailed studies with samples from unexposed individuals. The growth medium, type and amount of serum, source and amount of crude TCGF or IL-2, and the amount of PHA has been developed through cloning assays. The requirement for feeder cells has been demonstrated and the optimal number defined. Non-TG selected C.E.s have been shown to be proportional to the number of cells per well over the testable range of 0.1 to 16 cells per well and TG selected C.E. over the range of 0.1 to 4×10^4 cells per well. These results have led to the routine use of 1, 2, 5, and 10 cells per well in the absence of TG, and 2×10^4 cells per well in the presence of TG.[19]

Mutant frequency results with unexposed individuals have been analyzed for 238 samples from humans in the age range of 19 to 80 years. Significant effects of nonselected C.E. and of age were found. Multiple variable regression analysis yields the relationship:

$$\ln(\text{mutant frequency}) = 1.17 - 0.32 \ln(\text{unselected cloning efficiency}) + 0.02\,(\text{age in years})$$

This analysis of unexposed individuals provides the database for comparison with results from exposed individuals. At present, the effects of unselected C.E. and age are the only significant variables found for this assay in unexposed individuals.

hprt MUTANT FREQUENCY IN RADIOIMMUNOTHERAPY PATIENTS

Ionizing irradiation is a model mutagen and has been well studied with *in vitro* mutation induction assays. Studies of humans exposed to ionizing irradiation have shown increases in *in vivo hprt* mutant frequencies. These include atomic bomb survivors, individuals accidentally exposed through environmental contamination, radiotherapy technicians and patients, and individuals exposed to household radon.[4,21-29] We have studied patients treated for nonresectable hepatoma or cholangiocarcinoma with multiple cycles of radiolabeled anti-human ferritin antibody therapy.[30,31] Each cycle of treatment was normally at 8-week intervals and consisted of injections of 20 mCi of [^{131}I] on day 1 and 10 mCi on day 5. Each patient also received an induction therapy of 21 Gy external beam irradiation to the disease site, as well as concombinant chemotherapy. This patient group was employed to study the *in vivo* effects of chronic whole-body irradiation from an internal source. In this regard, it has similarities to irradiation received from "bone-seeking" environmental radioactive contaminants. The total body irradiation received from radioimmunotherapy (RIT) depends on the extent of antibody targeting, whole-body distribution, body mass, and the *in vivo* half-life of the antibody.

Calculation of radiation dose delivered to the blood for [^{131}I] is 7.1 cGy per mCi injected radioactivity. This radiation is delivered in an exponentially decaying activity ratio, with a mean absorbed rate estimated to be 0.059 cGy per hour for blood. Thus, the average treatment of 30 mCi would result in approximately 200 cGy of radiation to the blood.

The induction of somatic cell mutations involves two distinct steps which have been defined primarily through *in vitro* studies. The gene must first be damaged and a mutation fixed; then, the mutant phenotype must be expressed to allow mutant selection. In the assay for TG^r mutants, phenotypic expression requires the replacement of active HPRT enzyme with the mutated form. The kinetics of mutation induction and expression *in vivo* are largely unknown. The design of the RIT treatment procedure allowed us to obtain blood samples 2 months after a treatment; i.e., a 2-month *in vivo* expression time was the minimum time possible.

The study group consisted of 13 patients sampled before treatment and 32 samples obtained after treatment. The treatment samples can be subdivided into three groups: (1) one group of 11 samples obtained 2 months after the initial RIT treatment; (2) one group of 14 samples obtained 2 months after the most recent of multiple treatments; and (3) one group of 7 samples obtained 4 to 27 months after the last of multiple treatments. The data are summarized in Table 1. Because the mutant frequency values did not fit a normal distribution, the data were logarithmically (base e; ln) transformed in order to approximate a normal distribution. This transformation was identical to that employed with the values from unexposed humans described earlier. The analysis of the transformed values showed that, in each of the three groups, mean mutant frequency values are significantly different from the pretreatment mean value at the 0.05 level of significance. The three groups were not significantly different from each other in pairwise comparison. There was no correlation between mutant frequency and age or unselected C.E. in any of the four groups of data. The first group of samples drawn 2 months after the initial RIT treatment seemed to be the least complicated group with a single exposure to 29 to 51 mCi. Analysis of the 13 pretreatment values and these 11 single posttreatment values showed a significant linear regression of ln mutant frequency (ln Mf) on mCi of injected activity ($p < 0.05$, $r = 0.75$, slope = 0.05), demonstrating that the increased mutant frequency was proportional to the single dose received (Figure 1). These mutant frequency values were obtained 2 months after treatment, and we have no information on whether this is the optimal sampling time.

The remaining two groups of RIT patients received multiple treatments at 2-month intervals. One group of 14 samples were obtained 2 months after the last treatment, with a cumulative activity received of 58 to 314 mCi (mean ± SD = 141.0 ± 78.6 mCi). The second group of seven samples were obtained 4 to 27 months after the last treatment, with a cumulative activity received of 60 to 152 mCi (mean ± SD = 110.4 ± 32.9 mCi). The ln transformed mean mutant frequencies were not significantly different from the single treatment group, suggesting that the increased mutant frequency over untreated primarily reflects the effect of the last treatment. Linear regression analysis of each group showed a similar

Table 1
Mutant Frequency in RIT Patients

Group (N)	Sampling time after treatment	Mean MF $\times 10^{-6}$ (±SD)	Treatment (mCi) Last Mean ±SD	Last Range	Total Mean ±SD	Total Range
Pretreatment (13)	—	11.5 ± 5.1	—	—	—	—
Posttreatment						
Single (11)	2 months	82.7 ± 80.1	35.1 ± 8.4	29–51	35.1 ± 8.4	29–51
Multiple (14)	2 months	74.9 ± 61.6	32.9 ± 7.9	28–51	141.0 ± 78.6	58–314
Multiple (7)	4 to 27 months	32.3 ± 18.9	35.9 ± 9.3	29–51	110.4 ± 32.9	60–152

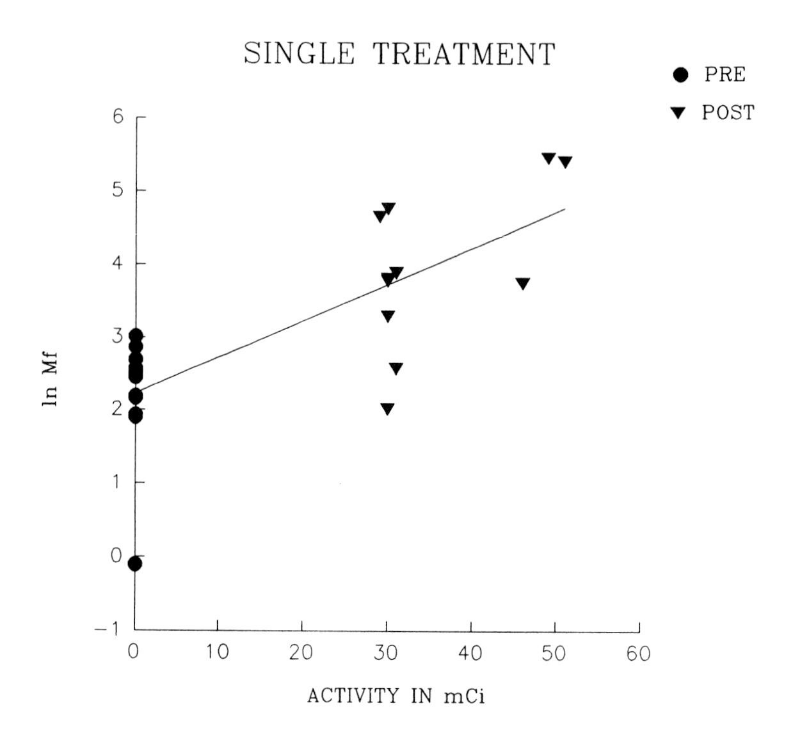

Figure 1. Relationship between ln Mf and mCi RIT (single treatment, 2-month sample). The results for 13 pretreatment samples (closed circles) and 11 posttreatment samples (closed inverted triangles) are presented. The line represents the linear regression fit (r = 0.75, slope = 0.05 = Δ ln Mf/mCi).

correlation of ln Mf with most recent dose of radioactivity received (Figure 2). Analysis of the total group of 45 samples showed a significant linear correlation of ln Mf with the most recent dose of radioactivity received ($p < 0.001$, r = 0.69, slope = 0.05). This is similar to that found with the single treatment samples. The correlation between ln Mf and total dose of radioactivity received was not as significant (Figure 3). The increased mutant frequency found in samples obtained 4 to 27 months after the last treatment suggests that the mutant cells persist *in vivo* in the absence of further treatment.

It is difficult to interpret the result that mutant frequency correlated better with the most recent dose of radioactivity received than with the total cumulative dose in the absence of a true, longitudinal study. One possibility is that this is the result of the cytotoxic nature of the RIT treatment, either directly through cell killing or indirectly through induced cell proliferation. The *hprt* mutant T-cells may be at selective disadvantage *in vivo,* especially during a proliferation response. In the absence of further treatment, elevated mutant frequencies are clearly maintained, as seen with the 4- to 27-month posttreatment samples. These results clearly demonstrate that exposure of humans to ionizing irradiation can be monitored by measurement of the *in vivo hprt* mutant frequency. The increased mutant fre-

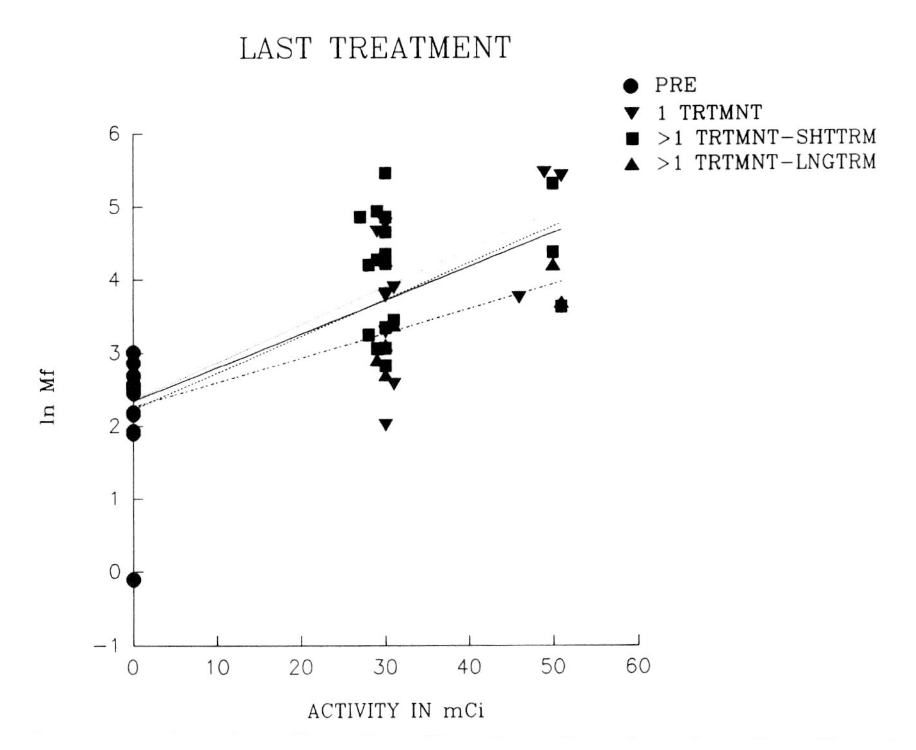

Figure 2. Relationship between ln MF and mCi RIT (single treatment or last treatment). The results for 13 pretreatment (closed circles), 11 single treatment (closed inverted triangles), 18 multiple treatment (2-month short-term sample, closed squares), and 7 multiple treatment (4- to 27-month long-term sample, closed triangles) are presented. The lines represent the linear regression fits for pretreatment plus single treatment (medium dashed line, r = 0.75, slope = 0.05); pretreatment plus multiple treatment (2-month sample, dotted line, r = 0.74, slope = 0.05); pretreatment plus multiple treatment (4- to 27-month sample, long dashed line, r = 0.67, slope = 0.03); and pretreatment plus the combined values for all treatments (solid line, r = 0.69, slope = 0.05).

quency appears to reflect recent exposures, although increases are still seen 2 years after exposure. Multiple exposures of the type involved in the RIT treatment do not appear to yield cumulative increases in the mutant frequency. However, the mutant frequency increases are proportional to the dose of radioactivity received in the last or only treatment.

MUTANT FREQUENCY VS MUTATION FREQUENCY

The T-cell cloning assay allows quantification of the TG^r *mutant* frequency occurring *in vivo* in humans. However, this does not necessarily provide a measure of the *in vivo hprt mutation* frequency. These terms are often used synonymously, although in cell populations capable of division, the frequency of mutants is not necessarily the same as the frequency of mutation events. If the *hprt*

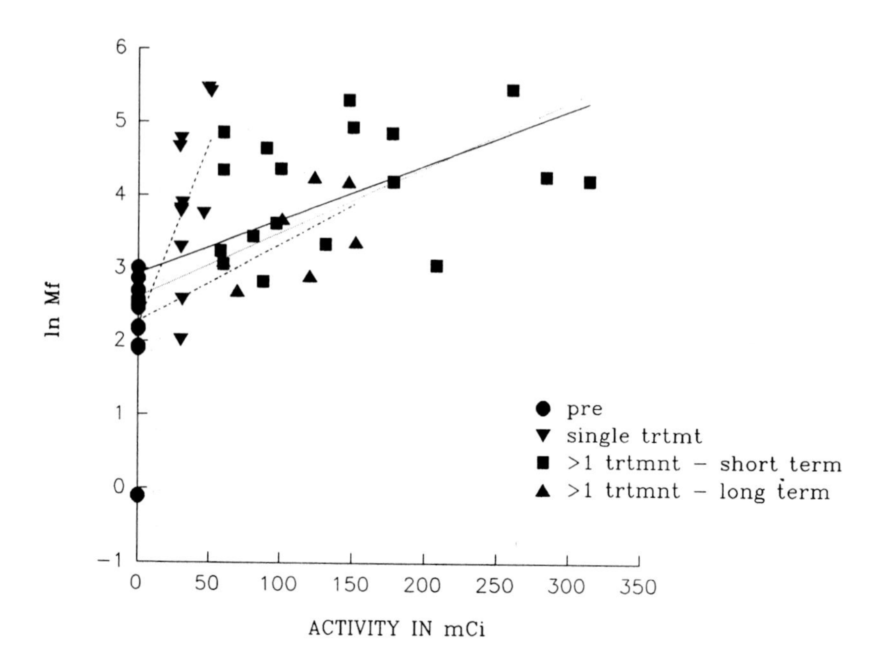

Figure 3. Relationship between ln Mf and mCi RIT (total or cumulative treatment). The results for the 49 samples described in Figure 2 are presented with the same symbols: closed inverted triangles (medium dashed line, r = 0.75, slope = 0.05); closed squares (dotted line, r = 0.68, slope = 0.009); and closed triangles (long dashed line, r = 0.67, slope = 0.01). The solid line represents the combined values fit (r = 0.51, slope = 0.007).

mutation itself does not inhibit cell division and if lymphocyte cell division occurs *in vivo*, the TG^r mutant frequency might be greater than the actual mutation frequency. However, in T-lymphocytes, the use of T-cell receptor (TcR) gene rearrangement pattern analyses can establish the *in vivo* relationship among TG^r mutant clones.

The T-cell population in humans displays an enormous heterogeneity *in vivo,* consistent with the needed heterogeneity of the immune response to antigenic challenge. There are different functional classes of T-cells; e.g., helper ($CD4^+$) and cytotoxic/suppressor ($CD8^+$) lymphocytes and subclasses within each class. There is heterogeneity in resting (G_0) vs proliferating T-cells, with the vast majority of adult cells being in the resting stage at any given time. Most extreme is the heterogeneity in the surface receptors present on mature T-cells, the TcRs which confer specific antigen reactivity.

G_0 T-lymphocytes are activated *in vivo* when they encounter antigen (presented by antigen presenting cells) which is recognized by the TcR proteins. These TcR proteins are composed of dimers of α and β or γ and Δ chains, and are associated with the CD3 antigen in a TcR-CD3 complex on the cell surface. The TcR genes are similar to those of the immunoglobulin genes in B-lymphocytes, in that they contain constant, variable, diversity, and joining segments which undergo rearrangement in the genomic DNA during T-cell differentiation. This process generates the diversity of the TcR proteins. The patterns of DNA rearrangement are readily recognized through Southern blot analysis of genomic DNA after appropriate restriction enzyme treatment.

We generally employ two restriction enzymes (HindIII and BamHI) and two TcR gene probes in order to define the TcR gene rearrangement pattern of a mutant clone.[32] The four TcR patterns of all mutant clones isolated from a single individual are then compared to define unique and shared patterns. If two or more mutant clones share the same TcR patterns, we consider these to have originated from the same *in vivo* precursor mature T-cell. This analysis can be combined with analyses of the *hprt* mutation in these clones to clearly define the relationship between mutant frequency and mutation frequency. If mutant clones show different TcR patterns (i.e., each has a unique pattern), each clone represents an independent mutation event arising in a mature T-cell. If mutant clones show different TcR patterns and the same *hprt* mutation, this may define a prethymic (i.e., pre-TcR gene rearrangement) mutation or a "hot spot" for mutation arising in mature, differentiated cells. If mutant clones show the same TcR pattern, they probably represent replicate isolates of the same *in vivo* mutation event. This is confirmed if they have the same *hprt* mutation and are thus true siblings. However, the possibility exists of multiple mutations occurring in an *in vivo* proliferating clone, which would result in mutant clones with the same TcR pattern but different *hprt* mutations. The combined TcR/*hprt* gene analyses can differentiate all these possibilities.

Our initial use of the TcR gene rearrangement patterns was to determine the number of unique mutant clones in order to correct the measured mutant frequency to calculate a mutation frequency. (We have previously described an example of an *in vivo* clonal expansion in which 94% of the mutants had the same TcR pattern. In this case, the measured mutant frequency of 470×10^{-6} was due to a mutation frequency of, conservatively, 28×10^{-6}.[33]) Our concern was that treatment-induced cell proliferation might lead to an elevated mutant frequency due to the clonal expansion of mutant cells with no increases in the actual mutation frequency. Our previous studies with unexposed, normal individuals have shown that 88% of the mutants have unique TcR patterns; the mutation frequency is, on average, 88% of the mutant frequency.[34,35] In the RIT study, pre- and posttreatment samples showed 86 and 89% of the mutants with unique TcR patterns, respectively (Table 2). Thus, as in unexposed individuals, the mean mutation frequency is not significantly different from the mean mutant frequency for these samples. Therefore, the increased mean Mf after RIT treatment is the result of an increased mean frequency of *mutations* in these patients.

Table 2
Mutation Frequency in RIT Patients

	"Normal"	Pretreatment	Posttreatment
No. of mutants	326	203	308
No. of unique TcR Patterns	287	175	274
Unique mutants (Mutations)	0.88	0.86	0.89
No. of TcR gene defined sets	17 doublets 2 triplets 3 quadruplets 1 nonamer	2 doublets 6 triplets 1 sextuplet 1 decamer	10 doublets 3 triplets 2 quadruplets 1 sextuplet 1octet

MOLECULAR ANALYSIS OF *hprt* MUTATIONS

A unique advantage of the T-cell cloning assay is the ability to propagate clones in order to analyze the mutations in the *hprt* gene. This allows the possibility of defining the spectrum of mutations induced by exposure to a given agent. Such a qualitative analysis of mutation events may be valuable for establishing the specific genetic effects of exposure to a mutagenic agent. In addition, defining the type of genetic damage in an indicator gene may allow predictions of genetic damage in suspected disease-related genes. Knowledge of expected damage could be important in approaches to the linkage between exposure and disease. In the specific case of the RIT patients, exposure to ionizing irradiation would be expected to result in mutations through large-scale deletions and rearrangements. These are the predominant type of mutations induced by *in vitro* irradiation of T-lymphocytes and other mammalian cells. In investigating this possibility, we performed Southern blot analyses of mutants from unexposed normals and pre- and posttreatment RIT patients (genomic DNA was digested with HindIII or PstI and hybridized with a full-length *hprt* cDNA probe). Our previous studies of unexposed normals showed that 14% (41/287) of the *hprt* mutations were the result of gross structural alterations detectable by Southern blot analysis (Table 3).[35] Of these 41 gross structural alterations, 9 were total *hprt* gene deletions, and 20 were partial gene deletions, for a total of 10% (29/287) deletion mutations. With the pretreatment RIT patients, 17% (20/118) of the mutants showed alterations, with 3 total and 10 partial *hprt* deletions, for a total of 11% (13/118) deletion mutations. In the posttreatment RIT patients, 38% (89/237) of the mutants showed alterations, with 37 total and 27 partial *hprt* deletions, for a total of 27% (64/237) deletion mutations. As a group, the posttreatment RIT patients show a significant increase ($p < 0.001$) in the proportion of mutants with structural alterations with 72% (64/89) of these being deletion mutations. The total deletion mutation is the predominant type, accounting for 16% (37/237) of the mutations,

Table 3
***hprt* Alterations in RIT Patients**

			Posttreatment		
	"Normal"	Pretreatment	Single (2 mo)	Multiple (2 mo)	Multiple (4–27 mo)
No. of mutations	287	118	12	89	136
No. with alterations	41 (0.14)	20 (0.17)	0	32 (0.36)	57 (0.42)
No. with partial deletions	20 (0.07)	10 (0.08)	0	11 (0.12)	16 (0.12)
No. with total deletions	9 (0.03)	3 (0.03)	0	12 (0.13)	25 (0.18)

as compared to 3% in both the normals and the pretreatment samples (9/287 and 3/118, respectively; Table 3). Therefore, not only does the proportion of deletion mutants arise after RIT, but the deletion sizes are larger.

The *hprt* alterations are presented for the three posttreatment groups in Table 3. We have analyzed too few mutants in the single-treatment group to draw any conclusions. However, in the multiple-treatment groups sampled, either 2 months (short term) or 4 to 27 months (long term) after the last treatment, the mutations show a similar fraction with *hprt* alterations, including the predominant total deletion mutations of 13 and 18%, respectively. These deletion mutations clearly persist in these patients after the cessation of treatment and appear to be a marker of *in vivo* exposure to ionizing irradiation.

We have begun a further analysis of *hprt* deletion mutations in order to eventually define the size of these deletions relative to the X-chromosome.[36] It has been suggested that an X-chromosome-linked gene like *hprt* will be insensitive to deletion events because of the haploid nature of such genes. The assumption is that large deletions will be lethal because of the concomitant loss of nearby "essential" genes. To investigate this possibility, we have examined the possible co-deletion of X-linked anonymous DNA sequences which have been mapped to the Xq26 region of the human genome. Six of these probes have been informative to date (DXS53, DXS79, DXS311, DXS86, DXS10, and DXS177). With total *hprt* deletions from unexposed individuals, 40% also show the loss of one or more of these linked markers. This co-deletion frequency rose to 75% with mutants from the RIT patients. This result suggests that ionizing irradiation induces larger deletions than those that occur "spontaneously". Consistent with this is the greater relative increase in total *hprt* deletions compared with partial deletions in these patients (Table 3).

The molecular analysis approach described here shows that *hprt* mutations induced *in vivo* in humans exposed to ionizing irradiation through RIT treatment are predominantly large structural alterations, especially total gene deletions which persist long after the radiation exposure has ended. Molecular analysis of human *in vivo* T-lymphocyte mutations can thus provide a molecular signature of exposure to a mutagenic agent.

CONCLUSIONS AND FUTURE DIRECTIONS

The use of a T-lymphocyte cloning assay allows quantification of the frequency of *in vivo* arising mutant somatic cells in humans. The further application of molecular analyses provides a measure of the *in vivo* mutation frequency and characterization of the specific mutation events. Studies with the *hprt* mutations are providing the spectrum of background mutations.[15] Southern blot studies define large gene alterations, and cDNA sequencing studies define small alterations, including single base changes. Similar approaches are defining the spectrum of induced mutations. The study described here demonstrates the induction of a specific type of *hprt* mutation in humans exposed to ionizing irradiation. Other studies have demonstrated the induction of mutations in humans exposed to chemical agents.[37-42] Characterization of these mutations is clearly the next step in this process of defining *in vivo* mutation spectra in humans.

The possible genetic effects of human exposure to genotoxic agents can be determined by measuring and analyzing somatic cell gene mutations. This is a crucial step in assessing human genetic risk from environmental exposure to genotoxic agents and for understanding the relevance of such mutations to disease.

ACKNOWLEDGMENTS

The authors' research is supported by NCI CA30688 and DOE FG028760502. This support does not constitute an endorsement of the views expressed. The authors thank L. M. Sullivan, T. C. Hunter, and M. T. Falta for their assistance in performing the RIT studies, and I. Gobel for manuscript preparation.

REFERENCES

1. **Albertini, R. J., and Robison, S. H.,** Human population monitoring, in *Genetic Toxicology,* CRC Press, Boca Raton, FL, 1991, 375.
2. **Stammatoyannopoulos, G., Nute, P., Lindsley, D., Farguhar, M. B., Nakamato, B., and Papayannepoulou**, Somatic-cell mutation monitoring system based on human hemoglobin mutants, in *Single Cell Monitoring System, Topics in Chemical Mutagenesis,* Plenum Press, New York, 1984, 1.
3. **Verwoerd, N. P., Bernini, L. F., Bonnet, J., Tanke, H. J., Natarajan, A. T., Tates, A. D., Sobels, F. H., and Ploem, J. S.,** Somatic cell mutations in humans detected by image analysis of immunofluorescently stained erythrocytes, in *Clinical Cytometry and Histometry,* Academic Press, San Diego, 1987, 465.
4. **Natarajan, A. T., Ramalho, A. T., Vyas, R. C., Bernini, L. F., Tates, A. D., Ploem, J. S., Nascimento, C. H., and Curado, M. P.,** Goiania radiation accident: results of initial dose estimation and follow up studies, in *New Horizons in Biological Dosimetry,* Wiley-Liss, New York, 1991, 145.
5. **Furthmayer, H.,** Structural analysis of a membrane glycoprotein: glycophorin A, *J. Supramol. Struct.,* 7, 121, 1977.

6. **Kudo, S., and Fukuda, M.,** Structural organization of glycophorin A and B genes: glycophorin B gene evolved by homologous recombination at *Alu* repeat sequences, *Proc. Natl. Acad. Sci. U.S.A.,* 86, 4619, 1989.
7. **Jensen, R. H., Bigbee, W. L., and Langlois, R. G.,** *In vivo* somatic mutations in the glycophorin A locus of human erythroid cells, in *Mammalian Cell Mutagenesis,* Cold Spring Harbor Laboratory, Cold Spring Harbor, NY, 1987, 149.
8. **Langlois, R. G., Nisbet, B., Bigbee, W. L., Ridinger, D. N., and Jensen, R. H.,** An improved flow cytometric assay for somatic mutations at the glycophorin A locus in humans, *Cytometry,* 11, 513, 1990.
9. **Strauss, G. H., and Albertini, R. J.,** Enumeration of 6-thioguanine resistant peripheral blood lymphocytes in man as a potential test for somatic cell mutations arising *in vivo, Mutation Res.,* 61, 353, 1979.
10. **Albertini, R. J., Castle, K. L., and Borcherding, W. R.,** T-cell cloning to detect the mutant 6-thioguanine-resistant lymphocytes present in human peripheral blood, *Proc. Natl. Acad. Sci. U.S.A.,* 79, 6617, 1982.
11. **Morley, A. A., Trainor, K. J., Seshadri, R., and Ryall, R. G.,** Measurement of *in vivo* mutations in human lymphocytes, *Nature,* 302, 155, 1983.
12. **Cole, J., Green, M. H. L., James, S. E., Henderson, L., and Cole, H.**, A further assessment of factors influencing measurements of thioguanine-resistant mutant frequency in circulating T-lymphocytes, *Mutation Res.,* 204, 493, 1988.
13. **Hakoda, M., Akiyama, M., Kyoizumi, S., Awa, A. A., Yamakido, M., and Otake, M.,** Increased somatic cell frequency in atomic bomb survivors, *Mutation Res.,* 201, 39, 1988.
14. **Janatipour, M., Trainor, K. J., Kutlaca, R., Bennett, G., Hay, J., Turner, D. R., and Morley, A. A.,** Mutations in human lymphocytes studied by an HLA selection system, *Mutation Res.,* 198, 221, 1988.
15. **Albertini, R. J., Nicklas, J. A., O'Neill, J. P., and Robison, S. H.,** *In vivo* somatic mutations in humans: measurement and analysis, *Annu. Rev. Genet.,* 24, 305, 1990.
16. **Patel, P. I., Nussbaum, R. I., Framson, P. E., Ledbetter, D. H., Caskey, C. T., and Chinault, A. C.,** Organization of the *hprt* gene and related sequences in the human geneome, *Som. Cell. Mol. Gen.,* 10, 483, 1986.
17. **Edwards, A., Voss, H., Rice, P., Civitello, A., Stegemann, J., Schwager, C., Zimmerman, J., Erfle, H., Caskey, C. T., and Ansorge, W.,** Automated DNA sequencing of the human HPRT locus, *Genomics,* 6, 593, 1990.
18. **Ostrosky-Wegman, P., Montero, R. M., Cortinas de Nova, C., Tice, R. R., and Albertini, R. J.,** The use of bromodeoxyuridine labeling in the human lymphocyte HGPRT somatic mutation assay, *Mutation Res.,* 191, 211, 1988.
19. **O'Neill, J. P., McGinniss, M. J., Berman, J. K., Sullivan, L. M., Nicklas, J. A., and Albertini, R. J.,** Refinement of a T-lymphocyte cloning assay to quantify the *in vivo* thioguanine-resistant mutant frequency in humans, *Mutagenesis,* 2, 87, 1987.
20. **O'Neill, J. P., Sullivan, L. M., Booker, J. K., Pornelos, B. S., Falta, M. T., Greene, C. J., and Albertini, R. J.,** Longitudinal study of the *in vivo hprt* mutant frequency in human T-lymphocytes as determined by a cell cloning assay, *Environ. Mol. Mutagenesis,* 13, 289, 1989.
21. **Ammenheuser, M. M., Au, W. W., Whorton, E. B., Jr., Belli, J. A., and Ward, J. B., Jr.,** Comparison of *hprt* variant frequencies and chromosome aberration frequencies in lymphocytes from radiotherapy and chemotherapy patients: a prospective study, *Environ. Mol. Mutagenesis,* 18, 126, 1991.

22. **Messing, K., Ferraris, J., Bradley, W. E. C., Swartz, J., and Seifert, A. M.,** Mutant frequency of radiotherapy technicians appears to be associated with recent dose of ionizing radiation, *Health Phys.,* 57, 537, 1989.
23. **Messing, K., Seifert, A. M., and Bradley, W. E. C.,** *In vivo* mutant frequency of technicians professionally exposed to ionizing radiation, in *Monitoring of Occupational Genotoxicants,* Alan R. Liss, New York, 1986, 87.
24. **Messing, K., and Bradley, W. E. C.,** *In vivo* mutant frequency rises among breast cancer patients after exposure to high doses of gamma radiation, *Mutation Res.,* 152, 107, 1985.
25. **Sala-Trepat, M., Cole, J., Green, M. H., Rigaud, O., et al.,** Genotoxic effects of radiotherapy and chemotherapy on the circulating lymphocytes of breast cancer. III. Measurement of mutant frequency to 6-thioguanine resistance, *Mutagenesis,* 5, 593, 1990.
26. **Hakoda, M., Akiyama, M., Hirai, Y., Kyoizumi, S., and Awa, A. A.,** *In vivo* mutant T cell frequency in atomic bomb survivors carrying outlying values of chromosome aberration frequencies, *Mutation Res.,* 202, 203, 1988.
27. **Ostrosky-Wegman, P., Montero, R., Palao, A., Cortinas de Nava, C., Hurtado, F., and Albertini, R. J.,** 6-Thioguanine-resistant T-lymphocyte autoradiographic assay. Determination of variation frequencies in individuals suspected of radiation exposure, *Mutation Res.,* 232, 49, 1990.
28. **Natarajan, A. T., Vyas, R. C., Wiegant, J., and Curado, M. P.,** A cytogenetic follow-up study of the victims of a radiation accident in Goiania (Brazil), *Mutation Res.,* 247, 103, 1991.
29. **Bridges, B., Cole, J., Arlett, C. F., Green, M. H. L., Waugh, A. P. W., Beare, D., Henshaw, D. L., and Last, R. D.,** Possible association between mutant frequency in peripheral lymphocytes and domestic radon concentrations, *Lancet,* 337 (May 18), 1187, 1991.
30. **Nicklas, J. A., O'Neill, J. P., Hunter, T. C., Falta, M. T., Lippert, M. J., Jacobson-Kram, D., Williams, J. R., and Albertini, R. J.,** *In vivo* ionizing irradiations produce deletions in the *hprt* gene of human T-lymphocytes, *Mutation Res.,* 250, 383, 1991.
31. **Nicklas, J. A., Falta, M. T., Hunter, T. C., O'Neill, J. P., Jacobson-Kram, D., Williams, J., and Albertini, R. J.,** Molecular analysis of *in vivo hprt* mutations in human lymphocytes. V. Effects of total body irradiation secondary to radioimmunoglobulin therapy (RIT), *Mutagenesis,* 5, 461, 1990.
32. **Nicklas, J. A., O'Neill, J. P., and Albertini, R. J.,** Use of T-cell receptor gene probes to quantify the *in vivo hprt* mutations in human T-lymphocytes, *Mutation Res.,* 173, 65, 1986.
33. **Nicklas, J. A., O'Neill, J. P., Sullivan, L. M., Hunter, T. C., Allegretta, M., Chastenay, B. F., Libbus, B. L., and Albertini, R. J.,** Molecular analyses of *in vivo* hypoxanthine-guanine phosphoribosyltransferase mutations in human T-lymphocytes. II. Demonstration of a clonal amplification of *hprt* mutant T-lymphocytes *in vivo, Environ. Mol. Mutagenesis,* 12, 271, 1988.
34. **Albertini, R. J., O'Neill, J. P., Nicklas, J. A., Allegretta, M., Recio, L., and Skopek, T. R.,** Molecular and clonal analysis of *in vivo hprt* mutations in human cells, in *Human Carcinogen Exposure: Biomonitoring and Risk Assessment,* IRL Press, Oxford, 1991, 103.
35. **Nicklas, J. A., Hunter, T. C., O'Neill, J. P., and Albertini, R. J.,** Molecular analyses of *in vivo hprt* mutations in human T-lymphocytes. III. Longitudinal study of *hprt* gene structural alterations and T-cell clonal origins, *Mutation Res.,* 215, 147, 1989.

36. **Nicklas, J. A., Hunter, T. C., O'Neill, J. P., and Albertini, R. J.,** Fine structure mapping of the *hprt* region of the human X chromosome (Xq26), *Am. J. Hum. Genet.,* 49, 267, 1991.
37. **Branda, R. F., O'Neill, J. P., Sullivan, L. M., and Albertini, R. J.,** Factors influencing mutation at the *hprt* locus in T-lymphocytes: women treated for breast cancer, *Cancer Res.,* 51, 6603, 1991.
38. **Dempsey, J. L., Seshadri, R. S., and Morley, A. A.,** Increased mutation frequency following treatment with cancer chemotherapy, *Cancer Res.,* 45, 2873, 1985.
39. **Huttner, E., Mergner, U., Braun, R., and Schoneich, J.,** Increased frequency of 6-thioguanine-resistant lymphocytes in peripheral blood of workers employed in cyclophosphamide production, *Mutation Res.,* 243, 101, 1990.
40. **Ammenheuser, M. M., Ward, J. B., Jr., Whorton, E. B., Jr., Killian, J. M., and Legator, M. S.,** Elevated frequencies of 6-thioguanine resistant lymphocytes in multiple sclerosis patients treated with cyclophosphamide: a prospective study, *Mutation Res.,* 204, 509, 1988.
41. **Bernini, L. F., Natarajan, A. T., Schreuder-Rotteveel, A. H. M., Giordano, P. C., Ploem, J. S., and Tates, A.,** Assay for somatic mutation of human hemoglobins, in *Mutation and the Environment. Part C: Somatic and Heritable Mutation, Adduction, and Epidemiology,* Wiley-Liss, New York, 1990, 57.
42. **Tates, A. D., Grummt, T., Tornqvist, M., Farmer, P. B., van Dam, F. J., van Mossel, H., Schoemaker, H. M., Osterman-Golkar, S., Uebel, C., Tang, Y. S., Zwinderman, A. H., Natarajan, A. T., and Ehrenberg, L.,** Biological and chemical monitoring of occupational exposure to ethylene oxide, *Mutation Res.,* 250, 483, 1991.

11

Human Gene Inducibility: A Marker of Exposure and Susceptibility to Environmental Toxicants

Greg N. Cosma and Seymour J. Garte

Nelson Institute of Environmental Medicine, New York University Medical Center, New York, NY

INTRODUCTION

Transcriptional regulation of mammalian genes is the subject of considerable research in molecular biology,[1] and elucidation of the mechanisms by which genes regulate their transcription has profound impact on the biological research of human diseases such as cancer.[2] In addition to this more traditional application of molecular biological research, the induction of target genes can be viewed as one event in a multiple-step pathway by which a cell responds to toxicant exposure, as demonstrated in Figure 1. Following tissue exposure, a xenobiotic may bind to critical intracellular macromolecules, resulting in cell injury or dysregulation of a cell function. In addition, the agent may bind to nuclear regulatory regions, either directly or via an intracellular receptor, thereby inducing transcription of a target gene. The subsequent synthesis of a protein product may then result in detoxification, for example, metabolism and excretion of the toxicant.[3] On the other hand, some other cell function may be affected that eventually culminates in further cell injury. Regardless of the sequalae, the measurement of target gene

0-87371-631-0/94/$0.00+$.50

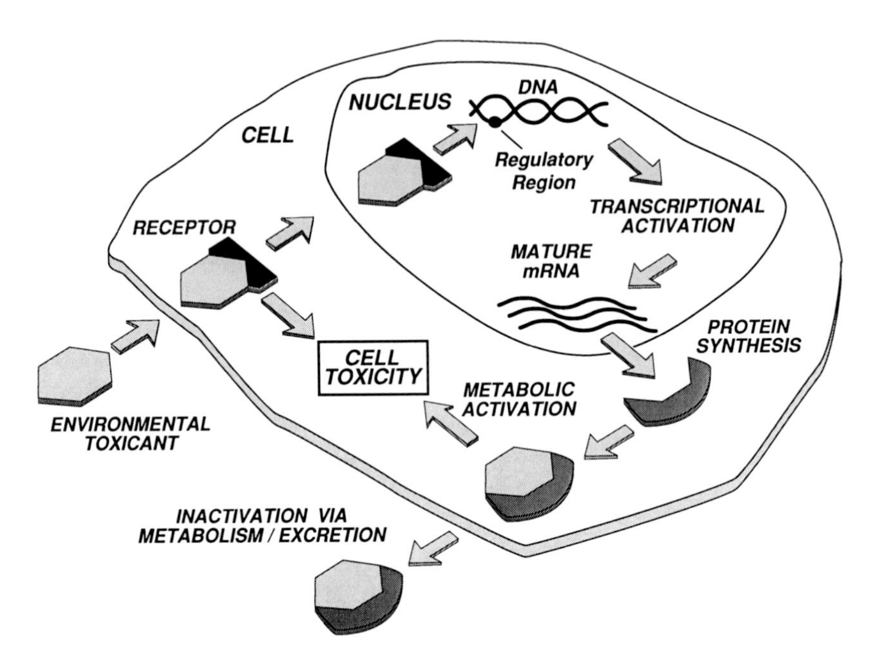

Figure 1. Scheme of molecular events that results from cellular exposure to a toxicant. Note that gene induction followed by protein synthesis can have opposing biological consequences to the organism.

inducibility can be regarded as a potential biological marker of a specific xenobiotic exposure. Furthermore, genetic control over this activity may influence individual susceptibility to disease after repeated exposures.

Modern epidemiologic research has begun to embrace these concepts of genetic regulation in order to develop biological markers of human exposure to environmental toxicants. Examples of molecular biomarkers that have been applied to human exposure assessment include: measurement of DNA-PAH adducts,[4] induction of chromosomal aberrations,[5] genetic mutations,[6] and DNA repair pathways.[7] The subject of this chapter is the use of human gene inducibility as a marker of toxicant exposure, and the identification of genetic factors that may affect transcriptional activity of inducible genes and disease susceptibility in exposed individuals. To highlight this rapidly evolving area of environmental research, we have focused on two principle xenobiotic-inducible human gene families: the cytochrome P450 and metallothionein genes and their responses to polycyclic aromatic hydrocarbons and heavy metals, respectively.

CYTOCHROME P450 GENES

The super gene family encoding the various members of the so-called "microsomal mixed function oxidase enzymes", or the cytochrome P450 proteins, has been studied extensively in the past decade.[8] Prior to these molecular investigations, the elucidation of substrate specificities of the P450 proteins occupied much of the toxicological research of xenobiotic metabolism.[3] These monooxygenase enzymes represent the first line of defense against toxic lipophilic chemicals. P450-mediated reactions involve the incorporation of atomic oxygen into the substrate, thereby allowing for their eventual excretion by making these compounds more hydrophilic. However, an undesirable consequence of these reactions causes certain chemicals to become activated into carcinogenic compounds. These P450-mediated pathways are the primary steps by which a majority of chemical carcinogens are activated into DNA binding mutagens.

It is clear that metabolism by cytochrome P450 enzymes can have opposing biological consequences. Perhaps because humans have been exposed to a wide array of diverse xenobiotics, we have evolved a large number of these proteins,[9] and it is likely that individual genetic control over their activities may influence susceptibility to diseases such as cancer.[10] With the availability of molecular biological probes for specific members of the human P450 genes, the study of the factors that control their inducibility at a transcriptional level is now possible.

CYP1A1 Gene

One member of the P450 family that has been studied extensively with respect to its transcriptional regulation by xenobiotics is the P450IA1, or CYP1A1, gene.[9] This P450 gene is responsible for the metabolism of commonly found environmental polycyclic aromatic hydrocarbons (PAHs): for example, the human carcinogen benzo[a]pyrene and other diverse toxicants such as TCDD or dioxin.[3] PAHs comprise a profound class of environmental toxicants, and their diversity and the extent to which they are found in the environment have made them a target of worldwide regulatory efforts. Besides their presence in tobacco smoke, PAHs are found in abundance in polluted ambient air as a result of motor vehicle exhaust, as well as from a variety of industrial sources (e.g., petrochemical refineries and steel foundries).[11] In addition, PAHs are a common source of aquatic pollution typically associated with industrial effluents. The effects of PAHs on human health have been well documented and recently reviewed.[12] They include carcinogenicity in lung, skin, scrotal, and various other human tissues, as well as immunotoxicity, reproductive toxicity, and teratogenicity. The ubiquity of PAHs in occupational and environmental settings, along with their highly toxic properties to human populations, has led to an intensive search for the factors responsible for mediating their biological effects.

The pioneering efforts in the laboratory of Nebert[13] first determined the molecular structure of the human CYP1A1 gene; and quickly following these insights, a search for the factors that affect its transcription began. It is now evident that CYP1A1 gene expression is the end result of a complex multistep pathway that begins after cell exposure to a PAH, and is mediated by the binding to an intracellular receptor, the Ah receptor.[14] It is believed that the Ah receptor belongs to a larger family of steroidal receptors, and that some of these other receptors may participate in the regulation of CYP1A1 gene activity. Following substrate binding, the Ah receptor complex travels to the nucleus where it binds to nuclear regulatory elements located in proximity to the structural CYP1A1 gene. The activation of these so-called "xenobiotic responsive elements" results in the induction of CYP1A1 gene transcription. In addition, a negative regulatory element exists that can potentially control transcriptional activity of the gene.[15] Furthermore, there exists a regulatory domain within intron 1 of the CYP1A1 gene that is controlled by the binding of glucocorticoid hormones. These endogenous agents can potentiate transcriptional activity induced by a PAH; however, glucocorticoids alone do not appear to regulate the gene in the absence of a PAH substrate.[16]

The apparent complexity of CYP1A1 gene regulation does not readily lend itself to a thorough evaluation of all the factors that control its transcription; however, there appear to be great differences in CYP1A1 activity between individuals.[17] Much of the recent research of human P450 metabolism has sought an explanation to this variability in CYP1A1 activity, and whether intrinsic factors, including genetic control, can affect gene expression, or whether exposure to toxicants can modulate the activity of the CYP1A1 gene. The first studies demonstrating differences in CYP1A1 activity between individuals were performed by measuring the catalytic activity of the CYP1A1 protein (aryl hydrocarbon hydroxylase) in peripheral blood lymphocytes.[17,18] These studies measured the inducibility of the protein in mitogen-stimulated cells exposed to a PAH *in vitro* from presumably unexposed individuals. As much as 100-fold differences in catalytic activity were found between individuals, which suggested the participation of intrinsic factors in regulation of human CYP1A1 activity. Furthermore, studies of identical vs fraternal twins revealed a greater variability in the latter group, which suggested a genetic component in the control of CYP1A1 gene activity.[19] These studies were followed by investigation of CYP1A1 activities in lung cancer patients, in which a higher catalytic activity of the AHH protein was found in lymphocytes from patients vs controls.[20] Subsequent studies were not able to confirm these initial results until methodologic difficulties were resolved, after which similar results were again reported in a larger population of lung cancer patients.[21] Since these patients had a history of tobacco use, it was difficult to ascertain whether the increase in CYP1A1 inducibility was a genetic susceptibility factor that predisposed individuals to lung cancer, or whether it was a result of their exposure to the PAH found in tobacco smoke.

Recent studies of CYP1A1 activity in lung tissue from smokers have confirmed the inducibility of the human gene by xenobiotic exposures.[22,23] Only normal lung tissue from active smokers had measurable levels of CYP1A1 mRNA which rapidly diminished after cessation of smoking. Furthermore, CYP1A1 gene expression in human lung tumors was more frequently elevated in smoking patients, although there appeared to be an altered control of transcription in many tumor samples.[24] In addition to these studies of lung cancer patients, CYP1A1 activity was assessed in a population of asbestos-exposed workers, and results of these studies demonstrated a higher level of AHH

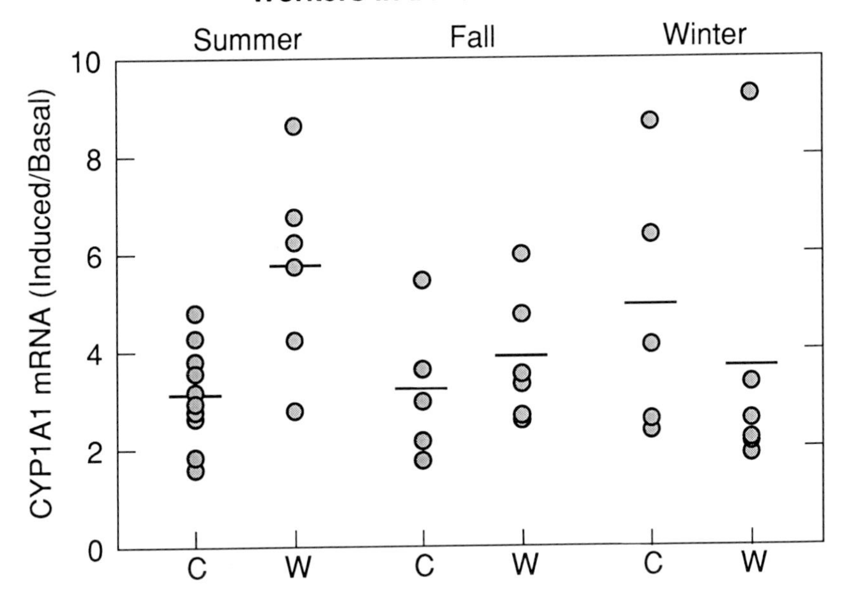

Figure 2. Seasonal study of individual responses of CYP1A1 gene expression from creosote-exposed workers (W) and unexposed individuals (C). Horizontal bars represent mean values of CYP1A1 mRNA levels.

inducibility in mitogen-stimulated lymphocytes from workers compared to unexposed individuals.[25]

The results of these previous studies of CYP1A1 inducibility suggested that the CYP1A1 gene may be under the control of both genetic factors and environmental toxicant exposures. What remained to be determined was whether transcriptional regulation of the CYP1A1 gene was under similar influences, or whether gene expression was independent of protein catalytic activity. Nebert et al. first showed a correlation between AHH activity and CYP1A1 mRNA levels in mitogen-stimulated human lymphocytes.[26] These results were later confirmed in a similar study of CYP1A1 gene inducibility and AHH activity in human lymphocytes.[27] More recently, our laboratory has studied CYP1A1 gene expression in stimulated lymphocytes from both unexposed individuals and PAH-exposed workers.[28] These studies were originally undertaken to determine whether the measurement of CYP1A1 gene inducibility may be a useful biomarker of PAH exposure in human populations. We chose to study a group of creosote-exposed railroad workers at seasonal intervals in order to assess whether changes in work conditions influenced expression of the CYP1A1 gene. These studies were performed by measuring basal and induced CYP1A1 mRNA levels in mitogen-stimulated lymphocytes. Figure 2 is a scatter plot of individual responses that were obtained from workers and controls when sampled in the summer, fall, and winter. Lymphocyte

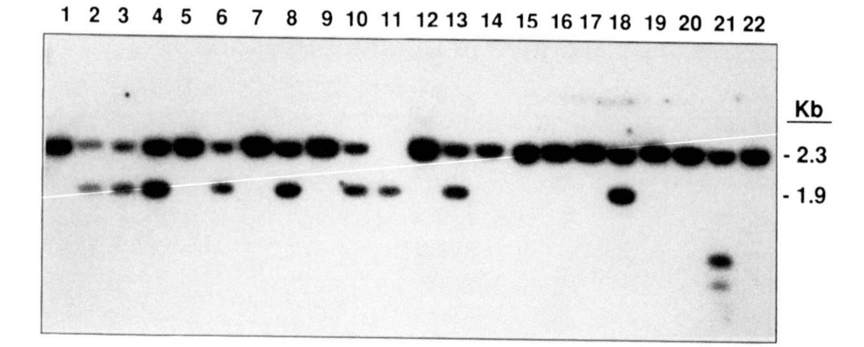

Figure 3. Restriction fragment length polymorphism (RFLP) analyses of CYP1A1 genotypes from 22 individuals. Genomic DNA was digested with Msp1 restriction enzyme; following gel electrophoresis and Southern transfer onto nylon membrane filters, DNA fragments were visualized by hybridization to a radiolabeled 3′ CYP1A1 cDNA probe. Individual 1 is an example of a normal homozygote, individual 2 is a heterozygote, and individual 11 is a homozygous variant.

CYP1A1 gene inducibility levels were significantly elevated in workers sampled in the summer when actual creosote exposures were greatest, but not from workers sampled in the colder seasons. The results of this pilot study suggested that CYP1A1 gene expression may be a useful indicator of PAH exposure, and that transcriptional activity of the gene can be modulated by environmental PAH exposure.

In addition to these studies that focused on the phenotypic differences in CYP1A1 gene activity between individuals, intensive interest has been paid to determining whether structural changes occur in the CYP1A1 gene that can influence its activity; for example, whether there exist DNA polymorphisms in individuals that affect either transcriptional activity or protein catalytic activity.[29] Recently, a polymorphism in the human CYP1A1 gene was described that appeared to be correlated to elevated lung cancer risk in a Japanese population.[30,31] Furthermore, this polymorphism appeared to be related to AHH enzyme inducibility, at least within one multigenerational family in which the CYP1A1 polymorphism was inherited in Mendelian fashion.[32] However, other studies failed to show any association between CYP1A1 genotype and lung cancer risk when studied in European populations.[33] Our laboratory has recently determined the racial distribution of this CYP1A1 polymorphism, and whether there exists a correlation between genotype and transcriptional activity of the gene.[34] Figure 3 is an example of the restriction fragment length polymorphism (RFLP) analyses that were performed on these individuals to determine their CYP1A1 genotypes. Our data demonstrated distinct racial differences in genotypic distributions of the CYP1A1 polymorphism, in which Asians exhibited a much higher frequency of the polymorphism compared to European- or African-Americans (Table 1). The second objective of these studies was to determine whether a correlation existed between transcriptional activity of the gene and DNA polymorphisms; when these two

Table 1
CYP1A1 Genotypic Frequencies in Three Racial Groups

Race	Homozygous normal	Heterozygous	Homozygous variant
European-American (33)	0.70 (23)	0.30 (10)	0
African-American (12)	0.58 (7)	0.42 (5)	0
Asian (23)	0.48 (11)	0.39 (9)	0.13 (3)

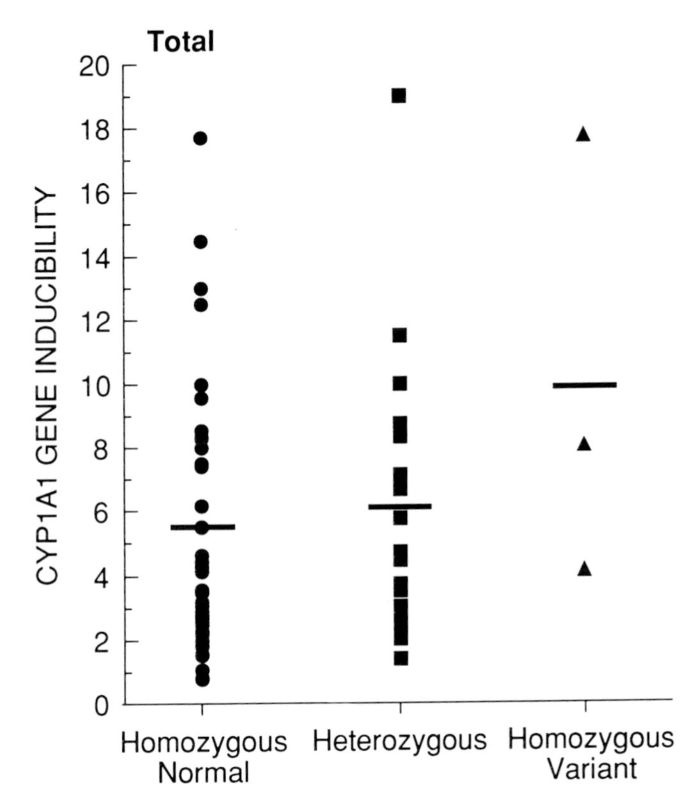

Figure 4. Comparison of CYP1A1 gene expression and CYP1A1 genotypes. Horizontal bars represent mean values of induced/basal CYP1A1 mRNA levels.

parameters were compared in these same individuals, we found no association between CYP1A1 gene inducibility and genotype (Figure 4).

Our racial distribution studies of CYP1A1 genotypes lent insight into the possible reasons why there appeared to be conflicting results in associating the Msp1 polymorphism with lung cancer risk in Asian vs European populations;[31,33] however, there remained to be determined the functional significance of the CYP1A1 polymorphism and how it could increase susceptibility to lung cancer. A recent

Table 2
Comparison of CYP1A1 Protein Activity and Msp1 Genetic Polymorphisms

	EROD Activity[a]	
Individual/genotype	**Basal (ave.)**	**Induced (ave.)**
Normal 1	0.048	0.210
Normal 2	0.027 (0.033)	0.400 (0.295)
Normal 3	0.023	0.275
Heterozyg 1	0.084	0.473
Heterozyg 2	0.075 (0.072)	0.164 (0.308)
Heterozyg 3	0.058	0.288
Homozyg 1	0.027	1.44
Homozyg 2	0.034 (0.045)	0.951 (0.999)[b]
Homozyg 3	0.074	0.607

[a] pmol resorufin/min/mg prot.
[b] $p<0.005$, compared to normal and heterozygote.

study by Hayashi et al.[35] suggested that the Msp1 polymorphism may be linked to another polymorphism in the CYP1A1 gene, and that this polymorphism occurs in a site that encodes for the catalytic subunit of the CYP1A1 protein. We have recently performed CYP1A1 enzymatic analyses in lymphocytes from Asian individuals of all three CYP1A1 genotypes and, as shown in Table 2, found a significant elevation in CYP1A1 protein activity in homozygous variant individuals. Interestingly, heterozygous individuals did not appear to display this change in CYP1A1 activity. If one compares the results of these studies to the studies of lung cancer risk in Japanese, a compelling association is apparent. Only homozygous variant individuals were at elevated risk to developing lung cancer as a result of tobacco usage,[31] and our studies have shown that these individuals also display a greater activity of the CYP1A1 enzyme, which is responsible for the metabolism of PAH procarcinogens found in tobacco smoke.[3] Thus, it would appear that the CYP1A1 polymorphism may be a genetic susceptibility factor in the development of PAH-related cancers, but that racial distributions of the polymorphism must be taken into account when determining disease susceptibility within different ethnic groups.

CYP2D6 Gene

Another cytochrome P450 gene that has been studied in human populations with respect to activity and DNA polymorphisms is the debrisoquine hydroxylase (CYP2D6) gene.[36] The genetic polymorphism in the CYP2D6 gene was first discovered by phenotypic differences in catalytic rates of debrisoquine metabolism in Caucasians.[37] Subsequent to these studies, reports of associations with lung cancer susceptibility surfaced,[38] but other studies could not confirm these initial findings.[39] These discrepancies were presumably due to the inadequacy of the phenotypic test which identifies individuals as either poor or extensive metabolizers.[40] Recently, a DNA RFLP in the CYP2D6 gene was discovered that appeared to be associated with differences in its activity,[41] although a less than perfect correlation exists between the RFLP and the debrisoquine poor metabolizer phenotype.[36] Using this genotype assay, a reaffirmation of the association between lung cancer risk and CYP2D6 polymorphism was demonstrated.[42] Unfortunately, a follow-up study by the same group within a larger case-control population did not find the association to be statistically significant.[43]

There remains an obvious need to determine whether a firm association exists between CYP2D6 polymorphisms and lung cancer risk. Furthermore, the potential procarcinogen substrates of this protein have yet to be identified. In this regard, it is interesting to note that individuals who are classified as poor metabolizer phenotypes appear to have a reduced cancer risk.[42] This would suggest that, in similar fashion to the CYP1A1 gene, an elevation in activity may be related to an enhanced production of carcinogenic metabolites. Finally, while there have been no studies conducted to determine CYP2D6 gene expression in diverse genotypic individuals, the recent cloning of the human gene and transcriptional start site should allow for a more thorough evaluation of genetic regulation of the CYP2D6 gene in human populations.[44]

CYP2E1 Gene

The human CYP2E1 gene is responsible for the metabolism of nitrosamines, many of which are human carcinogens and found in tobacco smoke.[45] The CYP2E1 protein can metabolically activate nitrosamines into DNA-binding mutagens, which is presumably responsible for their ultimate carcinogenicity.[46] A recent polymorphism in the human gene has been reported that appears to be related to an enhanced susceptibility to lung cancer among tobacco smokers.[47] While the human gene has been cloned and sequenced,[48] no studies to date have investigated the genetic regulation, including transcriptional control, of this P450 gene. The polymorphism reported by the Japanese group occurs within an intron sequence of the gene[47] and, like the CYP1A1 polymorphism, would not in itself be expected to give rise to a change in either transcription or catalytic activity of the gene product. However, these studies remain to be performed. Nevertheless, it would appear more likely that the polymorphism in the CYP2E1 gene may be linked to a mutation in another critical site of the gene, as has been demonstrated for the CYP1A1 gene,[35] or it is in linkage disequilibrium with another critical gene that is responsible for susceptibility to cancer, for example, an oncogene or tumor suppressor gene.

METALLOTHIONEIN GENES

The human health hazards associated with exposures to heavy metals are so numerous it has been estimated that more effort is spent in regulating levels of these toxic metals in the environment than any other class of pollutants.[12] There would appear to be just cause in this expenditure of both effort and fiscal resources. Their persistence in the environment, along with their ability to bioaccumulate in tissues, make metals a particular human health threat. For example, exposure to lead is associated with developmental neurological impairments in children and, apparently, can occur at very low levels of exposure as a result of ingestion of contaminated paint or inhalation of motor vehicle exhaust commonly found in urban air.[49] Other metals, such as cadmium and mercury, are typically associated with environmental exposures through contaminated drinking water or ingestion of contaminated foodstuffs.[50] These metals can elicit a wide range of adverse responses in exposed individuals, including kidney and reproductive toxicity as a result of cadmium exposure,[51] and peripheral and central nervous damage from mercury exposure.[52] In addition to these acutely toxic properties, heavy metals are also associated with the development of human cancers; for example, lung and possibly prostate tumors as a result of chronic cadmium exposures.[53] An even more disturbing potential health risk associated with low-level chronic exposures is the modulation of immunologic function that recent reports have described.[52]

The myriad of toxic properties that heavy metals exhibit toward humans is a logical explanation for the existence of a class of genes, the metallothionein (MT) genes, that function in part to bind and sequester these metals in the tissues in which they accumulate, and thus serve a protective role against heavy metal toxicity.[54,55] These genes code for low molecular weight, cysteine-rich metal binding proteins that bind avidly to heavy metals such as cadmium, mercury, and lead. Furthermore, MT genes serve to regulate tissue levels of essential metals such as zinc and copper.[56] MT genes perform this role in all mammalian species studied thus far, as well as in other eukaryotes; in humans, they belong to a large multigene family. So far, 14 individual human MT genes have been located in a locus on chromosome 16.[57] The transcriptional activation of the MT genes is one of the most thoroughly studied examples of environmental and physiologic regulation of gene expression that has been described to date. The following overview of MT gene regulation elucidates both the promise and potential hazards of utilizing transcriptional activation of target genes as biological markers of environmental toxicant exposures.

Molecular Biology of MT Gene Transcription

MT gene expression is the most studied and best understood model of metal-regulated transcription of a human gene. Even before the cloning of the human MT-II gene in 1982 in the laboratory of Karin,[57] molecular studies of MT had begun to search for its transcriptional factors. Karin et al.[58] first demonstrated that zinc and the glucocorticoid analog, dexamethasone, were primary transcriptional factors for the human MT gene. Following the cloning of the gene, studies began to map the structural regions surrounding the MT gene that were responsible for metal- and glucocorticoid-induced transcription. It is now apparent that a promoter region that is located directly upstream from the transcriptional initiation site of the MT gene contains both metal and glucocorticoid binding domains.[59] These cis-acting elements are termed MREs (metal responsive elements) and GREs (glucocorticoid responsive elements).

There are numerous MREs present in the human MT-II promoter, but only one GRE.[60] Glucocorticoids induce MT gene expression by binding first to an intracellular receptor, and then by a direct interaction of this trans-acting receptor with the GRE.[59] Metal induction of MT transcription is not as straightforward, but apparently is mediated by trans-acting metal-activated factors that recognize the MRE. Conceptually, the metal-activated factors behave like the glucocorticoid receptor, and presumably are activated by binding directly to metals. There is great interest in identifying the trans-acting metal elements of the MT gene, and recent studies have begun to characterize these DNA binding proteins,[61] including a human MT trans-acting metal binding protein.[62]

In addition to the GRE and MRE, there are other constitutive cis-elements in the MT promoter, and their presence most likely explains why other agents can transcriptionally activate the human MT gene. For example, cytokines such as interleukins-1, 6, and tumor necrosis factor can induce MT gene expression,[63] as well as bacterial endotoxin, or lipopolysaccharide.[64] The number of cis- and trans-acting elements that can potentially control transcription of the human MT gene allows for the regulation of its expression by both different classes of compounds and possibly by different members within a class of inducing agents, as is illustrated in the next section.

Metal-Regulated MT Gene Expression

The complexity of MT gene regulation, as was precedingly discussed, is best illustrated by metal-induced MT gene transcription. There appears to be a pleiotropic nature of action of metals on MT gene regulation, or, in other words, it appears that different metals affect the different MT genes in specific fashion. The studies of human MT gene regulation carried out by the laboratory of Gedamu are the most thorough investigations of metal-specific transcription to date. In one study, a comparison of MT-II, MT-IF, and MT-IG transcription revealed significant differences in gene inducibility values when human liver cells were exposed *in vitro* to zinc, copper, or cadmium.[65] For example, the MT-IIA gene was uniformly induced to very high levels by all three metals, but the MT-IF gene was poorly induced by these metals, particularly by copper. On the other hand, the MT-IG gene was well induced by copper and zinc, but very poorly induced by cadmium. Interestingly, the MT-IIA gene was the only MT gene that displayed a measurable level of basal gene expression in the absence of any inducing agent.

A subsequent investigation studied the time course of MT induction by heavy metals in which metal-specific differences on rates of transcription were observed, as well as differences in rates between MT genes.[66] These studies also demonstrated posttranscriptional control of MT mRNA levels with specific metals. Finally, in addition to metal- and gene-specific effects, another report demonstrated cell-type specific effects on MT gene inducibility.[67] A comparison between human liver, kidney, and lymphocytic cell lines revealed differences in mRNA levels for several MT genes following metal exposures *in vitro*. Furthermore, there appeared to be a relationship between DNA methylation status and gene expression in these cell lines. It was concluded that hypermethylation of the MT gene promoter may not allow access to the trans-activating factors necessary for initiation of transcription to occur.[67]

Environmental Modulation of Human MT Activity

The complexity of MT gene regulation may at first suggest that its use as a biological marker of metal exposures in human populations is conceptually sound but realistically improbable. However, the existence of such a highly evolved and sophisticated pathway of xenobiotic regulation indicates its integral role in the cellular response to metal exposure. Recent studies of human tissue MT levels add further credence to this view, and have included the measurement of MT protein in tissue and bodily fluids from unexposed and cadmium-exposed individuals. For example, a survey of autopsy tissues from a presumably unexposed population of Canadians revealed linear relationships between cadmium or zinc concentrations and MT protein levels in kidney, while liver samples demonstrated a linear relationship between zinc and MT levels.[68] A number of studies have determined excreted MT levels in a Japanese population environmentally exposed to cadmium, including the measurement of urinary MT protein by radioimmunoassay.[69] These studies demonstrated significantly elevated urinary MT levels in exposed individuals when compared to individuals living in an uncontaminated region. However, only cadmium-exposed women exhibited an elevated level of urinary MT, while among men living in this region, there was no apparent association between MT levels and cadmium exposure.[70] The biological mechanism behind the sex bias in MT response to environmental metal exposure remains unresolved at the present time.

MT levels have also been measured in individuals exposed to cadmium in the work place. A study of male workers employed in a cadmium production plant demonstrated higher levels of excreted MT in urine when compared to unexposed office workers, while longer-term employees displayed greater urinary MT levels than did new employees.[71] Furthermore, urinary MT levels were significantly correlated with liver and kidney cadmium burdens as measured by *in vivo* neutron activation. A study of Japanese men employed in a paint pigment factory also demonstrated higher urinary MT levels in workers exposed to relatively low-level cadmium dust (3 $\mu g/m^3$ for 20 yr).[72] While these studies have amply demonstrated the detection of excretable MT protein in cadmium-exposed human populations, there remains to be resolved the genetic basis for these observations. In other words, is the elevated level of urinary MT due to compromised renal filtration as a result of cadmium-induced nephrotoxicity, or is it caused by an increase in the circulating levels of MT protein due to systemic MT gene induction as a result of elevated tissue cadmium levels?

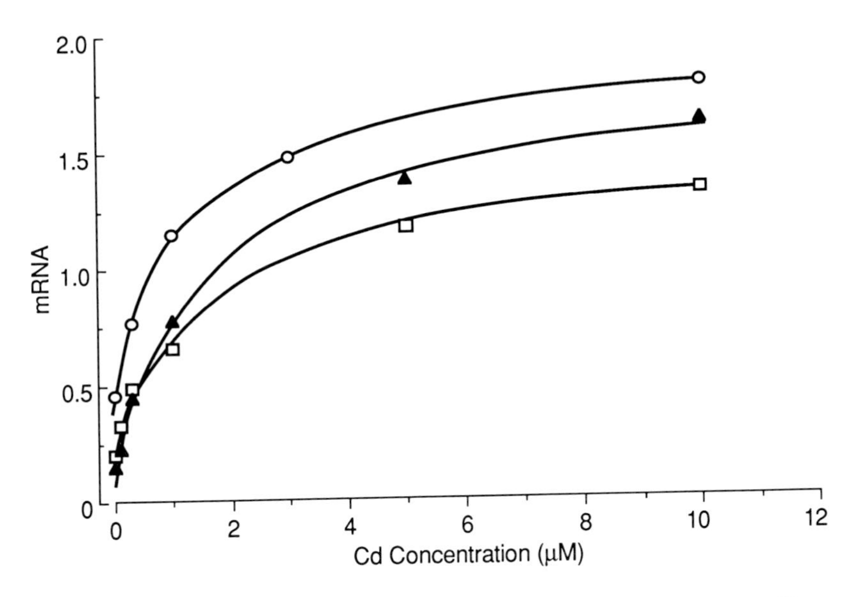

Figure 5. Dose-response relationship of MT gene expression in cadmium-exposed human lymphocytes. Depicted are results from lymphocytes of 3 separate individuals.

Human Lymphocyte MT Gene Inducibility

The preceding discussion pointed out a crucial lack of data to confirm the existence of human systemic MT gene induction following exposure to heavy metals. If one considers that external exposure to metals will eventually cause an elevation in peripheral blood concentrations,[73] then blood cells would appear to be a plausible surrogate cell type in which to measure MT gene induction. Their accessibility is also an advantage to their use for purposes of biological monitoring. The first studies of human MT induction in blood were performed with peripheral blood lymphocytes incubated *in vitro* with zinc, in which measurement of MT protein provided evidence for the inducibility of MT synthesis in these cells. To determine which human blood cells were capable of synthesizing MT in response to metal exposure, a study compared responses of mononuclear cells (monocytes and lymphocytes) to polymorphonuclear cells (neutrophils) following exposure *in vitro* to cadmium.[74] Results indicated that only mononuclear cells were capable of synthesizing MT protein in response to metal exposure, and that very little preexisting MT could be found in these cells; i.e. MT synthesis is a rapidly inducible process. Furthermore, treatment of cells with actinomycin prior to metal exposure indicated that the response to cadmium occurred at the level of gene transcription.[74]

The first studies to measure MT gene expression in human lymphocytes compared responses between protein and mRNA levels in cells exposed *in vitro* to cadmium.[75] A qualitative relationship between MT protein and mRNA expres-

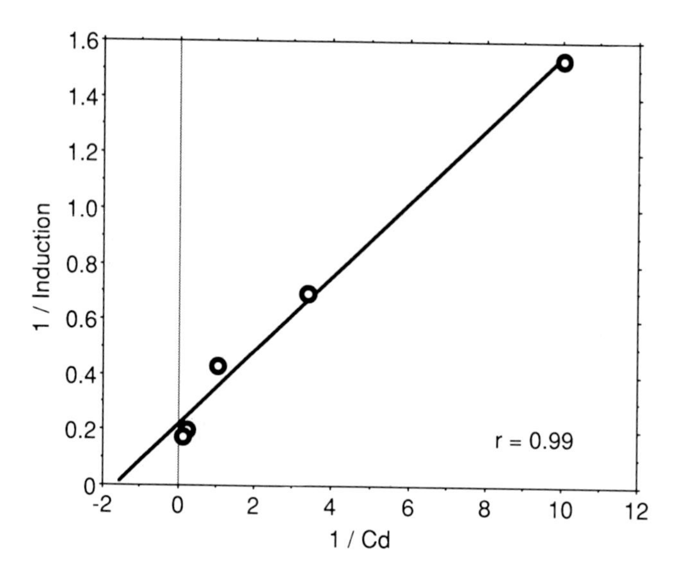

Figure 6. Double-reciprocal plot of data from Figure 5 for one individual.

sion levels was found, as well as a time course of induction that demonstrated peak mRNA levels by 12 h of exposure and reduced levels thereafter, whereas MT protein levels continued to rise as late as 4 days after continuous cadmium exposure. Interestingly, both lymphocyte basal and cadmium-induced MT mRNA values displayed considerable variation between healthy, nonsmoking individuals. When these data were calculated as induction ratios (induced/basal MT mRNA) and plotted on a histogram, the resulting frequency distribution of individual responses displayed a wide range of variability in human MT gene inducibility. Furthermore, a day-to-day variability was detected in individual gene induction values. The authors concluded that environmental and possibly genetic factors appear to regulate the inducibility of the human MT gene.

Our laboratory has conducted studies of MT gene expression in human peripheral lymphocytes to determine MT inducibility in unexposed individuals, as well as in experimentally exposed subjects. We have examined the kinetics of MT gene induction in lymphocytes exposed *in vitro* to cadmium, and compared responses between individuals. The plot shown in Figure 5 is a dose-response of $CdCl_2$-induced MT-II gene expression in lymphocytes from three individuals. The shape of the curve in all three cases suggested a saturable response of MT gene induction that occurred well below cytotoxic concentrations of cadmium.[74] The peak levels of MT mRNA that were achieved appeared to vary between individuals, and when these data were graphed on a double-reciprocal plot (as shown in Figure 6), the calculated values for the slopes and intercepts of these lines varied between individuals (Table 3). In addition, we sampled individuals repeatedly to determine intraindividual variability of MT gene inducibility; and as shown in Table 3, there appeared to be less intra- vs interindividual variability in MT responses, suggesting greater genetic control rather than environmental influence on MT gene activity.

Table 3
Slope Intercept Values Derived from MT Induction Curves from Figure 6

Individual[a]	Slope	Intercept
A_1	0.509	0.505
A_2	0.442	0.437
B_1	1.57	0.392
B_2	1.52	0.252
C_1	0.208	0.281
C_2	0.2216	0.243
D_1	0.357	0.299
D_2	0.594	0.250

[a] Each individual was sampled on two separate occasions.

Table 4
Metallothionein Gene Epression in Lymphocytes of Individuals Before and After Exposure to Zinc Oxide

Individual	Exposure	I/B	Fold increase
A	Before	0.71	1.42
	After	1.00	
B	Before	4.33	
	After	4.80	1.11
C	Before	1.56	
	After	1.90	1.22
D	Before	0.50	
	After	1.53	3.06
E	Before	1.47	
	After	1.90	1.29
F	Before	1.55	
	After	2.27	1.46
		1.59 ± 0.73 (SD)	

Note: $p < 0.01$, paired t-test.

In addition to our studies of human MT gene expression in unexposed individuals, we have conducted a pilot experiment in which we measured MT gene inducibility in volunteer subjects exposed to zinc oxide fumes at the current OSHA Permissible Exposure Level of 5 mg/m^3.[76] The experimental design of these studies allowed us to sample each individual before and after ZnO inhalation; thus, each individual served as his own control and we could compare responses before and after exposure. Table 4 provides induced/basal lymphocyte MT-II mRNA values for six individuals before and after exposure to ZnO. In each individual, there was a measurable elevation in MT gene inducibility after exposure and, although the increases were not large, they proved to be statistically significant. Furthermore, when these same individuals were sham-exposed to carrier gas, there were no elevations detected in MT gene inducibility. Interestingly, we found no consistent pattern of either basal or induced MT gene expression in these studies and, yet, external exposure to zinc appeared to modulate the inducibility of the MT gene when challenged *in vitro*. We do not have an explanation for this phenomenon at this time, but it is compelling to note that a similar gene induction response was found in our studies of CYP1A1 activity in PAH-exposed railroad workers.[28]

The complexity of MT gene induction, as was discussed in a preceding section, allows several possible mechanisms to be advanced concerning modulation of MT inducibility after external exposures to heavy metals. Interestingly, there is evidence in animal models of metal toxicity that demonstrate an enhanced synthesis of MT protein after challenging animals previously exposed to cadmium.[77] These studies also demonstrated protection against toxicity in animals previously exposed to the metal.[78] A plausible route by which MT gene inducibility could be modulated by prior metal exposure is via a change in activity of the trans-acting transcriptional factors of the gene.[59] For example, prior exposure to metals may increase the amount of these factors, or change the nature of their binding to the cis-acting elements of the MT gene, in such a fashion that upon reexposure there is an increase in the inducibility of the gene. Alternatively, prior exposure to metals may change the nature of the cis-acting DNA elements of the MT gene. A recent study demonstrated that by changing the number of MREs or GREs in the human MT-II promoter, a change in gene inducibility could be achieved.[60] A reduction in basal activity was accomplished by the removal of MREs, but inducible expression was also reduced. The addition of MREs increased inducible expression, but only by a modest amount. Alternatively, by introducing additional GREs into the promoter, a synergistic induction response was obtained without an increase in basal gene expression. The degree of transcriptional synergism was found to be dependent on the position and number of GREs introduced.

The implications of these effects of genetic manipulation on MT gene inducibility are twofold: (1) such changes in the human MT promoter may explain changes in MT gene inducibility after exposure to metals, and offers a mechanism by which environmental exposure to metals can modulate the activity of the gene; and (2) differences in the nature of the MT promoter may explain intrinsic differences in inducibility of the gene between individuals, and suggests a plausible route by which genetic modulation of MT activity occurs. The insights into MT gene regulation that these molecular studies have provided, along with our analyses of MT inducibility in human populations, illustrate the potential benefits that can be achieved by the use of molecular approaches to human exposure assessment.

CONCLUSIONS

This review of human gene inducibility and its use in assessment of toxicant exposure and susceptibility has pointed out both the potential benefits and inherent risks in applying molecular biological methods to human exposure assessment. The sensitivity and specificity of measuring transcriptional activity of target inducible genes following xenobiotic exposure was demonstrated by our studies of CYP1A1 activity in creosote-exposed railroad workers.[28] Our studies of MT gene inducibility following inhalation of zinc oxide fumes also illustrated the use of MT activity as a marker of heavy metal exposure in human populations. However, we have stressed as a precautionary note that multiple factors can affect the transcriptional activity of these genes, and that prudence must be exercised when drawing conclusions from these studies of human exposure assessment.

Perhaps as compelling as the use of gene inducibility to determine exposure is its use in determining disease susceptibility in affected individuals. The issue of individual susceptibility following toxicant exposure is central to the study of human health effects of environmental contamination. For example, it has long been recognized that the risk of developing lung cancer from tobacco use is much higher in approximately 10% of smokers.[10] Recently, more sophisticated statistical analyses of lung cancer risk in smokers have demonstrated that this elevated risk is due to a genetic factor, and that this factor is inheritable in a Mendelian fashion.[79]

Our studies of CYP1A1 polymorphisms and its relationship to functional activity of the gene provide strong evidence for the possible role of this gene in determining cancer susceptibility following PAH exposure in certain racial populations.[34] Of course, as was pointed out in the section on cytochrome P450 genes, there are a number of these genes in which polymorphisms have been discovered that appear to be related to cancer incidences in human populations.[38,47] It is highly likely that disease outcome in individuals following toxicant exposure is due to a combination of genetic factors, and that not one genetic alteration alone will dictate risk. It must be stressed that this is not to suggest that environmentally related human illnesses are genetic diseases. Exposure to a toxicant is the primary cause of the disease; however, in certain individuals with a genetic predisposition, the relative risk of developing the disease may be increased significantly.

The preceding discussion illustrates the dual nature of applying molecular biological techniques to the study of toxicant exposures in human populations. On the one hand, the measurement of genetic responses provides a potentially sensitive indicator of a biological effect due to the exposure. Alternatively, the response that is being measured may also be regulated in part by genetic factors. The advantage to this approach of exposure assessment is that it can provide potential information on the biological response and individual susceptibility to toxicant exposure. However, the disadvantage is that this approach will not yield a precise quantitative relationship between the level of exposure and the molecular biological response that is being measured. Inherent variability of responses between individuals because of genetic factors will not allow the type of statistical comparisons that are favored by regulatory agencies when analyzing dose-response relationships of exposure to biological response from laboratory studies of inbred animal strains.

There is an obvious need for independent verification of exposure levels when conducting molecular biomarker studies in human populations. Only then can the measurement of a molecular response in an individual be useful in assessing his relative risk to a particular exposure. When these responses are compared to the level of toxicant to which the population was exposed, then a more definitive conclusion can be reached as to the quantitative nature of the resulting dose-response relationship. Our laboratory has begun to perform such studies of exposure assessment in human populations and, as was described in this chapter, we have done so by measuring the inducibility of responsive genes as well as by determining genetic polymorphisms that may influence their activity. These studies depend entirely on an understanding of how transcriptional regulation is achieved by these genes. Furthermore, it is essential that the function of the protein products be understood for us to correctly interpret the significance that gene inducibility has on human health. The ultimate rewards of using molecular approaches in the study of human exposure assessment will only be accomplished by the continued application of fundamental molecular genetic research to molecular epidemiology studies in diverse human populations.

REFERENCES

1. **Lewin, B.,** *Gene Expression,* Vol. 2, John Wiley & Sons, New York, 1974, chap. 6.
2. **Franks, L. M., and Teich, N. M.,** *Introduction to the Cellular and Molecular Biology of Cancer,* Oxford University Press, New York, 1986, chap. 10.
3. **Pelkonen, O., and Nebert, D. W.,** Metabolism of polycyclic aromatic hydrocarbons: etiologic role in carcinogenesis, *Pharmacol. Rev.,* 34, 189, 1982.
4. **Jahnke, G. D., Thompson, C. L., Walker, M. P., Gallagher, J. E., Lucier, G. W., and DiAugustine, R. P.,** Multiple DNA adducts in lymphocytes of smokers and nonsmokers determined by ^{32}P-postlabeling analysis, *Carcinogenesis (London),* 11, 205, 1990.
5. **Perera, F. P., Santella, R. M., Brenner, D., Poirier, M. C., Munshi, A. A., Fischman, H. K., and Van Rysin, J.,** DNA adducts, protein adducts, and sister chromatid exchange in cigarette smokers and nonsmokers, *J. Natl. Cancer Inst.,* 79, 449, 1987.
6. **Recio, L., Cochrane, J., Simpson, D., Skopek, T. R., O'Neill, J. P., Nicklas, J. A., and Albertini, R. J.,** DNA sequence analysis of *in vivo hprt* mutation in human T lymphocytes, *Mutagenesis,* 5, 505, 1990.
7. **Harris, C. C.,** Commentary: interindividual variation among humans in carcinogen metabolism, DNA adduct formation and DNA repair, *Carcinogenesis (London),* 10, 1563, 1989.
8. **Gonzalez, F. J., Crespi, C. L., and Gelboin, H. V.,** cDNA-expressed human cytochrome P450s: a new age of molecular toxicology and human risk assessment, *Mutation Res.,* 247, 113, 1991.
9. **Nebert, D. W., and Gonzalez, F. J.,** P450 genes: structure, evolution, and regulation, *Annu. Rev. Biochem.,* 56, 945, 1987.
10. **Nebert, D. W.,** Role of genetics and drug metabolism in human cancer risk, *Mutation Res.,* 247, 267, 1991.
11. **Baum, E. J.,** Occurrence and surveillance of polycyclic aromatic hydrocarbons, in *Polycyclic Hydrocarbons and Cancer,* Vol. 1, Gelboin, H. V., and Ts'o, P. O., Eds., Academic Press, New York, 1979, 45.

12. **Lippmann, M.,** *Environmental Toxicants: Human Exposures and Their Health Effects,* Van Nostrand Reinhold, New York, 1992, chap. 6.
13. **Jaiswal, A. K., Gonzalez, F. J., and Nebert, D. W.,** Human dioxin-inducible cytochrome P450: complementary DNA and amino acid sequence, *Science,* 228, 80, 1985.
14. **Nebert, D. W.,** The *Ah* locus: genetic differences in toxicity cancer, mutation, and birth defects, *Crit. Rev. Toxicol.,* 20, 153, 1989.
15. **Hines, R. N., Mathis, J. M., and Jacob, C. S.,** Identification of multiple regulatory elements on the human cytochrome P450IA1 gene, *Carcinogenesis,* 9, 1599, 1988.
16. **Sherratt, A. J., Banet, D. E., and Prough, R. A.,** Glucocorticoid regulation of polycyclic aromatic hydrocarbon induction of cytochrome P450IA1, glutathione S-transferases, and NAD(P)H: quinone oxidoreductase in cultured fetal rat hepatocytes, *Mol. Pharmacol.,* 37, 198, 1990.
17. **Kouri, R. E., Oberdorf, J., Slomiany, D. J., and McKinney, C. E.,** Aryl hydrocarbon hydroxylase activities in cryopreserved human lymphocytes, *Cancer Lett.,* 14, 29, 1981.
18. **Whitlock, J. P., Jr., Cooper, H. L., and Gelboin, H. V.,** Aryl hydrocarbon (benzopyrene) hydroxylase is stimulated in human lymphocytes by mitogens and benz[a]anthracene, *Science,* 177, 618, 1972.
19. **Atlas, S. A., Vesell, E. S., and Nebert, D. W.,** Genetic control of interindividual variation in the inducibility of aryl hydrocarbon hydroxylase in cultured human lymphocytes, *Cancer Res.,* 36, 4619, 1976.
20. **Kellerman, G., Shaw, C. R., and Luyten-Kellerman, M.,** Aryl hydrocarbon hydroxylase inducibility and bronchogenic carcinoma, *N. Engl. J. Med.,* 289, 934, 1973.
21. **Kouri, R. E., McKinney, C. E., Slomiany, D. J., Snodgrass, D. R., Wray, N. P., and McLemore, T. L.,** Positive correlation between high aryl hydrocarbon hydroxylase activity and primary lung cancer as analyzed in cryopreserved lymphocytes, *Cancer Res.,* 42, 5030, 1982.
22. **Karki, N. T., Pokela, R., Nuutinen, L., and Pelkonen, O.,** Aryl hydrocarbon hydroxylase in lymphocytes and lung tissue from lung cancer patients and controls, *Int. J. Cancer,* 39, 565, 1987.
23. **Anttila, S., Hietanen, E., Vainio, H., Camus, A.-M., Gelboin, H. V., Park, S. S., Heikkila, L., Karjalainen, A., and Bartsch, H.,** Smoking and peripheral type of cancer are related to high levels of pulmonary cytochrome P450IA in lung cancer patients, *Int. J. Cancer,* 47, 681, 1991.
24. **McLemore, T. L., Adelberg, S., Liu, M. C., McMahon, N. A., Yu, S. J., Hubbard, W. C., Czerwinski, M., Wood, T. G., Storeng, R., Lubet, R. A., Eggleston, J. C., Boyd, M. R., and Hines, R. N.,** Expression of CYP1A1 gene in patients with lung cancer: evidence for cigarette smoke-induced gene expression in normal lung tissue and for altered gene regulation in primary pulmonary carcinomas, *J. Natl. Cancer Inst.,* 82, 1333, 1990.
25. **Naseem, S. M., Tishler, P. V., Anderson, H. A., and Selikoff, I. J.,** Aryl hydrocarbon hydroxylase in asbestos workers, *Am. Rev. Respir. Dis.,* 118, 693, 1978.
26. **Jaiswal, A. K., Gonzalez, F. J., and Nebert, D. W.,** Human P_1-450 gene sequence and correlation of mRNA with genetic differences in benzo[a]pyrene metabolism, *Nucl. Acids Res.,* 13, 4503, 1985.
27. **Amsbaugh, S. C., Ding, J.-H., Swan, D. C., Popescu, N. C., and Chen, Y.-T.,** Expression and chromosomal localization of the cytochrome P_1-450 gene in human mitogen-stimulated lymphocytes, *Cancer Res.,* 46, 2423, 1986.

28. **Cosma, G. N., Toniolo, P., Currie, D., Pasternack, B. S., and Garte, S. J.,** Expression of the CYP1A1 gene in peripheral lymphocytes as a marker of exposure to creosote in railroad workers, *Canc. Epid. Biomark. Prev.,* 1, 137, 1992.
29. **Nebert, D. W., and Jaiswal, A. K.,** Human drug metabolism polymorphisms: use of recombinant DNA techniques, *Pharmacol. Ther.,* 33, 11, 1987.
30. **Kawajiri, K., Nakachi, K., Imai, K., Yoshii, A., Shinoda, N., and Watanabe, J.,** Identification of genetically high risk individuals to lung cancer by DNA polymorphisms of the cytochrome P450IA1 gene, *FEBS Lett.,* 263, 131, 1990.
31. **Nakachi, K., Imai, K., Hayashi, S.-I., Watanabe, J., and Kawajiri, K.,** Genetic susceptibility to squamous cell carcinoma of the lung in relation to cigarette smoking dose, *Cancer Res.,* 51, 5177, 1991.
32. **Peterson, D. D., McKinney, C. E., Ikeya, K., Smith, H. H., Bale, A. E., McBride, O. W., and Nebert, D. W.,** Human CYP1A1 gene: cosegregation of the enzyme inducibility phenotype and an RFLP, *Am. J. Hum. Genet.,* 48, 720, 1991.
33. **Tefre, T., Ryberg, D., Haugen, A., Nebert, D. W., Skaug, V., Brogger, A., and Borresen, A.-L.,** Human CYP1A1 (cytochrome P_1450) gene: lack of association between the MspiI restriction fragment length polymorphism and incidence of lung cancer in a Norwegian population, *Pharmacogenetics,* 1, 20, 1991.
34. **Cosma, G., Crofts, F., Currie, D., Wirgin, I., Toniolo, P., and Garte, S. J.,** Racial differences in restriction fragment length polymorphisms and mRNA inducibility of the human CYP1A1 gene, *Canc. Epid. Biomark. Prev.,* 2, 53, 1993.
35. **Hayashi, S.-I., Watanabe, J., Nakachi, K., and Kawajiri, K.,** Genetic linkage of lung cancer — associated MspI polymorphisms with amino acid replacement in the heme binding region of the human cytochrome P450IA1 gene, *J. Biochem.,* 110, 407, 1991.
36. **Sugimura, H., Caporaso, N. E., Shaw, G. L., Modali, R. V., Gonzalez, F. J., Hoover, R. N., Resau, J. H., Trump, B. F., Weston, A., and Harris, C. C.,** Human debrisoquine hydroxylase gene polymorphisms in cancer patients and controls, *Carcinogenesis,* 11, 1527, 1990.
37. **Mahgoub, A., Idle, J. R., Dring, L. G., Lancaster, R., and Smith, R. L.,** Polymorphic hydroxylation of debrisoquine in man, *Lancet,* ii, 584, 1977.
38. **Ayesh, R., Idle, J. R., Ritchie, J. C., Crothers, M. J., and Hetzel, M. R.,** Metabolic oxidation phenotypes as markers for susceptibility to lung cancer, *Nature,* 312, 169, 1984.
39. **Speirs, C. J., Murray, S., Davies, D. S., Biola Mabadeje, A. F., and Boobis, A. R.,** Debrisoquine oxidation phenotype and susceptibility to lung cancer, *Br. J. Clin. Pharmacol.,* 29, 101, 1990.
40. **Henthorn, T. K., Benitez, J., Avram, M. J., Martinez, C., Llerena, A., Cobaleda, J., Krejcie, T. C., and Gibbons, R. D.,** Assessment of the debrisoquine and dextromethorphan phenotyping tests by gaussian mixture distributions analysis, *Clin. Pharmacol. Ther.,* 45, 328, 1989.
41. **Skoda, R. C., Gonzalez, F. J., Demierre, A., Meyer, U. A., and Gonzalez, F. J.,** Two mutant alleles of the human cytochrome P-450dbl gene (P450C2D1) associated with genetically deficient metabolism of debrisoquine and other drugs, *Proc. Natl. Acad. Sci. U.S.A.,* 85, 5240, 1988.
42. **Gough, A. C., Miles, J. S., Spurr, N. K., Moss, J. E., Gaedigk, A., Eichelbaum, M., and Wolf, C. R.,** Identification of the primary gene defect at the cytochrome P450 CYP2D locus, *Nature,* 347, 773, 1990.

43. **Wolf, C. R., Smith, C. A. D., Gough, A. C., Moss, J. E., Vallis, K. A., Howard, G., Carey, F. J., Mills, K., McNee, W., Carmichael, J., and Spurr, N. K.,** Relationship between the debrisoquine hydroxylase polymorphism and cancer susceptibility, *Carcinogenesis,* 13, 1035, 1992.
44. **Kimura, S., Umeno, M., Skoda, R. C., Meyer, U. A., and Gonzalez, F. J.,** The human debrisoquine 4-hydroxylase (CYP2D) locus: sequence and identification of the polymorphic CYP2D6 gene, a related gene, and pseudogene, *Am. J. Hum. Genet.,* 45, 889, 1989.
45. **Bartsch, H., and Montesano, R.,** Relevance of nitrosamines to human cancer, *Carcinogenesis,* 5, 1381, 1984.
46. **Hong, J., and Yang, C. S.,** The nature of microsomal N-nitrosodimethylamine demethylase and its role in carcinogen activation, *Carcinogenesis,* 6, 1805, 1985.
47. **Uematsu, F., Kikuchi, H., Motomiya, M., Abe, T., Sagami, I., Ohmachi, T., Wakui, A., Kanamaru, R., and Watanabe, M.,** Association between restriction fragment length polymorphism of the human cytochrome P450IIE1 gene and susceptibility to lung cancer, *Jpn. J. Cancer Res.,* 82, 254, 1991.
48. **Umeno, M., McBride, O. W., Yang, C. S., Gelboin, H. V., and Gonzalez, F. J.,** Human ethanol-inducible P450IIE1: complete gene sequence, promoter characterization, chromosome mapping, and cDNA-directed expression, *Biochemistry,* 27, 9006, 1989.
49. **Centers for Disease Control,** *Preventing Lead Poisoning in Young Children,* Atlanta, U.S. Department of Health and Human Services, No. 99-2230, 1985.
50. **Horn, N.,** in *Metabolism of Trace Metals in Man,* Vol. II, Rennert, O. M. and Chan, W. Y., Eds., CRC Press, Boca Raton, FL, 1984, 26-52.
51. **Friberg, L.,** *Cadmium in the Environment,* 2nd ed., CRC Press, Cleveland, 1974.
52. **World Health Organization,** *Inorganic Mercury,* Environmental Health Criteria, No. 119, 1991.
53. **International Agency for Research on Cancer,** *IARC Monograph on the Carcinogenic Risk of Chemicals to Man,* Vol. 11, 1976.
54. **Webb, M.,** Toxicological significance of metallothionein, *Experientia Suppl.,* 52, 109, 1987.
55. **Klaassen, C. D., and Lehman-McKeeman, L. D.,** Induction of metallothionein, *J. Am. Coll. Toxicol.,* 8, 1315, 1989.
56. **Cousins, R. J.,** Absorption, transport, and hepatic metabolism of copper and zinc: special reference to metallothionein and ceruloplasmin, *Physiol. Rev.,* 65, 238, 1985.
57. **Karin, M., and Richards, R. I.,** Human metallothionein genes — primary structure of the metallothionein-II gene and a related processed gene, *Nature,* 299, 191, 1982.
58. **Karin, M., Andersen, R. D., Slater, E., Smith, K., and Herschman, H. R.,** Metallothionein mRNA induction in HeLa cells in response to zinc or dexamethasone is a primary induction response, *Nature,* 286, 295, 1980.
59. **Thiele, D. J.,** Metal-regulated transcription in eukaryotes, *Nucl. Acid Res.,* 20, 1183, 1992.
60. **Filmus, J., Remani, J., and Klein, M. H.,** Synergistic induction of promoters containing metal- and glucocorticoid-responsive elements, *Nucl. Acids Res.,* 20, 2755, 1992.
61. **Westin, G., and Schaffner, W.,** A zinc-responsive factor interacts with a metal-regulated enhancer element (MRE) of the mouse metallothionein-I gene, *EMBO J.,* 7, 3763, 1988.

62. **Koizumi, S., Otsuka, F., and Yamada, H.,** A nuclear factor that interacts with metal responsive elements of a human metallothionein gene, *Chem.-Biol. Interactions,* 80, 145, 1991.
63. **Schroeder, J. J., and Cousins, R. J.,** Interleukin 6 regulates metallothionein gene expression and zinc metabolism in hepatocyte monolayer cultures, *Proc. Natl. Acad. Sci. U.S.A.,* 87, 3137, 1990.
64. **Durnam, D. M., Hoffman, J. S., Quaife, C. J., Benditt, E. P., Chen, H. Y., Brinster, R. L., and Palmiter, R. D.,** Induction of mouse metallothionein-I mRNA by bacterial endotoxin is independent of metals and glucocorticoid hormones, *Proc. Natl. Acad. Sci. U.S.A.,* 811, 1053, 1984.
65. **Sadhu, C., and Gedamu, L.,** Regulation of human metallothionein genes, *J. Biol. Chem.,* 263, 2679, 1988.
66. **Sadhu, C., and Gedamu, L.,** Metal-specific posttranslational control of human metallothionein genes, *Mol. Cell. Biol.,* 9, 5738, 1989.
67. **Jahroudi, N., Foster, R., Price-Haughey, J., Beitel, G., and Gedamu, L.,** Cell-type specific and differential regulation of the human metallothionein genes, *J. Biol. Chem.,* 265, 6506, 1990.
68. **Chung, J., Nartey, N. O., and Cherian, M. G.,** Metallothionein levels in liver and kidney of Canadians — a potential indicator of environmental exposure to cadmium, *Arch. Environ. Health,* 41, 319, 1986.
69. **Tohyama, C., Shaikh, Z. A., Nogawa, K., Kobayashi, E., and Honda, R.,** Elevated urinary excretion of metallothionein due to environmental cadmium exposure, *Toxicology,* 20, 289, 1981.
70. **Kido, T., Shaikh, Z., Kito, H., Honda, R., and Nogawa, K.,** Dose-response relationship between urinary cadmium and metallothionein in a Japanese population environmentally exposed to cadmium, *Toxicology,* 65, 325, 1991.
71. **Shaikh, Z. A., Ellis, K. J., Subramanian, K. S., and Greenberg, A.,** Biological monitoring for occupational cadmium exposure: the urinary metallothionein, *Toxicology,* 63, 53, 1990.
72. **Kawada, T., Tohyama, C., and Suzuki, S.,** Significance of the excretion of urinary indicator proteins for a low level of occupational exposure to cadmium, *Int. Arch. Occup. Health,* 62, 95, 1990.
73. **Kjellstrom, T., and Nordberg, G. F.,** A kinetic model of cadmium metabolism in the human being, *Environ. Res.,* 16, 248, 1978.
74. **Enger, M. D., Hildebrand, C. E., and Stewart, C. C.,** Cd^{2+} responses of cultured human blood, *Toxicol. Appl. Pharmacol.,* 69, 214, 1983.
75. **Harley, C. B., Menon, C. R., Rachubinski, R. A., and Nieboer, E.,** Metallothionein mRNA and protein induction by cadmium in peripheral-blood leucocytes, *Biochem. J.,* 262, 873, 1989.
76. **Gordon, T., Chen, L. C., Fine, J. M., Schlesinger, R. B., Su, W. Y., Kimmel, T. A., and Amdur, M. O.,** Pulmonary effects of inhaled zinc oxide in human subjects, guinea pigs, and rabbits, *Am. Ind. Hyg. Assoc. J.,* in press.
77. **Squibb, K. S., Cousins, R. J., Silbon, B. L., and Levin, S.,** Liver and intestinal metallothionein: function in acute cadmium toxicity, *Exp. Mol. Pathol.,* 25, 163, 1976.
78. **Feldman, S. L., Squibb, K. S., and Cousins, R. J.,** Degradation of cadmium-thionein in rat liver and kidney, *J. Toxicol. Environ. Health,* 4, 805, 1978.
79. **Sellers, T. A., Potter, J. D., Bailey-Wilson, J. E., Rich, S. S., Rothschild, H., and Elston, R. C.,** Lung cancer detection and prevention: evidence for an interaction between smoking and genetic predisposition, *Cancer Res. (Suppl.),* 52, 2694s, 1992.

INDEX

A

B

C

D

E

F

G

R

S

W

X

Y